# Structural Geology
and
# Rock Engineering

*The Carrara marble quarry in Italy where there is interaction between the tectonised marble properties and the extraction techniques.*

*The Marble Arch monument in London, UK, is a 19$^{th}$ century triumphal arch designed by John Nash in 1827. It is faced with Carrara marble and located at the west end of Oxford Street.*

# Structural Geology
## and
# Rock Engineering

John W Cosgrove • John A Hudson

Imperial College London, UK

Imperial College Press

ICP

*Published by*

Imperial College Press
57 Shelton Street
Covent Garden
London WC2H 9HE

*Distributed by*

World Scientific Publishing Co. Pte. Ltd.
5 Toh Tuck Link, Singapore 596224
*USA office:* 27 Warren Street, Suite 401-402, Hackensack, NJ 07601
*UK office:* 57 Shelton Street, Covent Garden, London WC2H 9HE

**Library of Congress Cataloging-in-Publication Data**
Names: Cosgrove, J. W. (John W.) | Hudson, J. A. (John A.), 1940–
Title: Structural geology and rock engineering / John W. Cosgrove
    (Imperial College London, UK) & John A. Hudson (Imperial College London, UK).
Description: New Jersey : Imperial College Press, 2016. | Includes bibliographical references.
Identifiers: LCCN 2016000965| ISBN 9781783269563 (hc : alk. paper) |
    ISBN 9781783269570 (pbk : alk. paper)
Subjects: LCSH: Rock mechanics. | Engineering geology. | Geology, Structural.
Classification: LCC TA706 .C67 2016 | DDC 624.1/51--dc23
LC record available at http://lccn.loc.gov/2016000965

**British Library Cataloguing-in-Publication Data**
A catalogue record for this book is available from the British Library.

Desk Editors: Suraj Kumar/Mary Simpson

Typeset by Stallion Press
Email: enquiries@stallionpress.com

This book is dedicated to Carol M. Hudson — with thanks for her 'forensic' talents in spotting grammatical and typing mistakes in the draft version of this book. The mistakes may not all have been eliminated but, thanks to Carol's help, there are now far fewer.

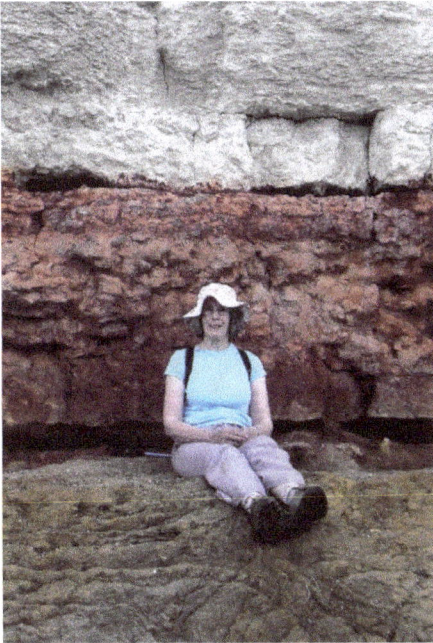

Carol Hudson

(sitting on Carstone stratum, a   bioturbated ferruginous sandstone conglomerate, leaning against the iron pigmented Red Chalk, and below the superincumbent grey-white Lower Chalk, UK)

*"The drive to connect.*
*The dream of a common language".*
Adrienne Rich

# PREFACE

The writing of this book was stimulated by four factors.

— Structural geologists have a great deal to contribute to rock engineering design and rock engineers need structural geology input for project design — which is now invariably computer based.
— The authors have consulted together on a variety of rock engineering projects in different countries.
— The authors have also given many Short Courses together on structural geology and rock engineering which has resulted in the development of a structured approach to the synthesis of the subjects.
— At Imperial College in London, the subjects of structural geology and rock engineering are both located in the Department of Earth Science and Engineering where the authors are based.

Although structural geology is concerned with interpreting past events, e.g., establishing how a particular rock formation was generated, and rock engineering design is concerned with future events, e.g., designing an underground repository for radioactive waste such that radionuclides will not migrate to the biosphere for many years, there is an intimate linkage between the two subjects. In particular, the idiosyncrasies of rock stress and rock fractures are common to the two subjects, as we outline in the introductory Chapter 1 overview and in the more detailed Chapters 2 and 3. The synthesis itself is demonstrated via the Chapter 4 case example of the Clifton Suspension Bridge in the UK. In the following Chapters 5–9, we provide overviews of a variety of engineering case examples, ranging through quarries, dams, opencast coal mining, underground rock engineering and historical monuments. Chapter 10 is a concluding discussion of the combined structural geology and rock engineering approach, and the Chapter 11 epilogue demonstrates how the approach even assists on the small scale.

*John W. Cosgrove*
*John A. Hudson*
*2015*

# ACKNOWLEDGEMENTS

The content of this book is based on our research, teaching and consulting experiences. We are grateful to the Department of Earth Science and Engineering, Imperial College London, and to our many students and colleagues over the years for enabling us, through the preparation and teaching of many lectures, to refine not only our understanding of structural geology and engineering rock mechanics but also the synthesis of the two subjects. Many of the examples presented in this book are based on our joint consulting experiences in the UK and overseas; thus, we are also grateful to our clients for enabling us to develop the synthesis approach and to present our experiences in this book.

We are also grateful to our 'local' publisher, Imperial College Press, and to their representatives Jane Sayers and Mary Simpson, for constant encouragement and to the copy editor and typesetter for transforming our Word files into this well-produced book.

# ABOUT THE AUTHORS

*John A. Hudson (left) and John W. Cosgrove (right) at a construction site in Shanghai, China*

## PROFESSOR JOHN W. COSGROVE

John W. Cosgrove is a Professor in the Department of Earth Sciences and Engineering, Imperial College London, UK. He obtained both his MSc (1969) and PhD (1972) from Imperial College. Following a two-year Post-doctoral Fellowship at McMaster University, Hamilton, Ontario, Canada, he returned in 1973 to Imperial College to take up a Lectureship and has since been promoted to Professor. He won the Paul Fourmarier Gold Medal, awarded by the Royal Academy of Belgium in 2005 for work on fluid induced failure in the crust and has also received awards for excellence in teaching from Imperial College. He was responsible for the M.Sc. course in Structural Geology & Rock Mechanics for the period (1978–1998). His present research interests relate to the interplay between stress, fractures and fluid flow in the Earth's crust. His earlier co-authored book (Price & Cosgrove 1990 "Analysis of Geological Structures") has been used worldwide. He has worked extensively on research for the hydrocarbon and mining industry (the dewatering of basins and the impact of the resulting fluids on mineralisation and hydrocarbon migration & concentration), radioactive waste disposal and other rock mechanics applications.

## EMERITUS PROFESSOR JOHN A. HUDSON

John A. Hudson graduated from the Heriot-Watt University, UK, and obtained his PhD at the University of Minnesota, USA. He has spent his professional career in consulting, research, teaching and publishing in engineering rock mechanics, and was awarded the DSc. degree by the Heriot-Watt University for his contributions to the subject. He has authored many scientific papers and books, and was the editor of the 1993 five-volume "Comprehensive Rock Engineering" compendium, and from 1983 to 2006 was editor of the International Journal of Rock Mechanics and Mining Sciences. Since 1983, he has been affiliated with Imperial College London as Reader, Professor and now Emeritus Professor. In 1998, he became a Fellow of the UK Royal Academy of Engineering and was President of the International Society for Rock Mechanics (ISRM) for the period 2007–2011. Additionally, he has completed consulting assignments in many countries. In 2015, the 7th ISRM Müller Award was conferred on Professor Hudson in recognition of "an outstanding career that combines theoretical and applied rock engineering with a profound understanding of the basic sciences of geology and mechanics".

# CONTENTS

# CHAPTER 1

# INTRODUCTION AND PURPOSE OF THE BOOK

*Structural Geology is the study of the natural processes that have deformed and fractured intact rock and rock masses.*

*Rock Mechanics is the study of mechanics applied to intact rock and rock masses.*

*Rock Engineering is the process of engineering on and in rock masses to create structures such as slopes, dam foundations, shafts, tunnels, caverns, mines and petroleum wellbores.*

*'Synthesis' is the combining of separate components to form a coherent whole.*

## 1.1 MOTIVATION

As intimated by the book's title, the purpose of the book is to

(a) Outline the principles of both structural geology and rock engineering (the latter via engineering rock mechanics) and
(b) to explain with illustrative examples why the synthesis of the two subjects is mutually beneficial.

1

The McGraw-Hill Dictionary of Scientific and Technical Terms defines structural geology as, "The branch of geology concerned with the form, arrangement and internal structure of the rocks" and rock mechanics as, "Application of the principles of mechanics and geology to quantify the response of rock when it is acted upon by environmental forces, particularly when human-induced factors alter the original ambient forces".

The evolutions of these two disciplines, structural geology and rock mechanics/rock engineering, have to a large extent developed separately. However, the two disciplines are in fact intimately related: the geological setting enables the necessary predictive capability for the engineering design of surface and underground structures constructed on and in rock masses, and rock mechanics knowledge supports structural geology understanding enabling the formation of the natural, ductile and brittle rock structures encountered in the Earth's crust to be understood.

## 1.2 CONTENT

In line with our motivation for writing the book, the structure and content of the book are as outlined in Table 1.1.

So, in the context of beginning by providing the necessary foundation material and as outlined in Table 1.1, following this introductory Chapter the basic principles of structural geology and rock mechanics for engineering are laid out in Chapters 2 and 3. In the case of the **structural geology** items in Chapter 2, there are 14 Sections outlining the principles of structural geology which include the brittle failure of rocks, fractures in rock masses and *in situ* rock stress. In the case of **rock mechanics for engineering** in Chapter 3, the subject is explained and summarised by outlining 50 rock mechanics principles.

## 1.3 AN INTRODUCTORY OVERVIEW OF THE SUBJECT

### 1.3.1 Structural geology and rock engineering

Before explaining the principles of both structural geology and engineering rock mechanics in Chapters 2 and 3, we now introduce the overall subject and describe why the synthesis of the two subjects is important.

Table 1.1. Structure and content of the book.

| Chpt. No. | Chapter title and content |
|---|---|
| 1 | **Introduction and Purpose of the Book**: Explanations of the structural geology and rock engineering subjects, and discussion of the pre-split blasting technique for engineering smooth rock faces — as one example of the links between the two subjects. |
| 2 | **Structural Geology Principles**: Coverage of the principles, beginning with an introduction to the subject, then explaining rock stress, brittle failure, stress in the Earth's crust, fluid-induced failure, fracture formation and propagation, recognition of fracture types, fracture geometry, the brittle-ductile transition, complex fracturing, fracture networks and their properties, and the interaction of *in situ* stress and fractures. |
| 3 | **Rock Mechanics Principles**: The basis of rock mechanics and rock engineering is presented in the form of 50 compact principles as a complement to the structural geology principles in Chapter 2. Thus, the information in Chapters 2 and 3 will provide the reader with a good foundation in both the structural geology and rock mechanics subjects and hence in using the integrated synthesis approach. |
| 4 | **Illustrative Synthesis Case Example, The Clifton Suspension Bridge, UK**: As an illustrative example of the structural geology–rock engineering combined approach to a rock stability assessment, the bridge foundation rock is described so that the engineering stability can be evaluated. |
| 5 | **Quarries**: The support that structural geology provides in the range of rock engineering applications begins with descriptions of four quarries and the interaction between the rock properties and the excavation process: a granite quarry in Poland, a limestone quarry in the UK, a granodiorite quarry in the UK and a marble quarry in Italy. |
| 6 | **Dams**: This Chapter continues the theme of linking the geological setting and the engineering objective by describing three dam stability case examples: two hydroelectric dams in Brazil and a water storage dam in Scotland. |
| 7 | **Opencast Coal Mining**: This UK example illustrates the significant effect of glacial disturbance on the near-surface geology and its influence on opencast coal mining operations. |
| 8 | **Underground Rock Engineering**: In this Chapter, we discuss the subject of risk (epistemic and aleatory) in the context of shale gas extraction, carbon dioxide storage and radioactive waste disposal. In all three cases, there is an intimate relation between the geological conditions and the engineering objectives. |

(*Continued*)

Table 1.1. (*Continued*)

| Chpt. No. | Chapter title and content |
| --- | --- |
| 9 | **Historical Monuments and Stone Buildings:** In this Chapter, we consider a range of historical monuments and stone buildings in the context of their geological origin and their engineering longevity. |
| 10 | **Concluding Discussion:** In the light of all the information in the preceding chapters, the text summarises some of the key factors and the importance of the combined structural geology–rock engineering approach. |
| 11 | **Epilogue:** As a final Chapter, the value of the combined approach is described as it applied, surprisingly, to establishing the authenticity of the UK Natural History Museum's fossil *Archaeopteryx lithographica*. |
|  | **References and Bibliography:** This Section contains the references used throughout the book's text, but there are also some bibliographic references included which readers might find useful. |

Consider the small tunnel in a limestone formation shown in Figure 1.1. This is a minor rock engineering project, but the photograph illustrates important points.

The tunnel has been driven through limestone beds that are gently inclined and contain strata-bound fractures perpendicular to the bedding — which break the rock mass into clearly defined blocks and which significantly affect the mechanical integrity of the rock mass. The enhanced water flow that results from this fracturing is indicated by the vegetation which is concentrated along the fractures. Failure of the limestone is visually evident at the tunnel portal where the existence of the bedding planes and cross-bedding fractures has allowed portions of the rock blocks to be released. Another aspect of the rock mechanics is illustrated by Figure 1.1: the rock stress component acting horizontally in the tunnel vicinity will be essentially zero because of the free surface formed by the cliff face to the right of the rock mass.

An example of the significant effect that such fractures can have on the stability of large-scale engineering projects is shown by the open-pit mining example in Figures 1.2a and 1.2b. Two large-scale, adversely orientated fractures/weakness zones have created a large rock wedge which has been able to slide when mining had proceeded to sufficient depth to release the rock block. The wedge failure was anticipated through measurements of

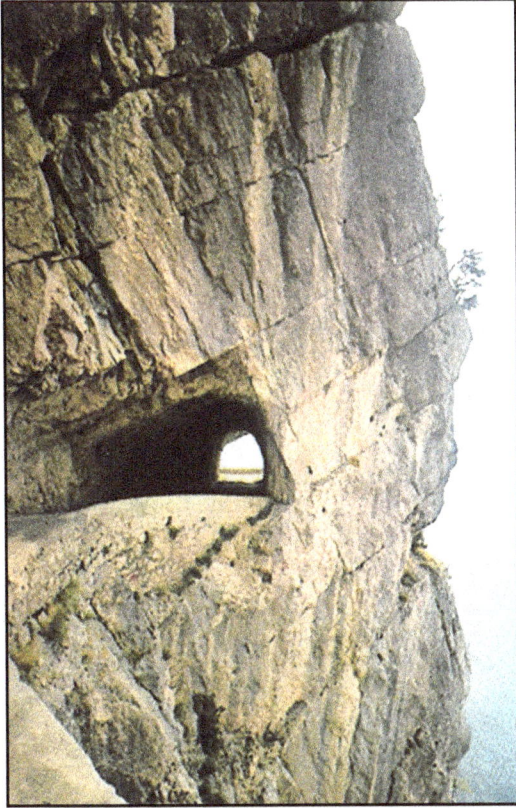

Figure 1.1.   Short tunnel in a limestone formation — illustrating the intimate relation between structural geology and rock engineering.

the mine wall movements, thus enabling the photographs to be taken when failure occurred. Such a major rock slide in an open-pit mine has significant risk and financial implications for the future; however, a fuller understanding of the structural geology of the site could possibly have resulted in a re-design of the open-pit geometry to minimise or eliminate such failure events.

In fact, the main aspect that is of concern in rock engineering is the possibility of collapse — which is primarily caused by two factors: the release of rock blocks and adversely high rock stress components. These are illustrated in Figure 1.3.

(a)

(b)

Figure 1.2.    Large wedge failure at the Teutonic Bore Mine, Western Australia, caused by two major fractures/weakness zones creating the wedge which slid into the open-pit mine. (a) The wedge beginning to form; (b) the wedge accelerating and creating dust which escapes along other pre-existing fractures. The scale of this event can be seen from the size of the haulage roads.

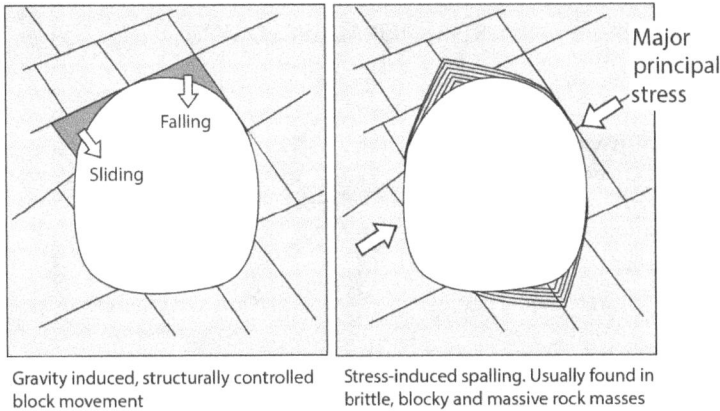

Gravity induced, structurally controlled block movement

Stress-induced spalling. Usually found in brittle, blocky and massive rock masses

Figure 1.3. The two main causes of failure in rock engineering: the movement of rock blocks and adversely high rock stresses (from D. Martin).

## 1.3.2 The dual roles of structural geology and rock engineering

The rock mechanics principles that have been developed in the two subjects of structural geology and rock engineering over the last 50 years or so serve to support two activities.

— The interpretation of past mechanisms that have led to the current rock mass deformations and fracturing, i.e., structural geology considerations; plus
— the ability to predict the rock mass behaviour when designing structures to be built on or in rock masses, i.e., rock engineering design.

This results in the different but complementary roles of structural geology and rock engineering in interpreting past events and in predicting future events, Figure 1.4.

The upper and lower rows in Figure 1.4 indicate the respective activities in structural geology and rock engineering for interpreting past events and predicting future events. The key difference between the two sets of activities is that structural geology is concerned with natural events, whereas rock engineering is concerned with the perturbations induced by engineering activity. The forces of nature have deformed and fractured the rock over

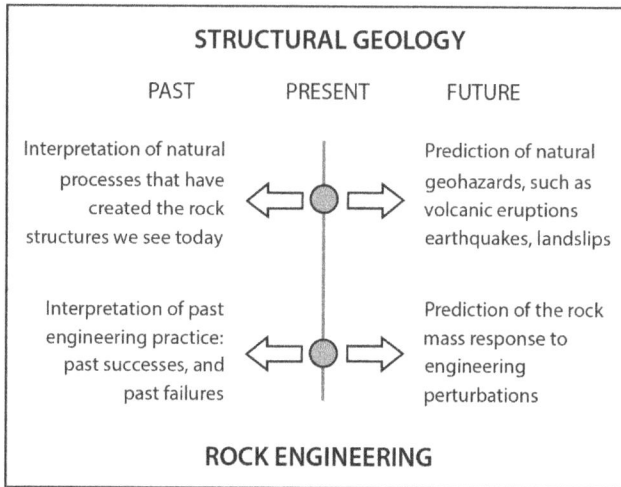

Figure 1.4. The dual roles of structural geology (upper row) and rock engineering (lower row) in terms of the interpretation of past events and the prediction of future events.

millions of years and the interpretation of the associated mechanisms through structural geology assists in the prediction of natural geohazards which may occur in the future. Similarly, the perturbations introduced by engineers as a result of drilling and blasting have enabled surface and underground structures, such as rock slopes and caverns, to be constructed.

With hindsight, we can establish whether such engineered structures have been successful and conduct forensic investigations if they have not been successful. This knowledge then enables better engineering design through the ability to predict the future: what will happen if a tunnel is driven in a particular rock mass in a particular direction at a particular depth? This ability to predict the future is crucial for rock engineering design; without it, there is no coherent basis for such design. So, it can be seen from Figure 1.4 that both the structural geology and rock engineering subjects have similar roles with regard to the interpretation of the past events and the prediction of future events and, most importantly, they both depend crucially on an understanding of the respective principles — which we outline in Chapters 2 and 3.

Despite these intimate links, the principles of rock mechanics for structural geology and the principles of rock mechanics for engineering are usually presented in separate books and via separate university courses. Also, the integration of the subjects may be lost during the site investigation for an engineering project where those involved will have different educational backgrounds. Furthermore, research journals (albeit with some exceptions) tend to be directed either towards the structural geology context or the engineering context. This book is an attempt to counter this tendency to separate the two topics by presenting a synthesis of the two subjects. The structural geologist has a great deal to contribute to the rock engineer; and the rock engineer cannot ensure safe design and construction without this knowledge. So, as Professors of structural geology and engineering rock mechanics in the (integrated) Department of Earth Science and Engineering at Imperial College London, UK, we have combined our experiences in teaching, research and consulting to provide this book describing the synthesis of structural geology and rock engineering.

### 1.3.3   An illustrative example of the link between structural geology and rock engineering: the pre-split blasting technique for creating a planar rock face

To provide an initial example of the intimate relation between geology and engineering, in this Section we describe the technique of pre-splitting and how this can be affected by the geological characteristics of the rock mass. (The use of pre-splitting in a granite quarry in Poland is also described in Chapter 5.)

#### 1.3.3.1 *Introduction to the pre-split blasting technique*

The interaction between structural geology and rock engineering is well illustrated by the blasting procedure known as 'pre-splitting'. The purpose of this blasting technique is to create a planar rock face, as might be required for a stable, permanent slope adjacent to a highway or in an open-pit mine. The technique consists of drilling a series of closely spaced, parallel blastholes, charging them with a low level of explosive, and then detonating all the explosives simultaneously, see Figure 1.5. A plan view of three pre-splitting blastholes within a larger set is shown in Figure 1.5(a). On detonation, high intensity compressional stress waves travel out from

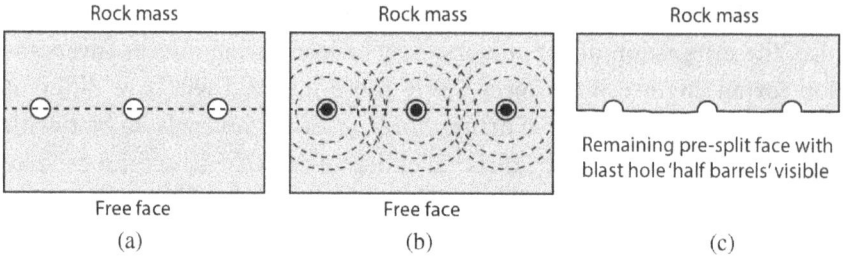

Figure 1.5.   (a) Overhead view of a horizontal rock face and the geometry of vertical blastholes (the circles along the dashed line) for a pre-splitting operation; (b) the action of the stress wave and gas pressure (dotted circles) in fracturing the rock and creating a new fracture between the blastholes along the dashed line; (c) the rock between the original free face and the new fracture along the line of the blastholes is then removed — leaving the resulting planar rock face with the 'half-barrels' of the blastholes visible.

the blastholes, Figure 1.5(b), and reflect off the adjacent blastholes as tensile waves — causing a crack to be formed along the line of the blastholes, which is further developed by the gas pressure from the explosive. After the broken rock is removed, Figure 1.5(c), assuming this is the lower part in the Figure, a planar face is produced with the 'half-barrels' of the blastholes visible on the face, see Figure 1.5(c).

Three aspects of the pre-splitting operation can cause irregularities in the resultant rock face. One is the accuracy of the borehole drilling, given that the boreholes should be parallel; the second is the use of a sufficient level of explosive to activate the pre-splitting failure mode, yet not so much that the rock would be shattered immediately adjacent to the borehole; the third is the geological nature of the rock mass, i.e., the characteristics of the intact rock and the pre-existing fractures. All these aspects are illustrated in the Figures in the following sub-sections.

### 1.3.3.2  *Examples of pre-split rock faces*

Four examples of pre-splitting the limestone rock faces along the A75 route north of Montpellier in the south of France are shown in Figures 1.6–1.9. An excellent example of successful pre-splitting in shown by the three rock benches in Figure 1.6. Note the tracery of fine vertical lines on the rock surfaces: these are the 'half-barrels' of the blastholes (Figure 1.5) which are

Figure 1.6.   Example of three roadside benches of planar rock faces created by successful pre-split blasting. The tracery of thin vertical lines is created by the half-barrels remaining from the explosive boreholes. (Limestone strata: A75 route north of Montpellier, south of France.)

now visible after the rock that was previously in front of them has been removed. The photograph clearly shows that the blastholes were drilled in a parallel pattern — which is essential for generating a planar, pre-split face. However, for the case in Figure 1.8 and despite the parallel blastholes, the resultant rock face is more irregular, especially in the upper bench. This has been caused by the increased pre-existing fracturing in these strata which has interacted with the pre-split blasting process.

This effect of the rock fracturing and its influence on the pre-split blasting is even more pronounced in Figure 1.9 in which it is difficult to even identify the remains of any blasthole half-barrels. If the strata are inclined

Figure 1.7. Roadside rock face created by pre-split blasting. Here, the parallel array of visible half-barrels indicates that these non-vertical blastholes have been angled to be perpendicular to the strata in order to avoid adverse blasting interactions with the strata boundaries. (Limestone strata: A75 route north of Montpellier, south of France.)

Figure 1.8. Roadside rock face created by pre-split blasting. Note how the fractured nature of the strata in the upper bench has caused the face to be more irregular than the faces in Figure 1.6. (Limestone strata: A75 route north of Montpellier, south of France.)

Figure 1.9.    Roadside rock face created by pre-split blasting. In this case, the dense fracturing of the strata has caused the face to be much more irregular — to such an extent that the half-barrels are hardly visible. (Limestone strata: A75 route north of Montpellier, south of France.)

to the horizontal, Figure 1.7, the pre-split blastholes should still be drilled perpendicular to the strata in order to optimise the blasting mechanism and avoid adverse additional fracturing induced by asymmetrical stress waves.

Another example is shown in Figure 1.10. This is a pre-split rock face in the Pre-Cambrian granitoid basement in Sweden where the pre-existing fractures, together with the newly created surfaces, have enabled tetrahedral rock blocks to form and fall or slide out from the face — leading to an irregular pre-split surface. In Figure 1.11, the effects of the bedrock characteristics on a pre-split rockface in Finland are clearly visible.

A similar example is evident in the underground Temppeliaukio Church (in the Töölö neighbourhood of Helsinki) designed by Timo and Tuomo Suomalainen and constructed in 1968 and 1969, see Figures 1.12(a)–1.12(c). The church was excavated in foliated and migmatised (partially melted) gneissic rock.

In Figure 1.12(b), notice the fact that some of the pre-split half-barrels are missing and that there are horizontal exfoliation joints present, see Figure 2.164. These exfoliation joints were caused by high horizontal stress components as the overlying rock was removed by erosion and tend to be sub-parallel to the ground surface, see Section 2.14.4. These joints have contributed to block fallout where the half-barrels are missing, Figure 1.12(c).

Figure 1.10. Interaction between blasting and pre-existing fractures at a pre-split rock face (Pre-Cambrian granitoid basement in Sweden). The well-developed pre-existing fractures (examples indicated by the white arrows) have caused local rock block fallout where three pre-existing fractures plus the new free face have caused tetrahedral rock blocks to form.

### 1.3.3.3 *Pre-splitting problems*

As illustrated already, the pre-splitting operation is adversely affected if the rock mass is significantly fractured and/or the pre-split blastholes are not drilled in a parallel array, either because of geological variations or simply because the drilling procedures were not adequate. Examples of pre-split faces in hard rock which have suffered these problems are illustrated in Figure 1.13.

### 1.3.3.4 *Understanding the mechanism of rock fracturing during the pre-splitting operation*

Research has been conducted for many years into the pre-splitting mechanism, i.e., to explain the fracturing caused by the stress wave, the gas pressure and the interaction between the two. As early as 1967, Kutter and

Figure 1.11.   Fractures in the rock mass affecting the pre-split blasting operation cause a rough, but acceptable, rock face in the Finnish bedrock (adjacent to Highway E18 west of Helsinki towards Turku, Finland).

Fairhurst (1967) discussed this subject in a paper presented at the 9th U.S. Symposium on Rock Mechanics and titled "The Roles of Stress Wave and Gas Pressure in Pre-splitting". They explained that, "The main role of the [stress] wave is to generate a densely radially fractured zone around the borehole. The subsequent gas pressure, exerted against the walls of a highly (radially) fractured cavity, generates a quasi-static stress field" and that, "under gas pressure, the longest of the radial cracks would extend first, and extension of two diametrically opposed radial cracks requires the least pressure."

Worsey (1981), Worsey *et al.* (1981) and Matheson (1983) further described and explained the pre-splitting mechanism, emphasising that, for a successful pre-split blasting operation, it is essential that the blast-holes are drilled parallel to each other, that the explosive is not coupled

(a)

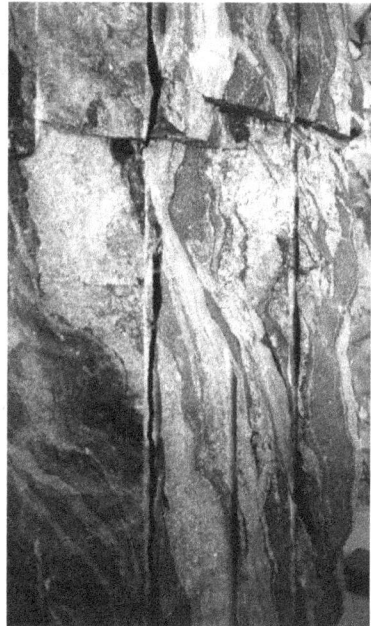

(b)                                          (c)

Figure 1.12. The underground Temppeliaukio Church in the Töölö neighbourhood of Helsinki. (a) View of the church walls in the foliated and fractured Pre-Cambrian bedrock; (b) the pre-splitting technique used to excavate the church; (c) evidence of block fallout (above the central pre-split blasting half-barrel) caused by pre-existing rock fractures.

directly to the rock (which would reduce the magnitude of the blast stress wave) and that the explosives in the holes are detonated simultaneously. Lopez Jimeno *et al.* (1995) in their book on "Drilling and Blasting of Rocks" provide detailed explanations of 'contour blasting' in their

(a)                          (b)                          (c)

Figure 1.13.   Examples of problems with pre-split faces in hard rock: (a) Rock wedge (top left of the photograph) formed by two pre-existing fractures; (b) the blasting process has been severely affected by insufficient explosive and significant fracturing; and (c) an extreme example of problems caused by non-parallel blastholes and major geological discontinuities.

Chapter 25, including the effects of 'rock stratification' on the geometry of the blasted surface.

In Figure 1.14, examples of the potential effects of strata boundaries and/or fractures on the geometry of the pre-split plane are shown in the left and central examples. If there is a high *in situ* stress field, the main blast induced fractures could all be developed parallel (Figure 1.14(c)) and adversely orientated in terms of the required pre-split plane when the fracturing occurs perpendicular to the minor principal stress, i.e., following the least work principle, see Section 2.3.3 and Figure 2.19.

### 1.3.3.5 *A final note on pre-splitting*

An interesting example of a pre-split plane using just an expanding cement (to protect adjacent buildings) in a line of boreholes is illustrated in Figure 1.15. In this case, there was no blasting *per se*, just the pressure caused by the expanding cement — yet a perfect pre-split plane was created.

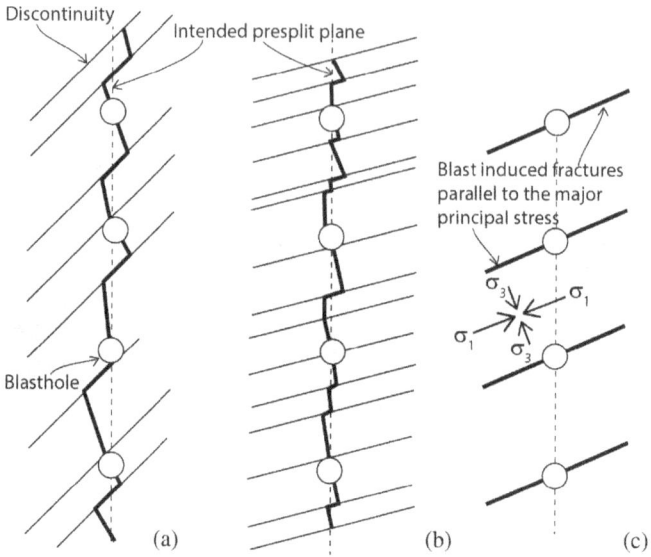

Figure 1.14. The influence of pre-existing fractures ((a) and (b)) and a strong *in situ* stress field (c) on the blasted pre-split face geometry, from Hudson and Harrison (1997).

Figure 1.15. Rock excavation using the pre-splitting technique but with just an expanding cement replacing the explosive, Scotland.

## 1.4 CHAPTER SUMMARY

The Chapter began with an explanation of the motivation for this book and its associated content, the purpose being to outline the principles of both structural geology and rock engineering (the latter via engineering rock mechanics) and to explain with illustrative examples why the synthesis of the two subjects is mutually beneficial — primarily because it is essential to understand the structural geology setting for rock engineering projects. To provide the necessary background, the principles of structural geology and engineering rock mechanics are provided next in Chapters 2 and 3, respectively. Chapter 4 then provides an example demonstrating why the synthesis of the two subjects is of mutual benefit. Chapters 5–9 include demonstration examples of engineering projects in rock masses in which a knowledge of structural geology is beneficial — if not essential. There is also a concluding discussion in Chapter 10, followed by an epilogue and the references/bibliography Section.

The introductory review of the synthesis of the structural geology and rock engineering was illustrated via a tunnel example and an open-pit slope failure example. The interpretation of past events and the prediction of future events for both subjects were highlighted. Finally, the specific example of pre-split blasting during engineering in rock masses was included and illustrated to demonstrate the interplay between geology and engineering through a series of photographs showing the increasing influence of rock fractures on the pre-split blasting operation.

The following two Chapters outline the principles of structural geology and rock mechanics. We have included these next because it is essential to have a background in these principles in order to understand and formulate the necessary synthesis for any given rock engineering project. For those readers who are new to the subjects, we have endeavoured to provide clear explanations so that the principles are directly understandable. Once these principles are absorbed, the content of the succeeding chapters concerning the engineering examples illustrating the synthesis between the two subjects will be readily comprehensible.

# CHAPTER 2

# STRUCTURAL GEOLOGY PRINCIPLES

## 2.1 INTRODUCTION

A large amount of structural geology research and field observations relates to the natural fractures in rock masses; also, many problems encountered in rock mechanics and rock engineering involve understanding and modelling rock fractures. The fractures are important because the properties of fractured rock masses can differ considerably from those of an intact rock sample. Because of this, it is essential that there is an understanding of:

(a) how fractured rock masses have been generated in response to the various stress fields that the Earth's crust has been subjected to over geological time. (as opposed to the fractures induced during the construction of an engineering project); and
(b) which properties the induced fracture network imparts to the rock (in terms of permeability, porosity, strength etc.).

With this in mind, the formation of fractures in the crust is now described in this Chapter, together with an explanation of the factors that control their type, regularity, and interaction to form fracture networks. Having described the fracturing of the crust and the generation of fracture networks through the superposition of individual fracture sets, the properties of these networks and their impact on the mechanical behaviour of the fractured rock mass is considered.

Fractures form in response to stress and so the chapter opens with a discussion of the factors that determine the stress field in the Earth's crust: namely the overburden stress and the stresses generated by plate tectonics. This is followed by a discussion of brittle failure which establishes the fundamental links between stress and fracture formation and enables one to predict not only the fractures that will occur naturally in response to the overburden and tectonic stresses but also the fractures induced by engineering activity, especially by the alteration of the natural stress field as a result of the removal of rock during engineering activities.

Finally, the way in which a series of fracture sets is linked to the succession of stress fields that affect the crust over geological time combine to form fracture networks is discussed, as well as their geometries and properties. Armed with this understanding, the rock engineer will be in a more powerful position to understand the likely behaviour of the fractured rock mass with which he or she is working and to use this in a predictive manner.

So the aim of this Chapter is to summarise the current understanding of the mechanisms and rock parameters that control brittle failure in rocks, see the overview in Figure 2.1. Particular attention is paid to the following topics:

— the link between stress and fracturing (i.e., the link between fracture type, orientation, regularity, etc., and stress);
— the types of fractures that can occur (extensional and shear);
— the relation between discrete fractures, diffuse fractures and fracture zones (the concept of a homogeneity index and brittleness);
— the small-scale processes linked to fracturing (process zones);
— fracture propagation (and the process of linking of fractures);
— the evolution of fracture sets (the concepts of under-saturation, saturation and super-saturation);
— fracture distribution (on one scale, i.e., controls on fracture clustering);

— fracture distribution (across all scales, i.e., fracture frequency/size distribution, scale dependency and self-similarity); and
— second-order structures linked to larger fractures (kinematic indicators).

Such a study needs to draw on the literature and understanding from a number of disciplines that relate to fracture development in rocks including:

— structural geology,
— tectonics,
— rock mechanics,
— fracture mechanics,
— engineering geology,
— fracture mineralisation,
— hydrology,
— hydrocarbon geology.

The hydrocarbon industry has decades of experience in obtaining information about fractures, including, *inter alia*, their geometry, orientation, frequency, length, aperture, distribution, conductivity, the information having been obtained from borehole cores, borehole scans and a variety of geophysical investigative techniques.

The remainder of this Chapter is divided into a series of short sections, each considering one of the topics listed earlier as follows.

— **Section 2.2**: The principal causes of stress in the crust are discussed and the link between stress, fracture type and fracture orientation is summarised.
— **Section 2.3**: The factors that control the expression of brittle failure in rocks are discussed.
— **Section 2.4**: The state of stress in the Earth's crust is described, together with its role in determining the distribution, orientation and type of fractures that form.
— **Section 2.5**: The important role of fluid pressure in fracture formation is described.

(a) $\sigma_1$  (b) $\sigma_1$  (c) $\sigma_1$  (d) $\sigma_1$  (e) $\sigma_1$  (f) $\sigma_1$

$\sigma_3 \rightarrow$   $\sigma_2$   $\leftarrow \sigma_3$

$\sigma_1$ | $\sigma_1$ | $\sigma_1$ | $\sigma_1$ | $\sigma_1$ | $\sigma_1$

| UNDEFORMED INTACT ROCK | TENSILE FRACTURES | SHEAR FRACTURES | BRITTLE SHEAR ZONES | DUCTILE SHEAR ZONES | PERVASIVE DUCTILE FABRIC |

Decreasing LOCALISATION of deformation →

Increasing PERVASIVENESS of deformation →

Increasing PRESSURE and TEMPERATURE →

Decreasing STRAIN RATE →

CONTINUITY (a)  DISCONTINUITY (b - d)  CONTINUITY (e - f) →

HOMOGENEOUS  INHOMOGENEOUS  HOMOGENEOUS →

Rock Mechanics Characterisation Capability

| Straight forward for small samples | Difficult requires bulk properties | Straight forward from small samples, rock is transversely anisotropic |

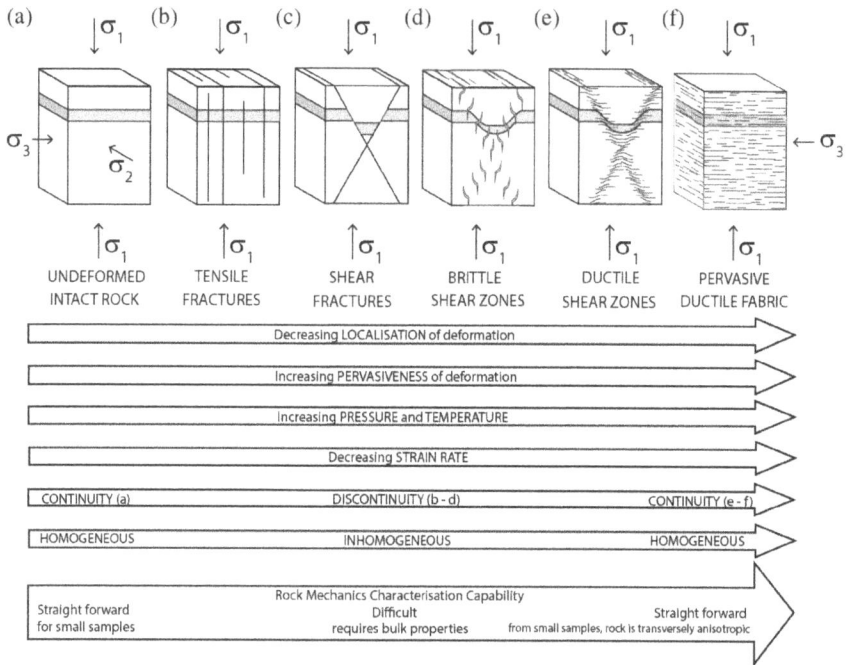

Figure 2.1.  The variety of deformation features, both brittle and ductile, that characterise many rocks. Each is generated during a single deformation event and the result of many such episodes of these types of deformation leads to the superposition of the structures and the formation of a fractured and deformed rock mass.

— **Section 2.6**: The small-scale processes linked to fracturing and the processes of fracture propagation are reviewed.

— **Section 2.7**: The features that enable one to differentiate between extensional fractures and shear fractures in the field are outlined.

— **Sections 2.8** and **2.9**: The controls on fracture spacing and on orientation and regularity are discussed.

— **Section 2.10**: The spectrum between brittle and ductile structures is described.

— **Section 2.11**: The way in which even the simplest stress regime in the crust can generate a complex fracture pattern is explained.

— **Section 2.12**: Noting that many rocks contain fracture networks which dominate the mechanical properties and deformational behaviour of the resulting fractured rock masses, the manner in which such fracture networks are built up by the superposition of individual fracture sets, each linked to a different geological stress regime, is explained. It is also shown how a detailed study of the networks can reveal the order in which the fracture sets were superimposed (a technique known as fracture analysis) and therefore reveal the chronology and orientation of stress regimes affecting the rock over geological time.

— **Section 2.13**: The properties of the fracture networks are also considered and it is shown that some of their most important mechanical properties, such as connectivity and conductivity, are scale-dependent.

— **Section 2.14**: The different scales of the various stress fields operating in the crust are discussed. They range from regional stresses linked to plate motion and which operate over thousands of square kilometres to highly localised stress fields formed at grain contacts. This Section also includes a discussion of the way in which ancient stress regimes can be locked into the rocks and released later as the rocks are brought to the surface during exhumation. This leads to the formation of exfoliation fractures which form at or very close to the Earth's surface and which can have a significant impact on the stability of the rock mass.

The structural geology principles described in the following sections of this Chapter thus provide the background to the influence that the geological setting can have on the design and construction stability of rock engineering projects. Following Chapter 3, which introduces the companion rock mechanics principles, in Chapter 4 and the succeeding chapters we present case examples explicitly illustrating the importance of understanding the geological setting when embarking on an engineering project.

## 2.2 UNDERSTANDING ROCK STRESS

Both structural geologists and rock engineers are concerned with the deformational structures (e.g., folds, rock fabrics, and fractures) that form in the Earth's crust. They form in response to **stress** which, as discussed in Section 2.14, can be generated by a variety of phenomena and can occur on a variety of scales. However, the state of stress in the crust is primarily the result of two factors: namely the overburden load and tectonic plate motion. Before discussing the state of stress and the related structures that form in response to it, the concept of stress is explained. The aim is to provide the reader with an insight into stress and some of the physical implications of the stress tensor[1] most relevant to both the structural geologist and the rock engineer.

Quantities observed in nature can be classified into categories depending on the number of parameters needed to define them. Thus, the simplest quantities (scalar quantities or zero rank tensors), such as mass or length, require just one parameter to completely define them. Other quantities, such as force, require three parameters (magnitude and direction); these quantities are termed vectors (first rank tensors). The more complex quantities of stress and strain require extra parameters to define them (second rank tensors). Because force is intimately related to stress but is a simpler quantity to understand, we will consider force first.

### 2.2.1 Types of forces

Forces can be divided into two types, **body forces** and **surface forces,** depending on their mode of origin. Body forces are forces that exist within a body as a consequence of it being situated in a potential field. Gravitational and electromagnetic forces are examples of this type. In contrast, surface forces are the result of forces applied to the boundaries of the body. Body forces act throughout the volume of a body simply because it is situated in a potential field. Figure 2.2 shows a body in a gravitational field. The intrinsic property of matter that interacts with gravity and determines the magnitude of the body force generated is mass, and a measure of the body force generated in the body is given by its weight. Unlike weight, mass is constant and independent of the magnitude of the gravitational field in which the body is situated. Thus, because of the different magnitudes of the gravitational fields on the Moon and the Earth,

---

[1] The word 'tensor' refers to a mathematical representation used to describe physical properties: a scalar is a zero rank tensor requiring one value for its specification (e.g., temperature), and a vector is a first rank tensor requiring three values for its specification (e.g., velocity in three dimensions). Stress is a second rank tensor requiring six components for its specification: three principal values and their directions.

There are two types of force (a) **BODY FORCE** & (b) **SURFACE FORCE**

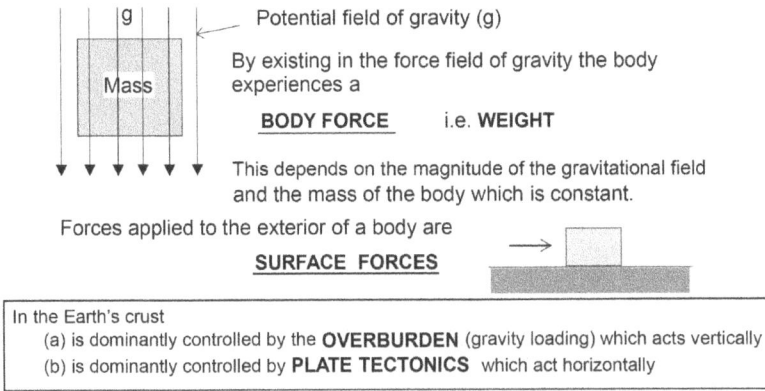

Potential field of gravity (g)

By existing in the force field of gravity the body experiences a

**BODY FORCE**    i.e. **WEIGHT**

This depends on the magnitude of the gravitational field and the mass of the body which is constant.

Forces applied to the exterior of a body are

**SURFACE FORCES**

In the Earth's crust
(a) is dominantly controlled by the **OVERBURDEN** (gravity loading) which acts vertically
(b) is dominantly controlled by **PLATE TECTONICS** which act horizontally

Figure. 2.2.   An illustration of body force and surface force.

the weight of an object will be different in the two locations, although the mass will be the same.

In contrast to body forces, contact or surface forces act at the point of contact between two objects. A surface force is a force exerted on the surface of an object and is a force that acts across an internal element in a material body or on an external surface (Figure 2.2).

Two of the most important mechanisms by which the Earth's crust is stressed are gravitational loading, (i.e., by the body force acting on a rock mass as the result of the overburden — which produces a vertical loading) and tectonic plate motion. The horizontal motion of plates causes collisions at plate margins and the generation of surface forces.

## 2.2.2 The resolution of forces

Force is a vector quantity, i.e., a quantity that requires two parameters in two dimensions for its definition (magnitude and direction). Thus, it is distinguished from scalar quantities such as length and mass which can be completely described by one parameter. Figure 2.3 shows a body subjected to a single force ($F$).

The state of force on any plane in the material, inclined at an angle $\theta$ to the applied force, can be determined by resolving the force into

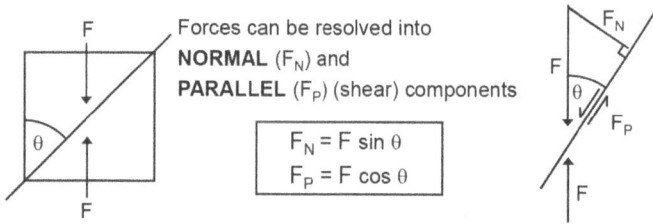

Figure 2.3.   The resolution of a force into its normal and shear force components.

components normal and parallel to the plane, i.e., $F_N$ (Equation (2.1)) and $F_P$ (Equation (2.2)) respectively as shown in Figure 2.3.

$$F_N = F \sin\theta, \tag{2.1}$$
$$F_P = F \cos\theta, \tag{2.2}$$

Thus, if the applied force $F$ is known, then it is a simple matter to determine the state of the normal and shear forces on any plane of interest. These equations apply to a uniaxial force field but can be easily extended for biaxial and triaxial force fields.

An analysis of the state of force within the Earth's crust caused by the overburden load is therefore relatively simple. Nevertheless, geologists and rock engineers almost invariably consider the state of stress in the crust rather than the state of force when they are attempting to understand past deformations or predict likely deformations in the future. The reason for considering stress rather than force, despite the former being a more complex quantity, is apparent from Figure 2.4 which shows two experiments, both involving the uniaxial loading of a rock. In both experiments the same rock type is subjected to the same force, but the response is very different. In Experiment A, the rock responds to the applied force by a small amount of elastic deformation. This deformation is reversible and the material returns to its pre-loading state once the force is removed. In Experiment B, the response of the rock to the applied force is different. The material is loaded beyond its elastic limit and fails catastrophically in a brittle manner. The deformation is permanent and is not recovered if the load is removed.

The only difference between the two experiments is the force per unit area, which is defined as the stress. Clearly, knowledge of the rock type and the applied force is not sufficient to enable the deformation to be predicted. It is because stress is a much more appropriate indicator of deformation than force that earth scientists focus on the former in their attempts to predict the deformation of the crust. With this explanation in mind, let us now develop the force analysis illustrated in Figure 2.3 into a stress analysis.

### 2.2.3 The resolution of stress and the stress equations

The force, $F$, representing the uniaxial force field and the components of normal force and shear force, $F_N$ and $F_P$ respectively, are re-expressed as stresses by dividing them by the area over which they act. The area of the top surface of the block shown in Figure 2.5 is ($a$) and the area of the surface of interest in the material inclined at an angle $\theta$ to the applied force and onto which the stress is being resolved is ($a'$).

The normal stress acting on the top surface of the block in the z direction is given by $F/a$ and the symbol used for normal stress is the Greek lower case sigma ($\sigma_z$) where the subscript indicates the direction in which the stress is acting. There is no applied component of shear force and therefore no shear stress parallel to the top surface of the block. By definition, any normal stress acting on a surface along which there is no shear stress, is a **principal stress** and any stress field can be defined by three mutually perpendicular principal stresses, termed $\sigma_1$, $\sigma_2$, and $\sigma_3$, and their directions. In the example in Figure 2.5 of a uniaxial stress field generated by the

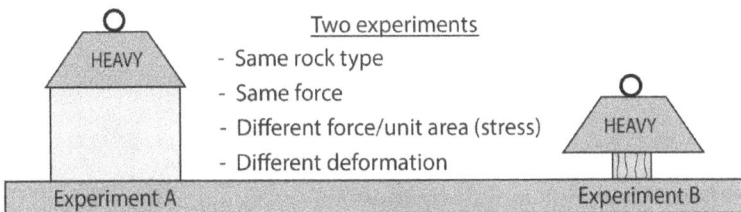

Figure 2.4. An illustration of the difference between force and stress. Both experiments utilise the same rock and the same force, but the type of deformation is different (A, small, elastic reversible deformation; B, large, irreversible deformation).

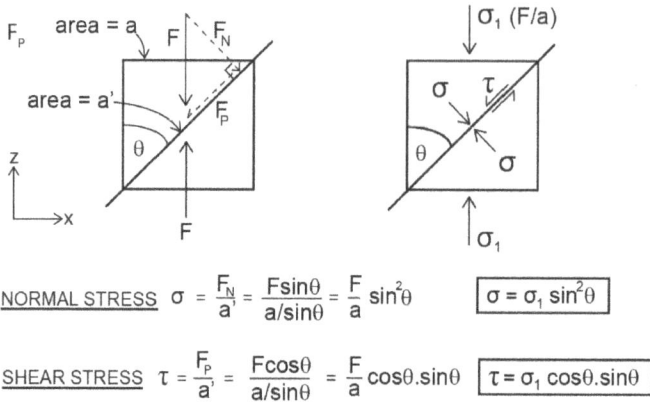

NORMAL STRESS $\quad \sigma = \dfrac{F_N}{a'} = \dfrac{F\sin\theta}{a/\sin\theta} = \dfrac{F}{a}\sin^2\theta \qquad \boxed{\sigma = \sigma_1 \sin^2\theta}$

SHEAR STRESS $\quad \tau = \dfrac{F_P}{a'} = \dfrac{F\cos\theta}{a/\sin\theta} = \dfrac{F}{a}\cos\theta.\sin\theta \qquad \boxed{\tau = \sigma_1 \cos\theta.\sin\theta}$

Figure 2.5.   The resolution of stress in a uniaxial stress field and the derivation of the uniaxial stress equations.

uniaxial force $F$, two of the principal stresses are zero. It follows that $\sigma_z$ is a principal stress and should be written as $\sigma_1$ as shown in Equation (2.3).

$$F/a = \sigma_z = \sigma_1. \tag{2.3}$$

The area $a'$ over which the components $F_N$ and $F_P$ act can be expressed in terms of the area of the top of the block $a$ and the angle $\theta$ of the plane of interest with respect to the applied principal stress (Equation (2.4)).

$$a' = a/\sin\theta. \tag{2.4}$$

To obtain the components of normal and shear stress on the plane of interest, the components of force (Equations (2.1) and (2.2)) are divided by $a'$ as shown by Equation (2.5).

$$\sigma = (F_N \sin\theta)/a' = (F \sin\theta)/(a/\sin\theta) = (F/a)\sin^2\theta. \tag{2.5}$$

Substituting Equation (2.3) into Equation (2.5), gives Equation (2.6).

$$\sigma = \sigma_1 \sin^2\theta. \tag{2.6}$$

Similarly the component of shear stress ($\tau$) acting along the plane of interest is given by the shear force $F_p$ (Equation (2.2)) divided by the area over which it acts $a'$ (Equation (2.4)) as shown in Equation (2.7).

$$\tau = (F_p \cos\theta)/a' = (F\cos\theta)/(a/\sin\theta) = (F/a)\cos\theta \cdot \sin\theta. \tag{2.7}$$

Substituting Equation (2.3) into Equation (2.7) gives Equation (2.8).

$$\tau = \sigma_1 \cos\theta \sin\theta. \tag{2.8}$$

Equations (2.6) and (2.8) are termed the uniaxial stress equations and they allow the state of normal and shear stress on any plane inclined at $\theta$ to the principal stress to be calculated.

Similar equations can be derived for a biaxial stress field defined by two principal stresses $\sigma_1$ and $\sigma_3$. These are Equations (2.9) and (2.10).

$$\sigma = \sigma_1 \sin^2\theta + \sigma_3 \cos^2\theta, \tag{2.9}$$

$$\tau = (\sigma_1 - \sigma_3) \cos\theta \sin\theta. \tag{2.10}$$

Having derived these equations, it is instructive to explore their implications as these provide an insight into the properties of stress.

## 2.2.4 The implications of the stress equations

Equation (2.9) gives the relation between the magnitude of the normal stress on a plane and the orientation $\theta$ of the plane. If the normal stress on planes oriented at various angles $\theta$ to $\sigma_1$ are represented by arrows whose orientations and lengths are proportional to the orientation and magnitude of the normal stress, an ellipse is delineated, see Figure 2.6. This is the 'stress ellipse' and it can be seen that the two principal stresses define the major and minor axes. Thus, the stress ellipse (stress ellipsoid if we were considering a triaxial stress field) springs naturally from the stress equation and describes the variation of normal stress associated with the stress field.

The physical implications of Equation (2.10), which describes the variation of shear stress with direction, can be conveniently demonstrated by plotting the equation graphically, Figure 2.6. As noted earlier, on planes normal to the principal stresses, $\theta = 0°$ and $90°$, the shear stress is zero. The maximum shear stresses occur on two planes, oriented at $\theta = 45°$ and $135°$, i.e., a conjugate set symmetrically arranged around the maximum principal stress, Figure 2.6. Note that the magnitudes of the maximum shear stresses on these two planes are the same, but the sign (i.e., the sense of shear) is different. For the plane $\theta = 45°$, the sense of shear is sinistral (i.e., looking across the plane the opposite side moves to the left) and, for the plane $\theta = 135°$, the sense of shear is dextral.

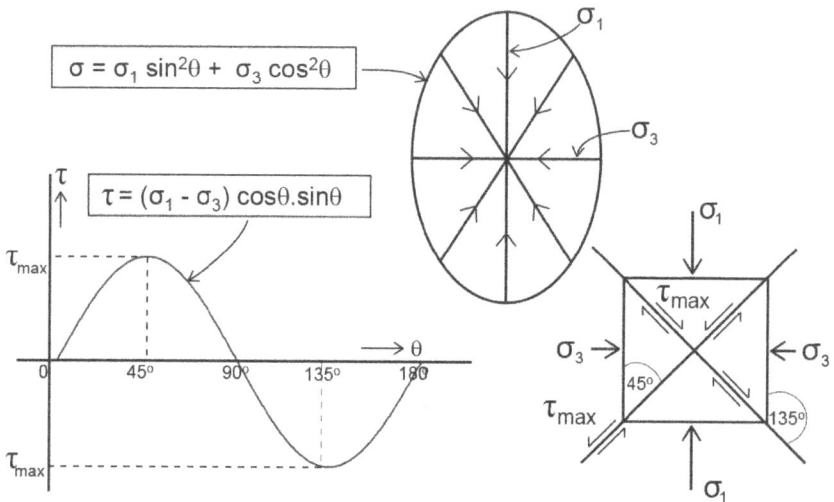

Figure 2.6. The graphical expression of the biaxial stress equations (Equations (2.9) and (2.10)) illustrating the physical expression of the stress ellipse and showing that the planes of maximum shear stress occur at $45°$ either side of the maximum principal compression $\sigma_1$.

As is noted in Section 2.3 on brittle failure, shear failure generally occurs along conjugate planes symmetrically arranged around the maximum principal compression. However, the shear fracture planes do not coincide exactly with the planes of maximum shear stress. Typically for rocks, the shear fractures form at an angle of approximately $30°$ either side of $\sigma_1$, rather than $45°$. The reason for this discrepancy is discussed in Section 2.3.2, Equation (2.15).

## 2.2.5 Mohr stress circle

Having discussed the physical implications of the stress equations individually to see how the normal and shear stress varies with direction, it is instructive to consider the two equations together and to plot these two stresses against each other for various values of $\theta$. The resulting plot is a circle, termed the Mohr stress circle after Mohr (1900), the German mathematician who was the first to express the stress equations in this helpful graphical form. Each point on the circle represents the state of shear stress ($\tau$) and normal stress ($\sigma$) on a particular plane within the stress field.

Plot of normal stress against shear stress on planes inclined at θ to σ

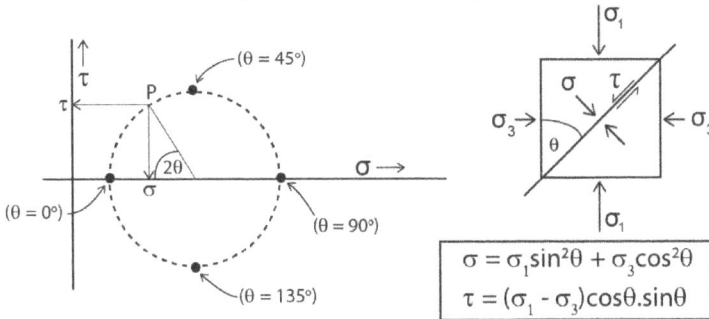

$$\sigma = \sigma_1\sin^2\theta + \sigma_3\cos^2\theta$$

$$\tau = (\sigma_1 - \sigma_3)\cos\theta.\sin\theta$$

Figure 2.7.   The plotting of the components of normal stress ($\sigma$) and shear stress ($\tau$) on a plane against each other using the biaxial stress equations. The stress state on each plane $\theta$ is represented by a point and the points for all planes for $\theta$ from $0°$ to $180°$ define a circle: the Mohr circle.

Consider the plane represented by point *P* in Figure 2.7. The value of shear stress and normal stress on this plane can be read off as shown and the orientation $\theta$ of the plane can be determined by joining point *P* to the centre of the Mohr circle. The angle this radius makes with the normal stress axis is $2\theta$. For example, the two planes on which the shear stress is a maximum, i.e., $\theta = 45°$ and $135°$, see Figure 2.6, have been marked on the Mohr circle. These planes have the maximum value of shear stress and, by joining the points on the circle representing these planes to the centre of the Mohr circle, we find that the planes have $\theta$ values of $45°$ and $135°$ (i.e., $2\theta = 90°$ and $270°$).

## 2.2.6 Different processes are controlled by different parts of the stress tensor

A schematic section through the upper part of the Earth's crust, a part characterised by brittle deformation (Section 2.3), is shown in Figure 2.8. Tectonic extension has caused brittle failure to occur which has resulted in the formation of two normal faults (see Figure 2.14(a) which define a fault-bounded valley known as a rift valley or graben (German for ditch). The black layer represents a salt horizon which has started to flow into

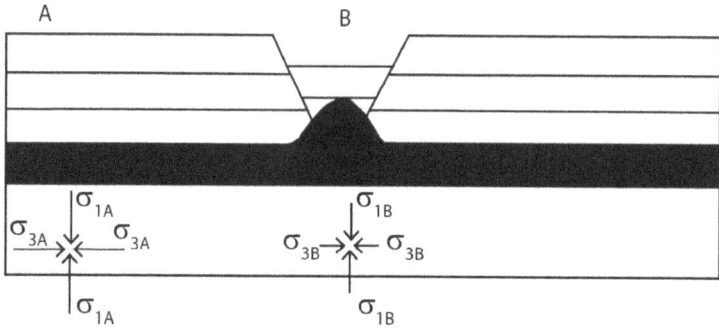

Figure 2.8. A schematic illustration of different stress states at the same depth but different positions in a layer resulting from differences in the overburden load and material properties. The diagram illustrates a rift valley and the black layer represents a salt horizon.

the graben. We can use this model to examine the link between stress, deformation and fluid flow.

Consider the stress state in the layer immediately beneath the salt after the crustal extension that generated the graben has been dissipated and the only factor stressing the crust is the overburden load. Let us assume that, because of differences in the density and/or thickness of the overburden, the overburden stresses, $\sigma_v$, in the layer immediately beneath the salt at positions A and B are different. It is shown in Section 2.4, see Equation 2.27, that the overburden stress will generate a horizontal stress within the layer — the magnitude of which is determined by the overburden stress and Poisson's ratio, a measure of the compressibility of the rock. Thus, if as a result of lateral variations in the lithology, the Poisson's ratio is different at points A and B then the horizontal stresses will also be different, even if the overburden stresses were the same.

For the sake of this discussion, let the vertical stress at A in an element of rock in the layer immediately beneath the salt be significantly larger than it is in an element at B and let the horizontal stress at A be close to the vertical stress and that at B significantly smaller, see Figure 2.8. Let us now ask the questions, which element of rock in the sub-salt, i.e., that beneath A or B is under the greater stress, which element will deform the most, and which way will fluids flow within this layer i.e., from element A to B or vice versa? The answer to the first question is relatively simple: the values of the principal stresses on element A are both greater than on element B and consequently the element at A will be under the greater stress. The answer to the second question is less obvious: although the stress state on element A is greater than element B, we have argued that the difference between the vertical and horizontal stresses at A is small. The state of stress is almost hydrostatic (i.e., with equal normal stress in all directions) and an imaginary circle stamped onto the layer at this position would remain almost a circle. In contrast, the vertical stress at B is significantly larger than the horizontal stress and, as a result, the imaginary circle would be deformed into an ellipse. Thus, although the absolute values of the stresses are smaller at B, the rock experiences a greater degree of distortion or deformation. This is because the ability of a stress field to deform a rock relates not to the absolute values of the principal stresses but rather to the difference between them, i.e., to the differential stress $(\sigma_1 - \sigma_3)$.

Let us now consider the answer to the third question, namely which way will fluids migrate within the sub-salt layer of Figure 2.8. The answer is revealed by the diagram which shows a layer of salt (black) within the succession. Salt behaves as a viscous fluid within the Earth's crust and, when extension of the crust results in the formation of rift valleys, as indicated in Figure 2.8, salt flows into the extended region. This is because the overburden stress in the salt at the sides of the rift valley, location A, is greater than that in the salt beneath the valley, location B. The salt responds to the resulting pressure gradient and flows from the high pressure regions to those of relatively low pressure. Because the salt acts as a fluid, the state of stress generated by the overburden will be hydrostatic and classical fluid dynamics can be used to determine the direction of fluid flow.

However, the state of hydrostatic stress which characterises salt in the crust is atypical of most rocks. Rocks in the upper part of the crust do not generally behave as fluids and can and do support a differential stress $(\sigma_1 - \sigma_3)$. The state of stress in a rock layer can vary laterally and is generally non-hydrostatic (see for example, the sub-salt layer in Figure 2.8) and it is often necessary to predict the migration of fluids within such layers which might act as water or hydrocarbon reservoirs. In such examples, the fluids migrate in response to gradients in the mean stress (i.e., $(\sigma_1 + \sigma_2 + \sigma_3)/3$). In these examples of fluid migration (salt, water, hydrocarbons), the important factor that controls flow is the absolute values of the principal stresses and not the differences between them.

Thus, we see that different processes, e.g., deformation and fluid migration, are controlled by different aspects of the stress tensor (in these examples the differential stress and the mean stress, respectively). We will see in the succeeding sections of this Chapter that other important features, such as the type of brittle failure that develops in a rock (shear failure, i.e., faults, or extensional failure, i.e., joints, see Section 2.3.4) and the regularity of fractures, see Sections 2.4.2 and 2.5.4, are also controlled by specific parts of the stress tensor (in these examples the differential stresses, $(\sigma_1 - \sigma_3)$ and $(\sigma_2 - \sigma_3)$.

## 2.3 BRITTLE FAILURE OF ROCKS

### 2.3.1 Background

Deformation within the Earth's crust takes on a variety of forms but they can be divided into two fundamentally different types: brittle and ductile. These are general terms which describe whether or not the deformation involves a loss of continuity of the rock mass (i.e., brittle failure which involves fracturing) or involves no loss of continuity (i.e., ductile failure, e.g., folding). The significance of these terms is illustrated by the fact that the Earth's crust can be divided into two parts — an upper part which is characterised by brittle deformation and a lower, relatively warm part characterised by ductile deformation, often termed the aseismic zone in which brittle deformation and the associated seismicity are relatively uncommon.

The upper, seismic zone is dominated by fractures and it is within this highly fractured, upper, seismic portion of the crust that rock engineers invariably work. Fractures form in response to stress and a particular stress field will give rise to a specific set of fractures and, because the Earth's crust has generally been subjected to several stress states over geological time, rocks commonly possess several fracture sets which combine to form a fracture network, see Figure 2.9. An understanding of how such fracture networks develop, and what their geometric properties are likely to be, is of first order importance to the rock engineer as it is often these fractures,

Figure 2.9.   Fractured limestone rock mass, South Wales, UK.

rather than the intrinsic properties of the rock, which control the bulk properties of the fractured rock mass, see Section 2.12.

Because fracturing (brittle failure) is so important in controlling rock properties and rock stability, the theory of brittle failure and the expression of this failure in the Earth's crust are discussed in this book. Brittle failure of the Earth's crust has been examined using a variety of techniques: the three most fruitful have been 'field observations', 'theoretical analysis' and 'modelling', the latter being either analogue or numerical. Field observations resulted in two different types of fractures being recognised: those with displacements parallel to the fractures (shear fractures, termed **faults** by geologists), Figure 2.10, and fractures on which the only movement is perpendicular to the fracture walls (extensional fractures, termed **joints**), Figure 2.11.

An understanding of the brittle deformation of the crust is essential in order to prepare for  major engineering projects in which it is necessary to determine the effect of applying new loads to the crust (e.g., the building of a dam) or removing loads from the crust by excavation (e.g., an open-pit mine). Considerable effort has therefore been applied to the task of obtaining a full understanding of the theory of brittle failure.

Figure 2.10. Conjugate shear fractures forming normal faults in the Carboniferous turbidites of Northcott Mouth, Bude, SW England.

Figure 2.11. Several sets of extensional fractures developed in a Liassic limestone bed at Lilstock on the southern margin of the Bristol Channel, UK. These sets combine to form a fracture network which dominates the bulk properties of the fractured rock mass.

## 2.3.2 Shear failure

Experimental work in which cylinders of rock were loaded to failure under a range of confining stresses (Figure 2.12(a)) has been used to generate large databases from which shear failure criteria for the rock could be determined. The cylinders were placed inside a confining jacket to which the confining stress was applied by means of fluid pressure, and then loaded axially to failure. The results from this work provided a range of stress conditions (i.e., values of the maximum compressive stress, $\sigma_1$) needed to cause shear failure under a variety of values of confining stress (i.e., $\sigma_3$). The results can be plotted in several ways. If the load needed to cause failure is plotted against the confining stress, many rocks produce an approximately straight line plot, Figure 2.12(b).

The intersection of the line with the $\sigma_1$ axis of maximum principal compression in Figure 2.12(b) gives the strength of the sample when the confining stress $\sigma_3$ is zero. This is termed the **uniaxial compressive strength**. Strength is defined as the maximum stress a material can sustain.

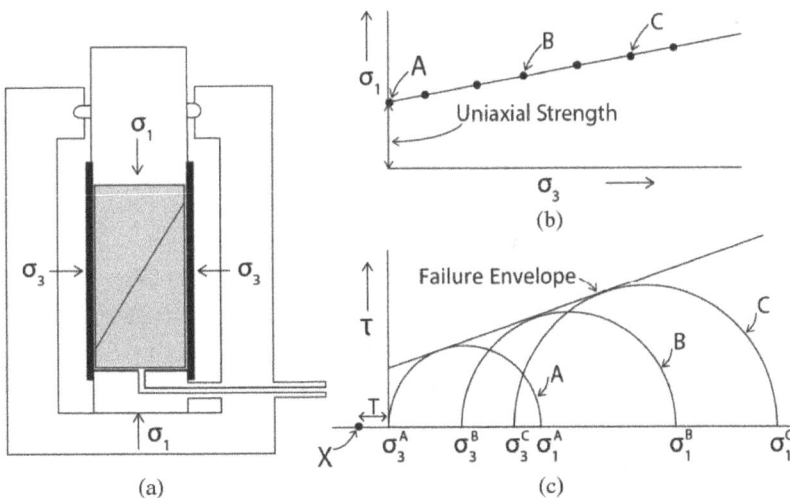

Figure 2.12. (a) Schematic representation of a triaxial deformation laboratory rig in which a confining stress is applied via a fluid pressure. The sample is then loaded axially to failure. (b) and (c) show plots of the experimental results. $\sigma$ is the normal stress and $\tau$ is the shear stress.

What is immediately apparent from the graph is that the 'strength' of a material is not a single quantity but depends (amongst other things) on the confining stress. It can be seen from the graph that the maximum compression the rock can sustain (i.e., its strength) increases with confining stress. Thus, it can be argued that the strength of a rock will increase with depth in the crust as the confining stress increases.

The data from the experiment shown in Figure 2.12 can be plotted in a different manner to that shown in Figure 2.12(b). Each experiment produces a value of $\sigma_1$ and $\sigma_3$ that caused the rock to fail. Thus, each pair of results defines a stress state which can be represented as a Mohr stress circle, Figure 2.7. So, the experimental results can be represented as a series of Mohr circles, each of which represents a stress state that causes the rock to fail by shear failure. This is shown schematically in Figure 2.12(c). The tangent to these Mohr circles, the failure envelope, defines the failure criterion for shear failure. For many rocks, the empirically derived failure criterion shown in Figure 2.12(c) approximates to a straight line with the equation $y = mx + c$, where '$m$' is the slope of the line, (tan $\phi$) and '$c$' its intersection with the $y$ ($\tau$) axis. Thus, this empirically derived failure criterion can be written as in Equation (2.11);

$$\tau = \sigma \tan \phi + c. \tag{2.11}$$

The criterion for shear failure was also derived theoretically by Coulomb (1776) and Navier (1833). Coulomb argued that, in order for shear failure to occur in a material, the shear stress acting along the potential fracture plane must be of sufficient magnitude to (i) cause a shear fracture to form and (ii) subsequently cause slip on that plane. He argued that the shear stress must reach the shear strength of the material (known as the cohesion, '$c$') to form a discrete fracture and overcome the frictional resistance to sliding on that fracture. Coulomb used Amonton's law of sliding friction — which states that the frictional resistance to sliding is determined by the normal stress, $\sigma$, acting across the plane multiplied by the coefficient of sliding friction, $\mu$, which is the tangent of the angle of friction ($\mu = \tan \phi$). Thus, his theoretically derived criterion for shear failure (Figure 2.13(a)) is given by Equation (2.12)

$$\tau = \mu\sigma + c \quad \text{or} \quad \tau = \sigma \tan \phi + c. \tag{2.12}$$

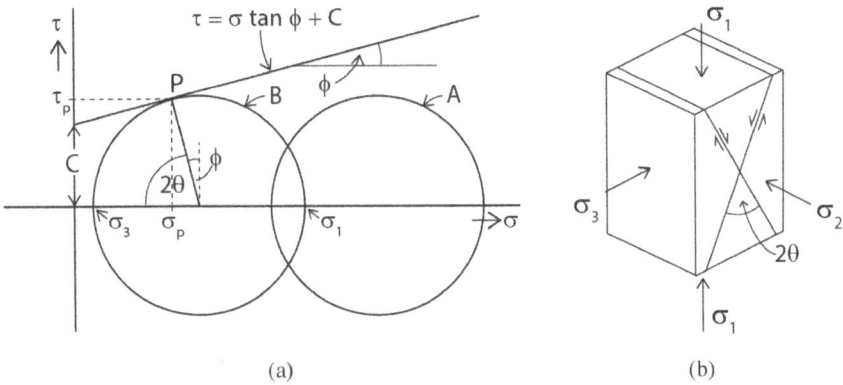

Figure 2.13. (a) The shear failure envelope and the two stress fields represented by two Mohr circles: stress state A is a 'stable' stress field, i.e., will not cause the rock to fail. Stress field B satisfies the failure criterion and will lead to shear failure. (b) The relation between a conjugate pair of shear failure planes and their causative stress field. The orientations of the shear fractures are determined by $\phi$ the angle of friction, see Equation (2.15).

The Navier–Coulomb criterion is identical to the empirically derived criterion, Equation (2.11) and Figure 2.12(c).

This remarkable correspondence between the two independently derived failure criteria provides great credibility. It can be seen from Equation (2.12) that the intersection of the empirically derived curve (Figure 2.13(a)) with the shear stress axis, $c$, gives the cohesion of the material (a measure of its ability to resist shear deformation, i.e., its shear strength) and that the slope of the line, $\phi$, is the angle of sliding friction which, as noted above, is related to the coefficient of friction, $\mu$. Armed with the shear failure criterion of a rock determined experimentally (Figure 2.12), it is a simple matter to determine whether or not a rock will be able to sustain a particular stress field imposed on it by the construction of a major building or dam.

It will be recalled from Figure 2.7 and the associated discussion that any 2D stress field can be represented by a Mohr circle which defines the magnitude of the normal and shear stress on a plane inclined at some angle $\theta$ to the maximum principal compression. Any stress state that falls below the failure envelope (e.g., stress state A in Figure 2.13(a)), will contain no plane on which the shear stress is of sufficient magnitude to cause shear failure. Such a stress state is referred to as a 'stable stress regime' for that material. Conversely, any stress state which touches or cuts the failure envelope (e.g., stress state B in Figure 2.13(a), will cause the material to fail. The stress state that satisfies the failure criterion is represented by point $P$ and can be read off from the two axes as shown (i.e., $\sigma_p$ and $\tau_p$). The orientation of the plane on which this stress state occurs and therefore along which shear failure will develop can be determined as shown in Figure 2.13(a), by joining point $P$ to the centre of the Mohr circle. The angle between this radius and the normal stress axis is $2\theta$, i.e., twice the angle between the failure plane and the maximum principal stress, $\sigma_1$, Figure 2.13(b).

From Section 2.2.3, the states of shear stress ($\tau$) and normal stress ($\sigma$) acting on any plane in a stressed material subjected to a biaxial stress field are given by the biaxial stress equations (Equations (2.13) and (2.14));

$$\sigma = \sigma_1 \sin^2\theta + \sigma_3 \cos^2\theta, \tag{2.13}$$

$$\tau = (\sigma_1 - \sigma_3) \sin\theta \cdot \cos\theta, \tag{2.14}$$

where $\theta$ is the angle between the maximum principal compression, $\sigma_1$, and the plane of interest. We can use these equations to determine the orientation of the failure planes (i.e., the orientation of the plane on which the failure criterion is first satisfied), by substituting for $\sigma$ and $\tau$ in Equation (2.11), using Equations (2.13) and (2.14) and differentiating the resulting expression with respect to $\theta$. The minimum value of this function, Equation (2.15), gives the value of $\theta$ along which the conditions for shear failure are first satisfied and has a value of

$$\theta = 45° - (\phi/2). \tag{2.15}$$

This relation can be derived from the geometry of Figure 2.13(a) using the Mohr circle B, i.e., the state of stress that caused failure, and the failure envelope. It can be seen by inspection that $2\theta = 90° - \phi$ and thus that $\theta = 45° - (\phi/2)$. The angle of internal friction, $\phi$, for most rocks is approximately 30°. It follows therefore from Equation (2.15) that the orientation of shear fractures in rock will be ~30° each side of the maximum principal compressive stress $\sigma_1$, Figure 2.13(b). These two sets of shear fractures are referred to as 'conjugate fractures' or 'conjugate faults'.

It follows from the above discussion and from Figure 2.13(b), both of which show that the orientation of a fault is determined by the orientation of the stress causing it, that faults in the Earth's crust could have any orientation. In fact, faults are found to fall into three types, namely normal, wrench (or strike-slip) and thrust, (Figures 2.14(a)–2.14(c)) respectively, corresponding to three specific stress configurations, i.e., those in which either $\sigma_1$, $\sigma_2$, or $\sigma_3$ is vertical. It is important to determine what causes the stress fields to be so constrained.

As noted briefly in the discussion relating to Figure 1.1 and as is discussed in Section 2.12.2 and Principle 31 in Chapter 3, a free surface cannot support a shear stress and as a result, in the vicinity of such a surface, the principal stresses must be both normal or parallel to it. Anderson (1942) argued that because the Earth's surface is a free surface it follows that the principal stresses must be constrained to one of the three orientations shown in Figures 2.14(a)–2.14(c)). It is interesting to note however that the formation of these three types of faults is not constrained to the upper part of the crust close to the Earth's surface. This may reflect the marked mechanical anisotropy which commonly characterises the crust as a result of the process of sedimentation. Weak bedding planes will be unable to sustain large shear stresses and, like the free surface, will have the effect of constraining the orientations of the principal stresses to being either approximately vertical or approximately horizontal. Other factors that tend to result in the principal stresses being either sub-vertical or horizontal are the processes by which the crust is stressed. The two principal causes of stress, as noted earlier in Figure 2.2, are the overburden which acts vertically and plate motion which is dominantly horizontal.

Thus, the reasons for the tendency for the stresses within the crust to be constrained into three main orientations, and therefore the reason for there being three major types of faults, relate to

— the intrinsic anisotropy of the crust,
— the free surface which defines the Earth's surface, and
— the principal processes which stress the crust, namely the overburden stress and plate motion.

We will return to the important concept of the influence of a free surface on the orientation of stress in Section 2.12 where we look at the build-up of fracture networks by the successive superposition of individual fracture sets and where we describe the technique of fracture analysis.

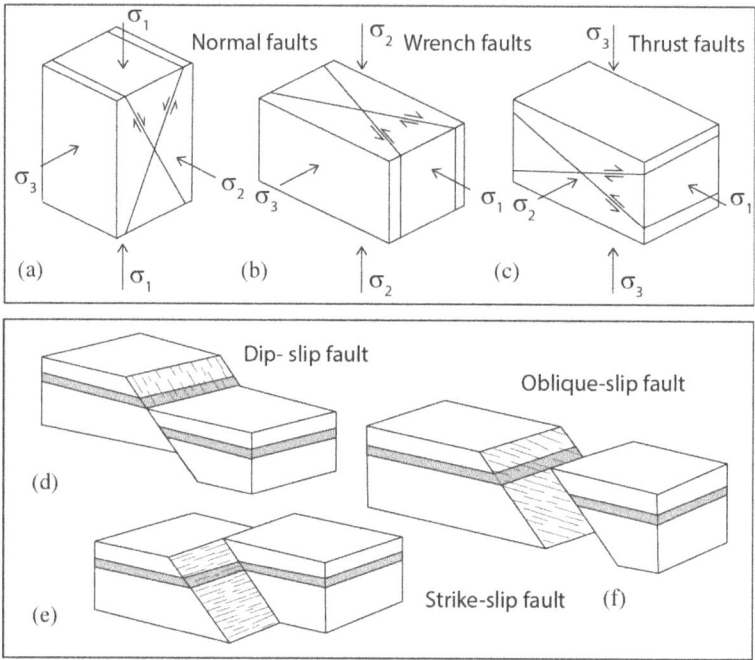

Figure 2.14. (a)–(c) show the orientations of the three major classes of faults with their associated stress fields. Note that for the formation of these faults the principal stresses are either horizontal or vertical, i.e., either parallel or normal to the Earth's surface. (d) A dip-slip fault, (e) a strike-slip fault and (f) an oblique slip fault. Note that the idealised faults (a)–(c), are either dip-slip or strike-slip. However, many natural faults have elements of both types of movement along them and are termed 'oblique-slip' faults.

### 2.3.3 Extensional failure

In addition to faults which occur commonly in rocks and which are recognised as the geological expression of shear failure, rocks frequently contain joints, i.e., extensional fractures. These fractures differ from shear fractures in two important ways. Firstly, there is no displacement parallel to the fracture: the only displacements are normal to the fracture and relate to its opening. Secondly, the fractures form normal to the minimum principal compression, $\sigma_3$, and thus lie in the $\sigma_1 - \sigma_2$ plane.

An experimental investigation of this type of failure is not as easy to carry out as it is for shear failure. Extensional failure occurs in the region where the principal stresses are tensile. In order to obtain data in this region, it is necessary to apply a uniform extensional stress to the sample rather than the compressive stress $\sigma_3$ shown in Figure 2.12(a), a task which is difficult to achieve, and then subject it to an axial extension. The only easily obtained data on extensional failure is the unconfined tensile strength of the material ($T$), and this is found to be approximately half the value of the shear strength (i.e., the cohesion, $c$), confirming that rocks (and other materials) are stronger in compression than in tension.

In contrast to the numerous experimental results obtained relating to shear failure, the single result for extensional failure (point $\sigma = T$, $\tau = 0$, point X in Figure 2.12(c) does not allow the form of the extensional failure criterion to be determined. Clearly it must link point X with the shear failure criterion in the region where it meets the shear stress axis (point 0, $c$). It was not until the work of Griffith (1920, 1924) that a theory of extensional failure was proposed that had the capability of predicting an extensional failure envelope. As will be seen, this envelope is compatible with the experimental results discussed above.

In 1920, Griffith proposed the energy-balance concept of fracture in his classic paper based on the principle of the energy conservation laws of mechanics and thermodynamics. His work laid the historical and scientific foundation for the science of Fracture Mechanics. He argued (Griffith, 1924) that the tensile strength of a material should be related to its inter-atomic bond strength. However, measured values of tensile strength were invariably several orders of magnitude lower than the predicted value based on this concept and Griffith suggested that this discrepancy could be accounted for by the existence within the material of microscopic flaws. He pointed out that stress magnification occurs at the flaw tips, which for the sake of his analysis were taken to be elliptical cracks, and demonstrated that the amount of magnification is controlled by the **eccentricity** (the ratio of the major to minor axis) and **orientation** of the microscopic crack with respect to the applied stress.

Stress magnification around a circular hole was reported by Kirsch (1898) and his results are illustrated diagrammatically in Figure 2.15 which shows a plate containing a circular hole, subjected to a biaxial stress field defined by the principal stresses $\sigma_1$ and $\sigma_3$. Stress redistribution around the hole results in a local reduction of $\sigma_3$ to ($3\sigma_3 - \sigma_1$) and local magnification of $\sigma_1$ to ($3\sigma_1 - \sigma_3$). It follows, therefore, that if $\sigma_1 > 3\sigma_3$, a

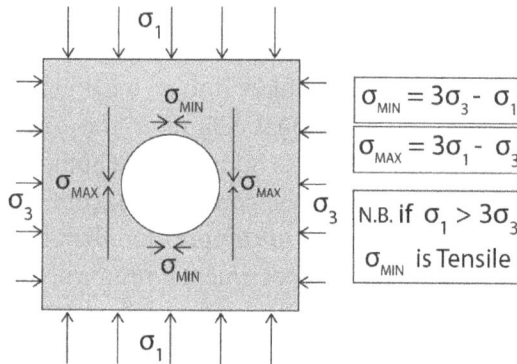

Figure 2.15. The generation of tensile stresses at points around a circular hole in a plate despite the fact that the applied stress field is compressive. The stress magnification is determined by the values of the applied stresses ($\sigma_1$ and $\sigma_3$) and, as shown in Figure 2.16, the eccentricity of the flaw.

tensile stress is generated at the edge of the hole, despite the fact that the applied principal stresses are both compressive.

Analytic solutions to the stress concentrations around elliptical holes were presented by Inglis (1913) who showed that the magnification factor was determined by the eccentricity of the ellipse. He showed that the greater the eccentricity the greater the magnification. This is illustrated in Figure 2.16. Around a circular hole the stress concentration is 3 (for uniaxial loading). If the eccentricity is increased from 1 to 2, the stress magnification factor increases to 5 and, for an ellipse with an eccentricity of 3, the stress magnification is 7. The general equation relating the eccentricity, $c/b$, and the stress magnification, $\sigma_m/\sigma_r$, where $\sigma_m$ is the magnified value of the stress at the crack tip and $\sigma_r$ the regionally applied uniaxial stress, is given by Equation (2.16).

$$\sigma_m/\sigma_r = 1 + (2c/b), \tag{2.16}$$

where $c$ and $b$ are the major and minor axes of the elliptical crack.

Griffith argued that it is this stress magnification at crack tips that enables local stresses to develop within the material which are of sufficient magnitude to overcome the inter-atomic bond strength and lead to tensile failure, even though the applied stresses are much lower.

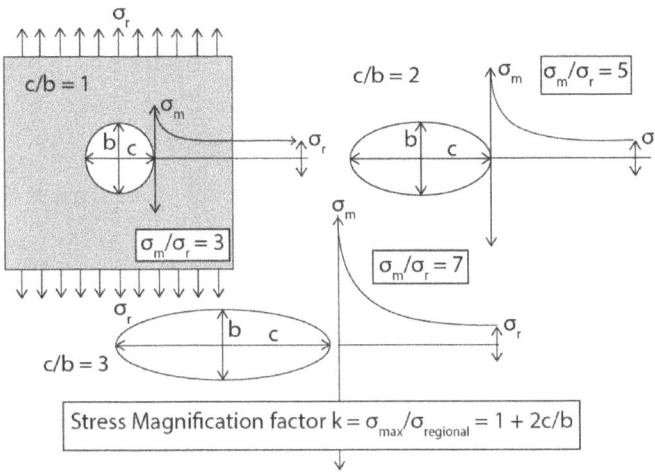

Figure 2.16. Stress magnification around holes in a uniaxially stressed plate. These holes represent elliptical fractures with different eccentricities. The background applied stress $\sigma_r$ is magnified close to the fracture tips to a value of $\sigma_m$. After Inglis (1913)

Using this concept, the conditions necessary for the initiation of a macroscopic fracture by microcrack growth were considered. He equated the surface energy required to increase the area of the fracture as the crack enlarged to the release in elastic strain energy in the material caused by the enlargement as a function of the increase in crack length, $c$. Figure 2.17 shows the balance between these two energy terms. The crack surface energy is given by Equation (2.17);

$$SE = 4c\lambda, \tag{2.17}$$

where $\lambda$ is the free surface energy per unit area, and the stored elastic energy (i.e., the stored strain energy, SSE) is given by Equation (2.18);

$$SSE = \pi\sigma^2 c^2 / 2E, \tag{2.18}$$

where $\sigma$ is the remote stress applied to the material and $E$ is its Young's modulus.

It can be seen from the graph that initially, when the crack is small, the energy required to lengthen the crack is greater than the energy released as

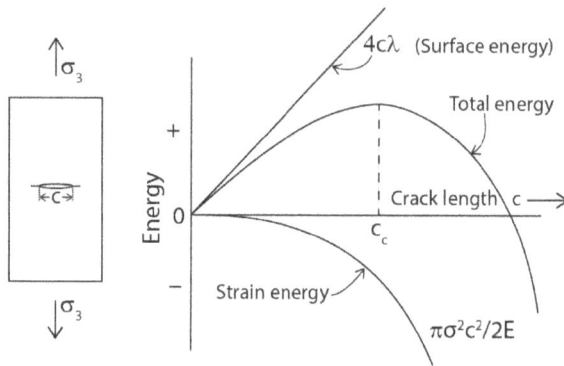

Figure 2.17.   (a) Griffith's model for a crack propagating in a rod and (b) the energy partitioning for the process. The graphs show the increase in surface energy ($4c\lambda$), the decrease in elastic strain energy ($\pi\sigma^2c^2/2E$) and the change in total energy as the crack length '$c$' increases.

a result of crack growth. Griffith argued that a crack is on the limit of propagation when these two quantities are equal. The crack size when the total energy reaches a maximum is termed the critical crack size, $c_c$. As the fracture propagates beyond this length, the input of energy needed to increase the surface area is less than that gained by the release of elastic strain energy. The fracture is then unstable and propagation accelerates, theoretically approaching the speed of sound in the medium in which the crack is situated.

The **Stress Intensity Factor**, $K$, is used in fracture mechanics to more accurately predict the stress state ('stress intensity') near the tip of a crack caused by a remote load. Griffith showed that for a mode 1 fracture (an extensional fracture) the stress intensity factor is given by Equation (2.19):

$$K_1 = \gamma\sigma(\pi c)^{1/2},\tag{2.19}$$

where $\gamma$ is a dimensionless parameter that depends on both the specimen and crack geometry, $\sigma$ is the remote applied stress, and '$c$' the fracture length. When this stress state becomes critical a small crack grows ('extends') and the material fails. The critical value for this quantity can be obtained by substituting the critical crack length and the associated remote stress into Equation (2.19) to give Equation (2.20):

$$K_{1c} = \gamma\sigma_c(\pi c_c)^{1/2}.\tag{2.20}$$

Note that the **fracture toughness** can be defined in terms of the stress intensity factor, $K$, but at a critical stress state.

Rearranging Equation (2.20) gives the critical applied stress for a given critical crack size, Equation (2.21):

$$\sigma_c = K_{1c}/\gamma(\pi c_c)^{1/2} \qquad (2.21)$$

and further rearrangement shows that the critical crack size, i.e., the largest flaw that can survive without propagating into a macrofracture for a given applied stress, Equation (2.22):

$$c_c = c_{max} = (K_{1c}/\gamma\sigma_c)^2/\pi. \qquad (2.22)$$

As noted above, once the critical stress is reached, fast fracture propagation occurs with the upper limit of the crack tip velocity being the speed of sound in that material.

Griffith also noted that, in addition to the eccentricity of the flaw influencing the stress magnification, the orientation of the flaw with respect to the applied stress field is also important. He showed that the crack tip stresses would be a maximum when Equation (2.23) is satisfied:

$$\cos 2\theta = (\sigma_1 - \sigma_3)/(\sigma_1 + \sigma_3), \qquad (2.23)$$

where $\theta$ is the angle between the long axis of the flaw and the maximum principal compressive stress direction, Figure 2.18(a). He also showed that, providing $\sigma_1$ is not equal to $\sigma_3$ and $(3\sigma_1 + \sigma_3) > 0$, the tensile stresses around the flaw tip reach the critical value necessary for the spreading of the fracture (i.e., for fracture propagation) when Equation (2.24) is satisfied:

$$(\sigma_1 - \sigma_3)^2 - 8T(\sigma_1 + \sigma_3) = 0. \qquad (2.24)$$

This relation between the principal stresses at failure can be re-expressed as Equation (2.25):

$$\tau^2 + 4T\sigma - 4T^2 = 0 \qquad (2.25)$$

and this shows that there is a parabolic relation between the normal stress and shear stress at failure, Figure 2.18(b), see Murrell (1958).

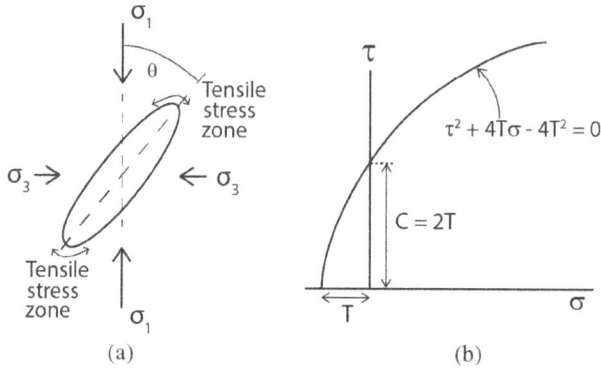

Figure 2.18. (a) The generation of tensile stresses at the tips of an elliptical crack — even though the stress state is compressive. The magnitude of the stress concentration at the crack tips is determined by the eccentricity of the ellipse and its orientation $\theta$ with respect to the applied stress. This concept was used by Griffith to determine his criterion of extensional failure, shown graphically in (b).

### 2.3.4 Combined shear-extension failure envelope for a brittle material

The two failure criteria (Figures 2.12(c) and 2.18(b) can be linked at the point where their slopes are the same to produce a complete brittle failure envelope, Figure 2.19(a). This Figure shows the two types of failure and the stress fields (represented as Mohr circles) that generated them. The diagrams (b) and (c) show the relation between the stress fields and the resulting fractures.

In order for failure to occur, the stress state has to satisfy the failure criteria, i.e., the Mohr stress circle must touch the failure envelope (see Figure 2.19). Inspection of this Figure shows that, in order to satisfy the shear failure criteria, the Mohr circle has to have a larger diameter (i.e., differential stress $(\sigma_1 - \sigma_3)$) than is required for extensional failure. These differences can be quantified.

It can be shown that the largest Mohr circle that can still touch the vertex of the parabola has a radius of $c$, i.e., the cohesive strength of the rock. It will be recalled that, for many materials, the cohesion, $c$, (the shear strength) is approximately twice the tensile strength ($T$). If the radius of the Mohr circle is less than $c$ (i.e., less than $2T$), then it is too small to

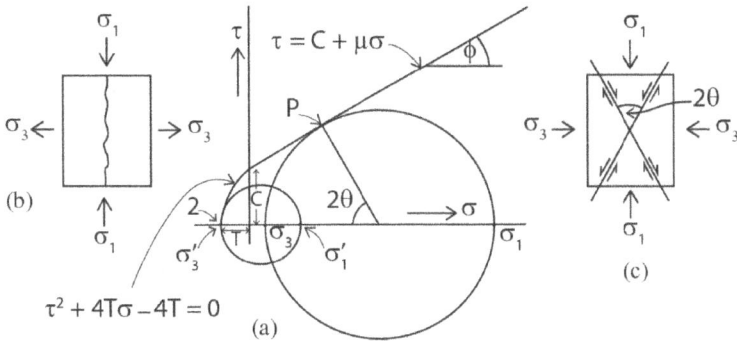

Figure 2.19.   (a) The combined brittle failure criterion derived by linking the linear Navier–Coulomb shear failure criterion with the parabolic extensional failure of Griffith. The stress conditions required for the two modes of failure are apparent. In order to satisfy the extensional failure criterion, a small differential stress $(\sigma_1 - \sigma_3)$ is required; whereas, shear failure requires a larger differential stress. (b) and (c) show the relation between the two modes of fracture and the causative stress fields.

touch the shear failure envelope. If such a stress field were to cause failure, it would always be by extensional failure. Conversely, if the radius of the Mohr circle is greater than $c$, (i.e., greater than $2T$), then the Mohr circle will be too large to satisfy the extensional failure criteria and, if such a stress state were to cause failure (i.e., touch the envelope), it would always be by shear failure. Thus, the stress conditions required for the two modes of failure are:

For shear failure $\qquad$ $(\sigma_1 - \sigma_3) > 4T$ and the Mohr circle must touch the failure envelope;

For extensional failure $\quad$ $(\sigma_1 - \sigma_3) < 4T$ and the Mohr circle must touch the failure envelope.

It was noted above in the discussion relating to Figure 2.19(a) that the stress state ($\sigma$ and $\tau$) on the plane along which shear failure occurs (point $P$) can be read off directly from the graph and the orientation $\theta$ of the failure plane with respect to $\sigma_1$ determined by joining point $P$ to the centre of the Mohr circle and measuring the angle $(2\theta)$ between it and the horizontal axis.

The orientation of the failure plane and the magnitude of the stress on it can also be determined for extensional fractures. Here, the Mohr circle

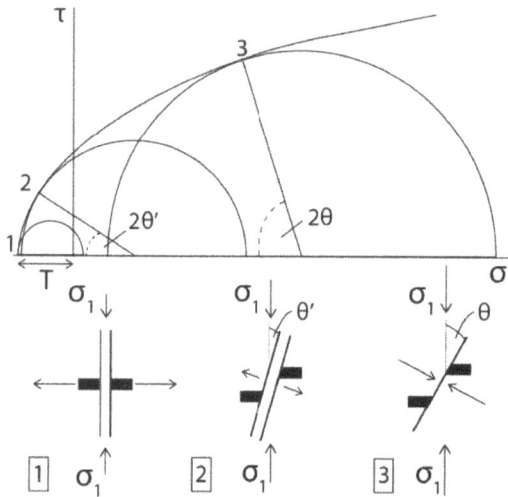

Figure 2.20.   Three states of stress represented by the Mohr stress circles, all of which touch the brittle failure envelope (i.e., at points 1, 2 and 3) and which will therefore lead to failure. The orientation of the fractures that result and the state of stress on the fracture planes are shown.

touches the failure envelope on the horizontal axis at point $(-\sigma_3, 0)$ and the state of stress on the failure plane can be read off as $(\tau = 0, \sigma = -T)$. The angle $(2\theta)$ between the line joining the point where the stress field touches the failure envelope and the horizontal axis is zero and therefore the angle between $\sigma_1$ and the fracture plane is zero, Figure 2.20, Case 1. Thus, for extensional failure, the fractures form parallel to $\sigma_1$, i.e., they open against the least principal stress $\sigma_3$, there is no shear stress acting along them, and the normal stress must be equal to $T$, the tensile strength of the rock.

We have considered the two principal modes of brittle failure, namely extensional failure which occurs when the differential stress is less than $4T$, i.e., when the Mohr circle representing the stress field touches the parabolic extensional failure envelope at its vertex (point 1 in Figure 2.20), and shear failure which forms when the differential stress reaches this value and the Mohr circle touches the shear failure envelope (point 3). However, it is possible for the Mohr stress circle to touch the extensional failure envelope at any point along the parabolic section between the vertex and the point where the parabola meets the shear failure envelope. It is interesting to consider the orientation of the resulting failure planes and the stress state acting along them.

Consider the stress state which touches the failure envelope at point 2, Figure 2.20. Using the procedure outlined above, it can be seen that the fractures will have characteristics of both extensional failure (the normal stress is negative) and shear failure (there is a shear stress acting along the plane) and that the angle $2\theta$ is considerably less than for classical shear failure (~60°), Figure 2.13. Because these fractures display features of both extensional and shear failure, they have been termed hybrid fractures. The portion of the failure envelope relating to **hybrid fractures** extends from $(\tau = 0, \sigma = -T)$ to $(\tau = \tau_0, \sigma = 0)$ beyond which the normal stress on the failure plane becomes compressive.

Natural examples of conjugate hybrid fractures are shown in Figure 2.21. Note the small value of $2\theta$ and that, in addition to the shear

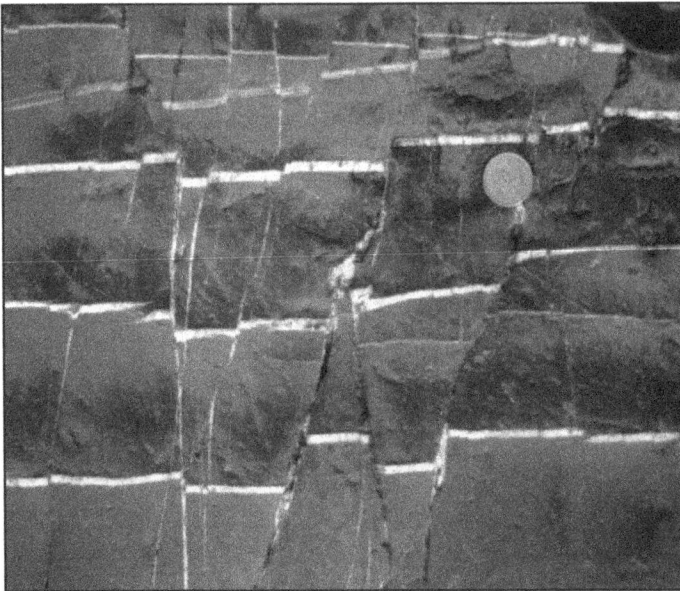

Figure 2.21.   Examples of hybrid extensional/shear fractures. Note the small angle $2\theta$ between the conjugate fractures and the clear evidence for movement both along the fracture and across it.

displacement along these fractures, there has been fracture normal opening which has resulted in the precipitation of quartz along the fractures.

These fractures are uncommon in nature and it is interesting to try to establish why this is so. It is possible to determine the range of differential stresses (i.e., the range of diameters of the Mohr circles) that can contact the portion of the failure envelope that gives rise to hybrid fractures. As noted, the Mohr circles will have a diameter greater than $4T$ and will contact the extensional failure envelope in the extensional regime, i.e., to the left of the shear stress axis. This range is shown in Figure 2.22 which shows the range of differential stress under which these three modes of failure form, together with the corresponding values of $2\theta$. Compared to the other modes of failure, the range over which hybrid failure will occur is small (between the two dashed lines representing $4T$ and $5.66T$, the differential stress value at which the normal stress on the fracture plane becomes compressive). It is also noted that the nonlinear dependence of $2\theta$ on the differential stress for hybrid failure (note the rapid increase in $2\theta$

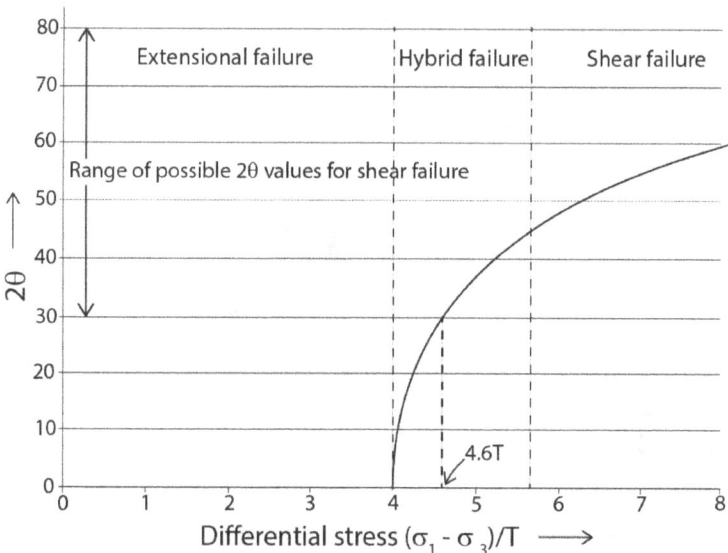

Figure 2.22.   Plot of differential stress against $2\theta$ for the three modes of failure. Extensional failure occurs in the range $0T$–$4T$, Hybrid failure in the range $4T$–$5.66T$ and Shear failure for all values of differential stress greater than $5.66T$. $T$ is the tensile strength of the rock.

as the differential stress just exceeds $4T$) results in the range of differential stress values where this angle is less than 30°, i.e., when the hybrid nature of the failure planes will be apparent, being even smaller.

In addition, although many rocks have an angle of sliding friction of ~30° and therefore a value of $2\theta$ of around 60° (see Equation (2.15)), the value of $\phi$ for rocks can vary greatly, less than 10° for some very soft rocks and more than 50° for some very hard rocks. It follows that the values of $2\theta$ between conjugate fractures that are formed by 100% shear failure can range from 80° to as low as 30°, see Figure 2.22. Thus a small value of $2\theta$ does not necessarily imply that hybrid failure has occurred. If we allow that this range of values can be generated as 100% shear fractures, then the range of differential stress over which hybrid failure is likely to occur is between $4T$ and $4.6T$, i.e., the value of differential stress above which $2\theta$ is greater than 30° (see Figure 2.22). These observations provide a possible explanation as to why hybrid failure planes appear so seldom in nature. One way of convincingly demonstrating the occurrence of true hybrid fractures is to show that there was extension across the fractures as well as shear along them at the time of their formation. The precipitation of minerals along the conjugate fractures shown in Figure 2.21, which have a $2\theta$ value of around 20°, shows clearly that both types of displacement occurred during their formation and that they are therefore examples of true, hybrid fractures.

## 2.4 THE STATE OF STRESS IN THE EARTH'S CRUST

### 2.4.1 Introduction

In Section 2.3, it was noted that faults fall into three main categories (normal, strike–slip and thrust) and that these structures correspond to three specific stress states in the crust: namely, conditions when one of the principal stresses, $\sigma_1$, $\sigma_2$ or $\sigma_3$ is vertical, Figures 2.14(a)–2.14(c). It was argued that the reasons that the principal stresses are constrained to these three orientations is a combination of the following:

1. The free surface represented by the Earth's surface (i.e., the principal stresses have to be either normal or parallel to a free surface because it cannot sustain a shear stress).

2. The intrinsic planar anisotropy of the crust resulting from a variety of processes, including that of sedimentation.
3. The two main mechanisms by which the crust is stressed, namely the overburden which imparts a vertical load and plate tectonics which tend to impose horizontal principal stresses.

Because of the intimate relation between stress and fracturing discussed in the previous Section, it is useful to determine the state of stress in the crust — as this will enable the type of fractures that can form and their orientation to be determined.

### 2.4.2 Overburden stress

For the sake of simplicity, let us consider a portion of the crust unaffected by plate tectonics, i.e., a 'tectonically relaxed' region in which the state of stress is controlled dominantly by the weight of the overburden. The vertical stress $\sigma_v$ in such a regime is given by Equation (2.26);

$$\sigma_v = z\rho g, \tag{2.26}$$

where $z$ is the depth in the crust, $\rho$ the mean density of the overburden and $g$ the acceleration due to gravity. This is the force per unit area acting on a stratum at depth $z$ and derives directly from Newton's definition of force (force = mass × acceleration). This overburden stress can generate a horizontal stress, the magnitude of which is determined by the boundary conditions and the intrinsic properties of the rock. If it is argued that the rock cannot expand laterally in response to the overburden load (because of the confining effect of the surrounding material), then the horizontal strain will be zero. However, in order to prevent the horizontal expansion of the rock which would occur if the constraints of the surrounding rock were to be removed, there is a horizontal confining stress in the surrounding rock. This stress is controlled by the compressibility of the rock (its Poisson's ratio) and its value is given by Equation (2.27);

$$\sigma_h = \sigma_v/(m-1), \tag{2.27}$$

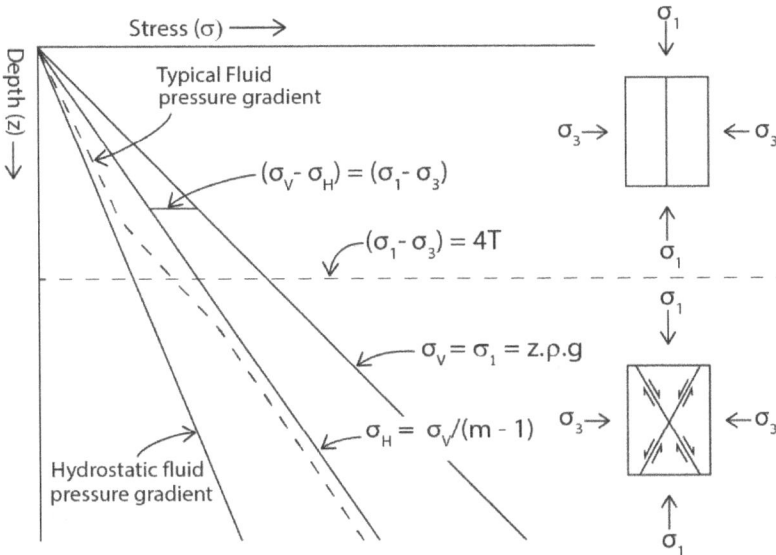

Figure 2.23. Graph showing the increase of the vertical stress, horizontal stress and fluid pressure with depth in the crust when the stress is the result of the overburden load. As the differential stress increases with depth, so the mode of failure changes from extensional failure (joints) to shear failure (faults).

where $m$ is the Poisson number (the reciprocal of Poisson's ratio which is the ratio of the lateral expansion and the axial contraction caused by a load acting on an unconfined sample of the rock).

The vertical and horizontal stress can be plotted against depth and assuming that the overburden density remains constant it can be seen from Equations (2.26) and (2.27) that they both increase in a linear manner, Figure 2.23. The Earth's crust below the water table is saturated with water and, assuming that the pores in the rocks are connected to each other, then there will be a uniform increase in fluid pressure with depth. The fluid pressure at any depth can be determined from Equation (2.26) by inserting the appropriate density. This pressure increase (the hydrostatic fluid pressure gradient) is shown diagrammatically in Figure 2.23. However, if at some depth, the pores become isolated as a result of compaction and cementation, then the fluid pressure increases more rapidly and the sediments become over-pressured. This effect is shown by the dashed line of Figure 2.23.

Knowing (i) the stress state responsible for the stress profile (i.e., a vertical $\sigma_1$ resulting from the overburden and an induced horizontal stress, $\sigma_3$) and (ii) the magnitude of the differential stress, it is possible to determine the type and orientation of the fractures that might form. It is apparent from Figure 2.23 that the differential stress $(\sigma_v - \sigma_h = \sigma_1 - \sigma_3)$ increases linearly with depth. Thus, in the upper part of the crust where the differential stress is less than $4T$, vertical extensional fractures will form and, at a depth where the differential stress reaches $4T$, shear fractures dipping at around 60° will form. However, in order to fully describe the geometry of the fractures that form as a result of the overburden stress, the impact of the intermediate stress $(\sigma_2)$ needs to be considered. This is shown in Figure 2.24 and discussed in the following paragraph.

The 2D stress state represented in Figure 2.23 is in the $\sigma_1/\sigma_3$ vertical plane; $\sigma_2$ acts normal to this plane and controls the orientation (strike) of the

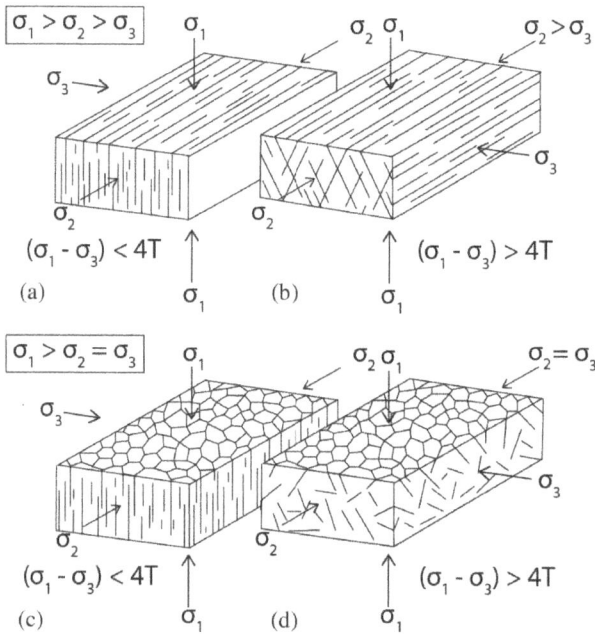

Figure 2.24.   The impact of the intermediate principal stress $(\sigma_2)$ on the strike direction of extensional and shear fractures. In (a) and (b), $\sigma_2 > \sigma_3$ and the strike orientation of the resulting fractures is consistent and regular. In (c) and (d) $\sigma_2 = \sigma_3$ and the fracture strike directions are random, resulting in polygonal arrays.

Figure 2.25. A fracture network in a limestone bed. All three fracture sets making up the network have a markedly consistent strike (ENE–WSW, NE–SW and NW–SE), even though they have different degrees of continuity. This indicates a significant difference between $\sigma_2$ and $\sigma_3$ at the time of fracture formation. North is taken to be parallel to the vertical edge of the photograph.

fractures. If there is a significant difference between $\sigma_2$ and $\sigma_3$ (e.g., Figure 2.24(a), then there is a direction in which the fractures can open most easily defined by the least principal stress, $\sigma_3$. All the fractures will open in this direction (following the least work principle) resulting in an array of fractures parallel to the $\sigma_1-\sigma_2$ plane and striking parallel to $\sigma_2$. Each of the three fracture sets making up the fracture network shown in Figure 2.25 has a constant strike despite their different degrees of continuity, (see Section 2.12) and it can be argued that there was a significant differential stress $(\sigma_2-\sigma_3)$ albeit less than $4T$ at the time of their formation. In a similar manner, at depths where the differential stress is sufficiently high to generate shear fractures (Figure 2.24(b)), the shear fractures will form as a conjugate set, both of which strike consistently parallel to $\sigma_2$. Conversely, if $\sigma_2 = \sigma_3$, then it is equally easy for the fractures to open in any direction, (e.g., Figures 2.24(c) and 2.24(d), and there will be no tendency for their strike directions to align. This results in a polygonal array of extensional fractures in the upper part of the crust and randomly striking shear fractures at depths where the differential stress $(\sigma_1 - \sigma_3)$ exceeds $4T$. Under this stress regime, the strike of shear fractures will also be random, Figure 2.24(d).

Thus, there is a range of possible geometries for the extensional and shear fractures that form in the crust as a result of the overburden stress. Depending on the relative values of $\sigma_2$ and $\sigma_3$, the fractures may display a consistent strike direction and produce a clearly defined fracture set (i.e., when $\sigma_2$ and $\sigma_3$ are significantly different), as shown in Figures 2.24(a) and (b), or they may be randomly oriented (i.e., when these two principal stresses are the same).

All intermediate forms of fracture arrays can occur and the regularity of the strike of the fractures can be used as an indicator of the relative magnitudes of $\sigma_2$ and $\sigma_3$ at the time of fracture formation. Natural examples of polygonal columns of extensional fractures of the type shown in Figure 2.24(c) are shown in Figure 2.26. The left hand photograph shows the plan view of the bed, which reveals the polygonal pattern of the fractures, and the right hand photograph shows a joint face at right angles to the bedding, which reveals the vertical orientation of the fractures. The scale of these polygonal arrays of fractures can vary: inspection of seismic images from the North Sea reveals polygonal arrays of normal faults (cf. Figure 2.24(d)) where the diameter of the polygons is around 500 m, Figure 2.27.

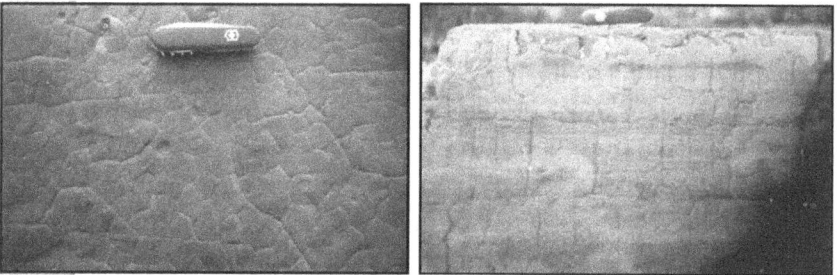

Figure 2.26.   Polygonal fractures of the type shown in Figure 2.24(c). The left hand photograph shows a view of the bedding surface and the right hand photograph shows the fractures exposed on a plane normal to bedding. From Kimmeridge Bay, Dorset, southern England.

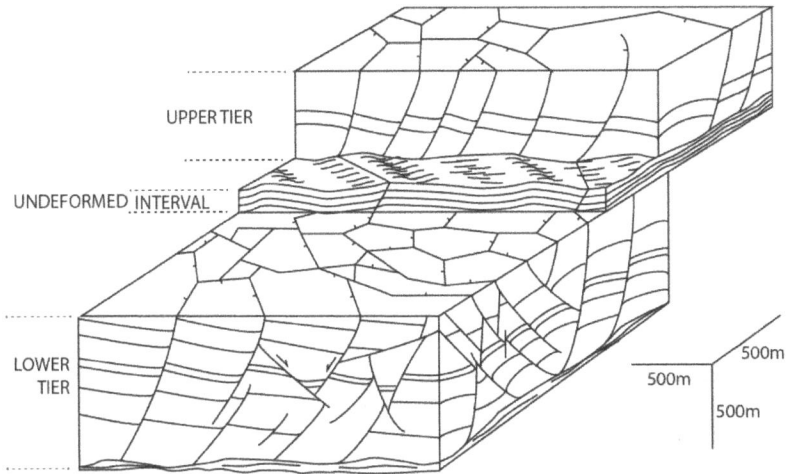

Figure 2.27. A large-scale, polygonal array of shear fractures from the North Sea. This indicates that, in the plane of the bedding (the $\sigma_2$–$\sigma_3$ plane), there was little difference between the two principal stresses during fracture formation. After Cartwright *et al.* (2003).

## 2.5 FLUID INDUCED FAILURE IN ROCK

### 2.5.1 Introduction

The discussion of brittle failure in Section 2.3 assumed that the deformation was occurring in a dry environment. However, rocks below the water-table are saturated with fluids, the two main types being **meteoric** fluids, mainly water recently derived from rain and snow, and **connate** fluids, i.e., liquids that were trapped in the pores of sedimentary rocks as they were deposited, mainly brines. It is important to consider the impact of these fluids on fracturing.

### 2.5.2 The problem of forming extensional fractures in the crust

In the Section on brittle failure, two important modes of brittle failure were identified, namely extensional fracturing which produced joints and shear fracturing which generated faults. The summary diagram showing the two failure criteria and the relation between the stress causing fracture formation and the resulting fractures is also included in this Section as Figure 2.28.

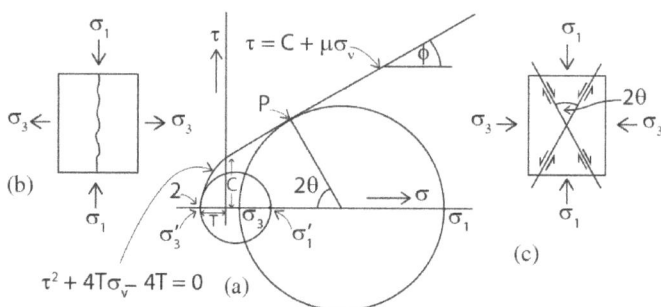

Figure 2.28.   (a) The combined brittle failure criteria derived by linking the linear Navier–Coulomb shear criterion with the parabolic extensional failure of Griffith. (b) and (c) show the relation between the orientations of the extensional and shear fractures and the stresses that cause them.

It can be seen that shear failure forms when the principal stresses are compressive, i.e., to the left of the shear stress axis and that extensional failure characterises the extensional regime to the right of this axis. It is also clear from this diagram that, in order for extensional failure to occur, at least one of the principal stresses ($\sigma_3$) must be extensional ($\sigma_3 > T$, the tensile strength of the rock). However the state of stress in the crust is generally compressive (see Figure 2.23) and it is therefore surprising to find that the upper part of the crust is dominated by extensional fractures, Figure 2.29. In order to resolve this apparent contradiction, it is necessary to consider the impact of fluid pressure on rock failure.

### 2.5.3 The impact of fluid pressure on the formation of fractures in rocks

Below the water table, the Earth's crust is saturated with water and the water pressure increases with depth as a consequence of the increasing hydrostatic head. However, this hydrostatic fluid gradient is only applicable to a system where the water in the rock pores is in contact and can move freely. If, as a consequence of burial and compaction or cementation the water in the pores becomes isolated (the depth at which this occurs is known as the fluid isolation depth), then the hydrostatic gradient can no longer be used to predict the water pressure at depth. Fluid pressures can begin to exceed the hydrostatic; as the rocks compact, the sediments become

Figure 2.29.   A network of extensional fractures (joints) cutting a Liassic limestone bed. Lilstock, north Somerset, UK.

overpressured, see Figure 2.23. Extremely high pressures can be generated in this way — as is dramatically illustrated in blowouts which occur when oil drillers drill into an overpressured horizon.

It was noted in the previous Section that the parameter which controls when brittle failure in the crust occurs and the resultant type of fracturing that forms is the state of stress in the crust. At any particular depth, the fluid in the crust will generate a hydrostatic stress, i.e., a pressure, and in order to determine the total state of stress in the crust and thus be in a position to predict whether or not failure will occur, this fluid pressure must be added to the lithostatic stress. The effect of a fluid pressure on the lithostatic stress is shown diagrammatically in Figure 2.30. It should be noted that stresses are additive and cannot exist in the same location independently. Thus, the state of stress in the crust in a tectonically relaxed region, when there is a fluid pressure in the rock, is obtained by adding the lithostatic stress to the fluid pressure. This is shown in Figure 2.30(a) where the lithostatic stress is represented by the stress ellipse whose axes are the horizontal and vertical stresses, both of which are compressive, and the fluid pressure is represented by a circle which shows the hydrostatic pressure, $p$, acting so as to oppose the lithostatic stresses. The resulting state of stress is represented by the stress ellipse with axes $(\sigma_1 - p)$ and $(\sigma_3 - p)$. The impact of a fluid pressure on the state of stress along a plane is shown in

(a)

(b)

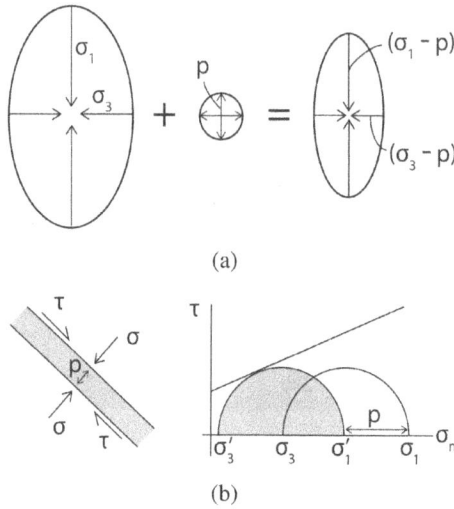

Figure 2.30. (a) Schematic representation of the effect of adding a fluid pressure to a lithostatic stress (represented by the circle of diameter $p$ and the ellipse with axes $\sigma_1$ and $\sigma_3$, respectively). The resulting effective stress state is shown by the ellipse with axes $(\sigma_1-p)$ and $(\sigma_3-p)$. (b) Fluid pressure along a fracture reducing the lithostatic stress $(\sigma)$ to an effective stress $(\sigma-p)$. The effect of a fluid pressure $p$ on the lithostatic stress state $\sigma_1$ and $\sigma_3$, is shown on the Mohr diagram.

Figure 2.30(b). It can be seen that the pressure, $p$, opposes the normal stress acting across the plane reducing it to $(\sigma_n-p)$ but has no effect on the shear stress. All the normal stresses are reduced and so, as shown in Figure 2.30(a), $\sigma_1$ becomes $(\sigma_1-p)$ and $\sigma_3$ becomes $(\sigma_3-p)$ and the impact on the Mohr circle is to move it to the left by an amount $p$, Figure 2.30(b).

The fluid pressure reduces the effect of the normal stress to an **effective stress** of $(\sigma-p=\sigma')$. It is clear from Figure 2.30(b) that the fluid pressure will drive a stable stress regime (the left-hand Mohr circle) towards the failure envelope. If the fluid pressure is sufficiently high to cause the Mohr circle to touch the envelope, fracturing will occur. This is termed fluid-induced failure or hydraulic fracturing.

### 2.5.4 The expression of fluid-induced failure in rocks

Although many geologists are familiar with the concept of hydraulic fracturing, the expression of hydraulic fracturing in rock can be confusing.

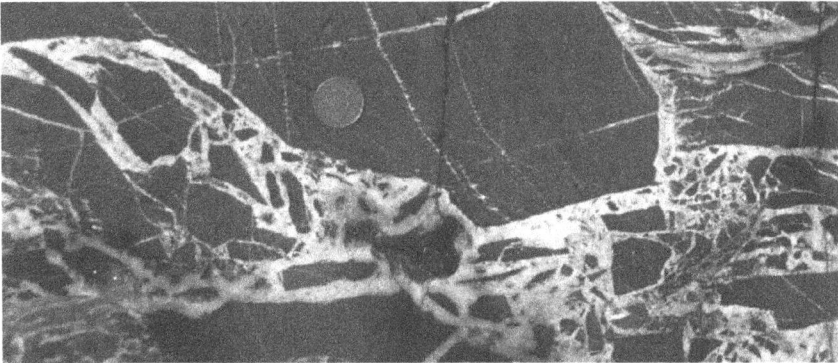

Figure 2.31.    An example of fluid induced failure from the Rusey fault zone SW England. The disorganised array of extensional fractures indicates that the differential stress $(\sigma_1 - \sigma_3)$ at the time of fracturing was low. The coin is 2 cm in diameter.

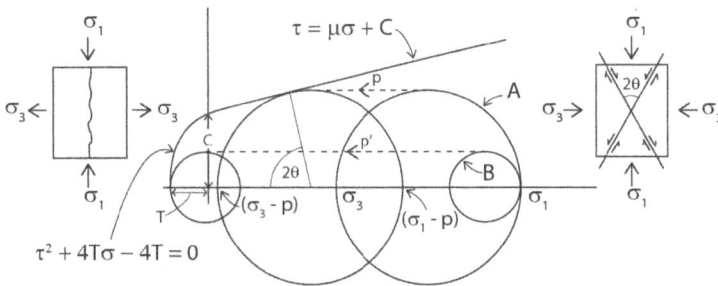

Figure 2.32.    Two stable lithostatic stress states represented by Mohr circles A and B with a high and low differential stress respectively. The lithostatic stresses are changed to effective stresses, $(\sigma - p)$, by a fluid pressure which moves the Mohr circles to the left where they come into contact with the shear failure and extensional failure envelopes resulting in the formation of shear fractures and extensional fractures.

Commonly, the impact of a high fluid pressure is assumed to 'blow the rock apart' and produce a breccia, Figure 2.31.

However, it is apparent from Figure 2.32 that this is not generally the case. The expression of brittle failure induced by an increase in fluid pressure will be determined by the magnitude of the differential stress. The two stable stress states represented by the Mohr circles A and B shown in Figure 2.32 have both been pushed into contact with the failure envelope by an increase in fluid pressure. Stress state A, with the large differential stress, generates shear failure, and stress state B, with a differential stress less than $4T$, generates extensional failure. This diagram resolves the apparent contradiction of the formation of numerous extensional fractures (joints) in the upper crust which is generally under a compressive stress regime, i.e., an increase in fluid pressure can cause a compressive lithostatic stress ($\sigma_1$, $\sigma_2$ and $\sigma_3$) to become effective stresses (($\sigma_1-p$), ($\sigma_2-p$) and ($\sigma_3-p$)) driving the compressive stress regime into the extensional regime where it can impact with the extensional failure envelope and generate extensional fracture, Figure 2.32.

The stress states A and B in in Figure 2.32 have the same value of $\sigma_1$ which commonly occurs in two adjacent layers at some depth, $z$, in the crust when the environment is such that the stress in the rock is dominated by the weight of the overburden. In this situation both layers have the same vertical stress, $\sigma_1$, but the horizontal stresses are determined by the material properties (primarily the Poisson's ratios, Equation (2.27)) and are therefore different. It can also be seen from Figure 2.32 that the fluid pressure required to cause failure is different in the two rocks, i.e., $p$ and $p'$. A higher pressure ($p'$) is required to cause failure in the rock with the lower differential stress, stress state B.

It was noted earlier, Section 2.4.2, that extensional fractures open up against the least principal stress (i.e., according to a least work principle) and that extensional failure can occur under a range of differential stress values ranging between zero and $4T$, where $T$ is the tensile strength of the rock. Figure 2.33 shows four stress states with differential stresses ranging from just under $4T$ (Mohr circle i) to zero (Mohr circle iv). Note that the stress state of zero differential stress represents a hydrostatic stress condition, i.e., $\sigma_1 = \sigma_2 = \sigma_3$. All four stress states will give rise to extensional failure, but their expression will be varied. Stress state i has a significant differential stress and so there is a clear direction of relatively easy opening of the fractures — which therefore display a marked preferred orientation, Figure 2.33(i). In contrast, under the hydrostatic state of stress represented by Mohr circle iv, the normal stress is the same in all directions. There is therefore no tendency for the fractures to be aligned and they form in all directions with equal ease. The result is the formation of fractures in all directions which generates a polygonal fracture network, Figure 2.33(iv), and can lead to the formation of a typical breccia texture. As the differential stress of the stress field reduces (Mohr's circles ii and iii in Figure 2.33), so

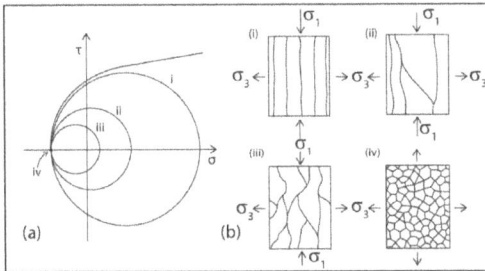

Figure 2.33.  (a) Shows four stress states (i–iv) represented by Mohr circles, all of which touch the brittle failure envelope and have a differential stress less than $4T$. They will all therefore result in the formation of extensional fractures. (b) Shows the expression of these failures. Natural examples in sandstones are shown in the photographs. The fracture regularity in the upper photograph indicates a relatively large differential stress (see b(i)) at Millook, North Cornwall. The lower photograph shows irregular quartz veins in the Devonian sandstone of St Anne's Head, Pembrokeshire, South Wales, corresponding to a lower differential stress (cf. b(ii) and (iii)).

the tendency for the fractures to align decreases, fractures (ii) and (iii). Thus, and depending on the value of the differential stress, extensional fractures can range from randomly orientated fractures to a well organised fracture set in which all the fractures are aligned. The photographs in Figure 2.33 show examples of extensional failure under a relatively high differential stress, upper photograph, and low differential stress, lower photograph.

This discussion of the effect of the differential stress $(\sigma_1 - \sigma_3)$ on fracture regularity in the $\sigma_1 - \sigma_3$ plane complements that in Section 2.4 where the impact of the differential stress $(\sigma_2 - \sigma_3)$ on fracture regularity in the $\sigma_2 - \sigma_3$ plane was examined. We can now consider how the magnitude of $(\sigma_1 - \sigma_3)$ places limits on the geometry of the fractures when viewed in the $\sigma_2 - \sigma_3$ plane. Because $(\sigma_2 - \sigma_3)$ can never be greater than $(\sigma_1 - \sigma_3)$, the alignment of the fractures in the $\sigma_2 - \sigma_3$ plane can never be better than the alignment in the $\sigma_1 - \sigma_3$ plane. It follows that the fractures shown in Figure 2.33(b) may have strikes as regular in orientation as the alignment of the fractures on the $\sigma_1 - \sigma_3$ planes illustrated, but can never be more regular.

However, as the $(\sigma_2-\sigma_3)$ value decreases, the fracture strikes will become progressively more random eventually generating a polygonal array when $(\sigma_2-\sigma_3)$ = zero, Figure 2.24(c). The fracture patterns shown in Figure 2.33(b) (i–iv) which represent a progressive reduction in the value of $(\sigma_1-\sigma_3)$ will have strikes that are progressively less regular until, in the example of Figure 2.33(b)(iv), the stress state will be hydrostatic and the organisation of the fracture strikes in the horizontal plane will be random. A polygonal array of fractures will therefore form both on this plane and on the vertical plane, resulting in the formation of a breccia texture in which the rock is broken into randomly shaped fragments, see Figure 2.31.

In addition, if the differential stress exceeds $4T$, then the resulting hydraulic fractures will be either hybrid or shear fractures. It should be noted that it is generally not possible to determine by inspection whether the fractures were caused by hydraulic fracturing or simply by a lithostatic state of stress.

### 2.5.5 Fluid induced failure in anisotropic rocks

It can be seen from the previous text and Figures that the orientation of fractures, be they extensional or shear, is determined by the orientation of the causative stress field. The failure criterion for the formation of extensional fractures by high fluid pressure is that the fluid pressure ($p$) must equal the sum of the tensile strength ($T$) of the rock plus the normal stress acting across the potential fracture, ($p = T + \sigma_n$). This is shown diagrammatically in Figure 2.34(a). As the fluid pressure builds up, this condition is first met along planes normal to the least principal compression, Figure 2.34(b).

However, an implicit assumption built into this argument is that the rock is homogeneous and isotropic. If the rock is anisotropic, and many rocks do possess a planar anisotropy either as a result of the processes involved in their formation, such as the bedding anisotropy in sedimentary rocks, or as a result of subsequent deformation, such as the cleavage and schistosity that characterises many metamorphic rocks, then the tensile strength of the rock is not a constant value. In the case of an isotropic rock, the tensile strength is constant in all directions and the material properties have no influence on the orientation of the fractures — which is controlled completely by the stress. However, when the rock is

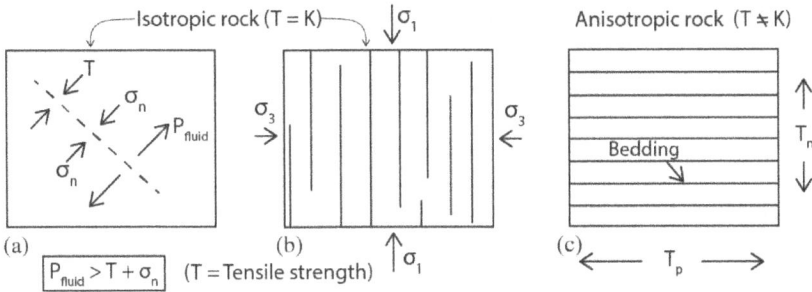

Figure 2.34. (a) A block of isotropic rock and the conditions necessary to induce a hydraulic fracture parallel to the dashed line. (b) The block in (a) subjected to an overburden load and an induced horizontal stress. Conditions for hydraulic fracturing are first met along the vertical planes normal to $\sigma_3$. (c). An anisotropic (bedded) rock subjected to the same stress field as the rock in (b). $T_p$ is the tensile strength parallel to bedding and $T_n$ is the tensile strength normal to the bedding.

anisotropic there is an orientation within the rock along which fracturing will occur more easily. Thus, when the fluid pressure builds up in such a rock, the anisotropy of the stress field and of the material compete to control the orientation of the fractures that form.

In Figure 2.34(b), the least principal stress $\sigma_3$ is horizontal. So, in an isotropic rock, the fluid-induced extensional fractures will be vertical, as illustrated. However, in an anisotropic material, Figure 2.34(c), the material properties as well as the stress have an influence on the orientation of the resulting fractures. The rock has two distinct tensile strengths: one large, parallel to the bedding ($T_p$); and the other smaller, normal to the bedding ($T_n$). If the stress regime shown in Figure 2.34(b) is superimposed on Figure 2.34(c), the stress will favour the formation of vertical fractures (normal to bedding) and the material properties will favour the formation of horizontal, i.e., bedding-parallel, fractures. For any particular stress condition, it is possible to determine which set of fractures will form.

The condition for fluid induced vertical fractures to form is that the fluid pressure should equal or exceed the normal stress ($\sigma_3$) and the tensile strength parallel to bedding ($T_p$) (Equation (2.28));

$$p_{\text{fluid}} = \sigma_3 + T_p. \tag{2.28}$$

The condition for fluid induced horizontal fractures to form is that the fluid pressure should equal or exceed the normal stress ($\sigma_1$) and the tensile strength normal to bedding ($T_n$), (Equation (2.29));

$$p_{\text{fluid}} = \sigma_1 + T_n. \tag{2.29}$$

In order for bedding-parallel fractures to form, i.e., for the material properties to dominate and for the fractures to open against $\sigma_1$, the condition in Equation (2.29) should be reached before the condition in Equation (2.28) as shown in Equation (2.30);

$$\sigma_1 + T_n < \sigma_3 + T_p,$$

i.e.,

$$(\sigma_1 - \sigma_3) < (T_p - T_n). \tag{2.30}$$

If the difference between the tensile strengths is greater than the differential stress, the material properties determine the orientation of the fractures; and, if it is less, the stress controls fracture orientation. If $(\sigma_1 - \sigma_3) = (T_p - T_n)$, it is equally easy for fractures to form normal to both the minimum and maximum principal stresses. A natural example where this has occurred is shown in Figure 2.35.

Figure 2.35. A thin section of a shale showing the synchronous formation of quartz veins, both normal to and parallel to the bedding which is horizontal. This indicates that, at the time of vein formation, the differential stress ($\sigma_1 - \sigma_3$) was approximately equal to the difference between the tensile strength of the shale parallel and normal to the bedding ($T_p - T_n$).

## 2.5.6 The impact of stress on the flow of fluids through a fracture network

Many rock masses are characterised by fracture networks caused by the superposition of a number of fracture sets. In such a rock mass, any increase in fluid pressure is likely to open up appropriately orientated fractures rather than generate new hydraulic fractures. If we consider a simple example in which all the fractures have zero tensile strength, then it can be seen that at fluid pressures below the least principal stress, $\sigma_3$, none of the fractures will open. When the fluid pressure reaches $\sigma_3$, fractures normal to $\sigma_3$ will open but all other fractures will remain closed. As the fluid pressure increases, so the range of orientation of fractures that can open will increase — until the point when the fluid pressure is equal to the maximum compression, $\sigma_1$. At this time, all fractures will be open. This is illustrated diagrammatically in Figure 2.36(a) and quantified in the graph of Figure 2.36(b).

Evidence for the selective opening of fractures when the fluid pressure lies between $\sigma_1$ and $\sigma_3$ can be found in nature when only certain orientations of fractures within a fracture network are mineralised, the remainder being barren (see Cosgrove, 2004). This phenomenon records the

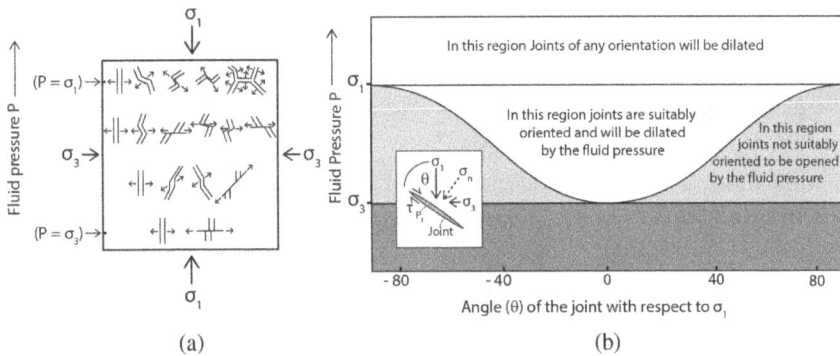

(a)          (b)

Figure 2.36.   (a) Fractures at various orientations with respect to the applied stress field. As the fluid pressure increases, the first fractures to open are those normal to $\sigma_3$. As the pressure continues to increase, so a greater range of fracture orientations can open, until, when the pressure equals $\sigma_1$, all fractures are open. (b) A graph showing the range of orientations of fractures that are opened by the fluid pressure as the pressure rises from $\sigma_3$ to $\sigma_1$. Based on Delayney *et al.* (1986).

influence of ancient stress fields on the response of a fracture network to fluid pressure and the preferential opening of fractures in a particular orientation along which fluids subsequently migrated.

Related to this discussion is the present day movement of fluids through fracture networks subjected to a differential stress. Most rocks in the crust and the fracture networks they contain are currently under such a stress regime caused by a combination of the overburden load and plate motion. This tends to close fractures normal to the maximum compression and open those parallel to it. The result is that, all things being equal, fluids tend to move along the fractures sub-parallel to the maximum compression direction and little or no flow occurs along fractures at a high angle to it. This has been observed in the Scandinavian shield where the highly conductive fractures trend NW–SE, i.e., parallel to the maximum compressive stress, ~25 MPa compression, linked to the opening of the Atlantic to the west, and the formation of the Alps to the south.

## 2.6 SMALL-SCALE PROCESSES LINKED TO FRACTURE FORMATION AND FRACTURE PROPAGATION

### 2.6.1 Introduction

In Section 2.3, the two fundamental types of fractures, extensional fractures and shear fractures, have been discussed. These are shown diagrammatically in Figure 2.37, the extensional mode (Mode I) in (a) and the shear failure in (b) and (c), the latter two modes being the in-plane or sliding mode (Mode II) where the slip direction is normal to the fracture tip, (b), and the anti-plane shear or tearing mode (Mode III) where the slip direction is parallel to the fracture tip, (c).

Any blind shear fracture, i.e., a fracture that is contained within a rock and which does not crop out on a rock face, will contain areas where the displacement at the fracture tip is Mode II and areas where it is Mode III. It can be seen from Figure 2.37 that the movement linked to extensional failure is normal to the fracture surface and that linked to shear failure is parallel to the fracture surface. In this Section, the small-scale processes associated with the formation and propagation of these fractures is considered.

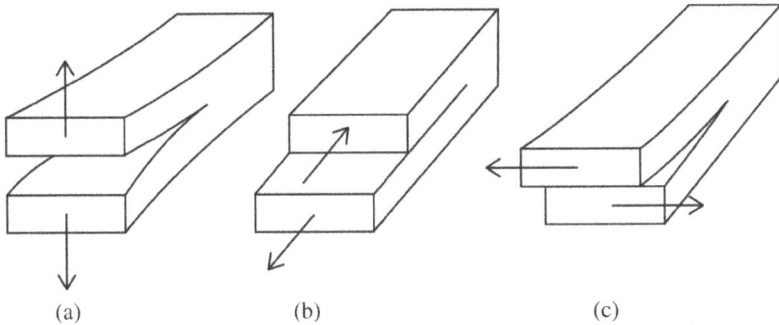

|   (a)   |   (b)   |   (c)   |

Figure 2.37. The fundamental fracture modes: (a) Opening extensional mode, Mode I; (b) In-plane shear or sliding mode, Mode II and (c) Anti-plane shear or tearing mode, Mode III.

## 2.6.2 Small-scale processes linked to fracturing

A study of fracture development on a microscopic scale has shown that a **deformation zone** is generated in front of the advancing fracture tip. This is often represented as a zone of microfracturing, also known as a **process zone,** which has been linked directly to the zone of stress concentration that forms around a fracture tip. This phenomenon of stress concentration or magnification around fracture tips was used by Griffith in his theory of extensional failure, Section 2.3.3. The detailed distribution of stress in this zone around the tip of an extensional fracture subjected to a far–field stress in which the greatest tensile stress is normal to the fracture is shown in Figure 2.38.

Concentrations of both shear and normal stresses occur in the vicinity of the tip. The contours of shear stress, Figure 2.38(a), define two distinct lobes that are oblique to the fracture tips and are separated by a local minimum along the fracture axis. In contrast, the normal stress is characterised by a single concentration ahead of the fracture tip with two maxima at 60° with respect to the fracture axis. A similar stress concentration can result from a build-up of fluid pressure in the fractures without the need for the far-field stress, Figure 2.39(a). The fluid pressure ($P_i$) and the far field stress ($\sigma_3$) together define the 'opening' or 'driving stress'. If the far–field stress is tensile (−ve) the two stresses act together ($-\sigma_3 - P_i$). If the far-field is compressive, the fluid pressure in the fracture would have to exceed the

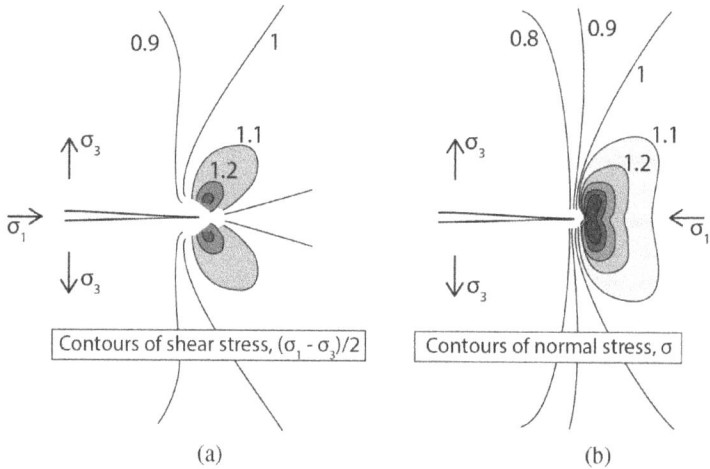

Figure 2.38.   Stress concentrations at a fracture tip caused by a remote stress field $\sigma_1$ and $\sigma_3$. (a) Contours of normalised maximum shear stress, and (b) of normalised normal stress acting on planes parallel to the fracture. From Eichhubl (2004).

remote stress before it could exert an effective tensile stress on the fracture walls leading to a stress concentration at the fracture tip.

This stress concentration can be quantified, Equation (2.31). The magnitude of both the normal and the shear stress at any point within the zone of stress concentration is determined by the position ($\theta$) and distance ($r$) of that point from the fracture tip (see Figure 2.39(a)) and by the 'Stress Intensity Factor' ($K$), Equations (2.19) and (2.20).

$$\sigma \propto \frac{K\theta}{\sqrt{2\pi r}},$$
(2.31)

where $K$ is determined by the fracture geometry '$L$' (the distance from the fracture tip to the point where the fracture width reaches the mean aperture '$a$', see Figure 2.39(a), the driving stress ($\sigma_3^R - P_i$) where $P_i$ is the fluid pressure inside the fracture, and a dimensionless 'modification factor' ($m$) which has a value of 1 for a straight internal crack some distance from the remote boundaries of the elastic body, Lawn and Wilshaw (1975) and Mandl (2005).

The localised nature of this stress concentration is apparent from Equation (2.31) which indicates that the stress concentration is inversely

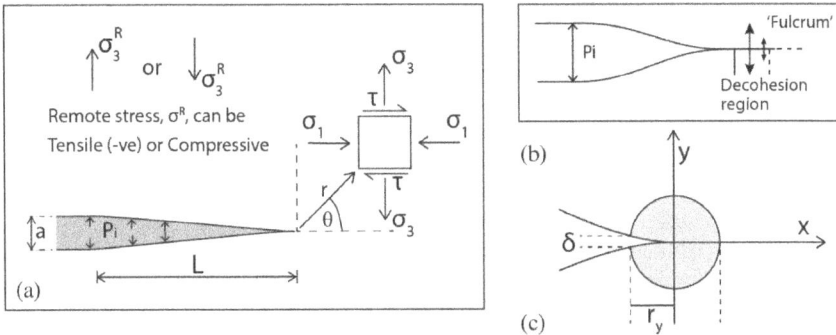

Figure 2.39. (a) The stress concentration in the near-tip region of a tensile fracture with an internal fluid pressure. (b) The leverage action of the fracture walls on the near-tip region. (c) Schematic drawing of the nominal plastic zone, nonlinear zone or process zone. From Atkinson (1987).

proportional to the square root of $r$, the distance from the fracture tip, and therefore drops off quickly away from the fracture.

This zone of stress concentration in which the fracture propagation occurs is often referred to as a 'plastic' zone and is a region in which nonlinear deformation occurs. It is also known as the 'decohesion' region, Figure 2.39(b). This Figure shows the leverage action of the fracture walls on the near-tip region. Figure 2.39(c) is a schematic drawing of the nominal plastic zone, nonlinear zone or process zone. For simplicity, it is assumed that this zone is circular in the $x$–$y$ plane, although in practice this will depend upon the state of stress. The intersection of the zone with the crack surface defines the point at which the crack tip opening displacement $\delta$ is measured. The drawing shows simple tension deformation where the principal tensional stress is parallel to the $y$-axis. Experimental work has indicated that the process zone is characterised by a cloud of microfractures, some of which link to allow the main fracture to propagate. This is discussed in the following Section.

## 2.6.3 The process of fracture propagation

Two mechanisms have been proposed to explain the process of fracture propagation, namely 'Growth through a process zone' and 'Growth by fracture linkage', Figures 2.40(a) and 2.40(b), respectively.

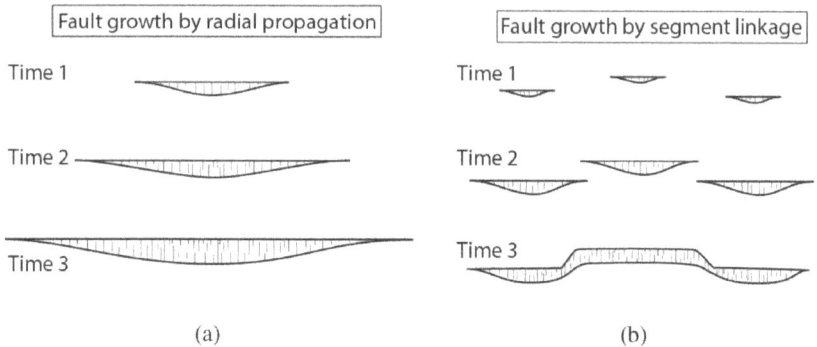

Figure 2.40.   Two modes of fracture growth. (a) Radial fault propagation and (b) Fault segment linkage.

### 2.6.3.1 *Growth of a fracture by propagation through a process zone*

The growth of a fracture by the development of a process zone in front of the advancing fracture tip has been studied experimentally and the results of two such studies are shown in Figure 2.41. This work has indicated that the process zone is characterised by a cloud of microfractures, some of which link to allow the main fracture to propagate. The left-hand diagram in this Figure is a schematic representation of the influence of the process zone on macrocrack growth. Tensile deformation across the plane $y = 0$ increases through sketches (i)–(v). Sketch (ii) shows the microcracks mostly isolated; at this stage, linear elastic behaviour is still observed. In (iii), the microcracking becomes more intense; some microcracks link up and nonlinear behaviour begins to characterise the deformation. In (iv), the macrocrack extends by the linking of microcracks within the now fully developed zone of nonlinear elasticity. In (v), further macrocrack extension occurs by migration of the process zone through the material ahead of the microcrack tip.

A similar increase in the density of microfractures in the process zone at the tip of a growing tensile fracture is shown in Figure 2.41(b), which schematically illustrates the different stages in fracture propagation and the development of the microcrack damage zone in the 'double cantilever beam' testing procedure for opening fracture propagation in a specimen of Salem limestone.

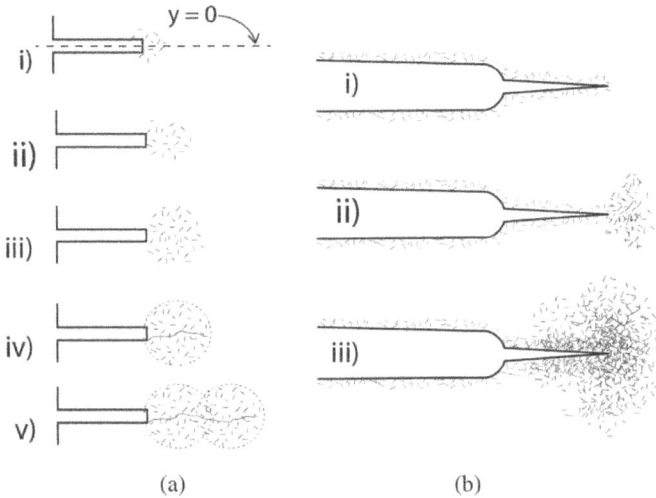

Figure 2.41.   (a) Schematic drawing illustrating the development of a process zone and its influence on macrocrack growth. Tensile deformation across the plane $y = 0$ increases through sketches (i)–(v), (from Atkinson, 1987). (b) 'Double cantilever beam' testing procedure for opening fracture propagation in Salem limestone. Sketches (i)–(iii) show the evolution of the microfracturing in the damage zone at various stages of fracture propagation, from Hoagland *et al.* (1973).

A study of the process zone around the tips of stressed fractures by Evans and Blumenthal (1984) and Eichhubl (2004); Figure 2.42 shows that the damage at fracture tips is sometimes concentrated in side lobes, Figure 2.42(a), and sometimes as frontal damage, Figure 2.42(b), see also Figure 2.38(a) and 2.38(b). These observations are clearly relevant to the formation of shear and extensional fractures. From the earlier discussion of brittle failure where the magnitude of differential stress required for the formation of extensional and shear fractures was quantified, it can be argued that, under low differential stress, the frontal damage pattern is the more efficient, encouraging the formation of extensional fractures and, under high differential stress, the side lobe pattern is more efficient, encouraging the formation of shear fractures. Diagrams (c)–(f) show a sequence of fracture growth by pore growth and coalescence. Fractures are inferred to grow with their long axis perpendicular to the least principal effective compressive stress, $\sigma_3$.

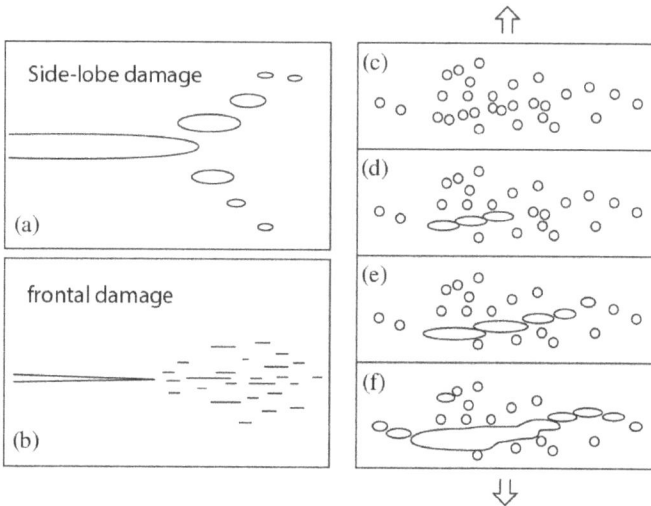

Figure 2.42. Diagram showing damage distribution at fracture tips, loaded in tension perpendicular to the fracture: (a) side-lobe damage, after Evans and Blumenthal (1984); (b) frontal damage, from Eichhubl (2004). Diagrams (c)–(f) show a sequence of fracture growth by pore growth and coalescence. Fractures are inferred to grow with their long axis perpendicular to the least principal effective compressive stress, $\sigma_3$. From Eichhubl (2004).

It might be argued that, as the fracture develops and passes through the process zone, evidence of this zone would be left along the margins of the fractures and that the thickness of this zone of microfracturing would reflect the size of the fracture. However, thin section study indicates that in most rocks these predicted marginal deformation zones cannot be detected. In addition to the microfracturing that occurs in the process zone of both extensional and shear fractures, other fractures are likely to develop along the walls of shear fractures as a result of continued movement along the fracture. This has the effect of generating a fracture zone, rather than a discrete fracture — a topic we will return to when the brittle–ductile transition is discussed in Section 2.10.

## 2.6.3.2 *Fracture propagation in granular materials*

In the preceding Sections on tensile failure and fracture propagation, emphasis has been focussed on the stress concentration that occurs in material at fracture tips when the material is subjected to either a compressional or extensional stress. However, in granular materials such as rock, there are other factors that lead to local stress concentration where various types of deformation such as pressure solution or fracture initiation and propagation are likely to occur. When a granular material is loaded, stress is transmitted through the material across grain contacts resulting in local intense stress concentrations, Figures 2.43(a) and 2.43(b). This causes large and heterogeneous strains to develop and cracking can occur.

In nature, sediments become buried in sedimentary basins and are subjected to overburden stresses that generate stress concentrations at the grain contacts. As lithification occurs by the various processes linked to diagenesis, material is precipitated in the pores and the sediments become cemented. This locks the elastic strains generated by the stress concentrations into the rock so that, on the removal of the overburden load during exhumation, the elastic strains are not released. Figure 2.43 shows an experiment designed to illustrate this process. It exploits the photoelastic properties of perspex which when deformed and viewed through polarized sheets reveals the distribution of strain within the individual grains and the concentration of deformation at the grain contacts. After vertical loading the model, epoxy resin was poured into the void space. The vertical load was maintained whilst the resin set mimicking the effect of a cementation. On removal of the load the 'cement' prevented the expansion of the grains locking in the stored strain energy. This is shown diagrammatically in the sketches in Figure 2.43(a). The possible impact of this stored strain energy on fracturing is considered in Section 2.14.4 where it is proposed that residual stresses in the rock may be responsible for the formation of exfoliation fractures during the process of rock exhumation.

In Figure 2.43(b), the Chinese finite element computer program RFPA has been used to simulate the progressive failure of rock particles when a steadily increasing compressive vertical stress is applied to an assemblage of particles in a crushing chamber, Tang and Hudson (2010). The fundamental and progressive nature of brittle failure of a single particle when diametrically loaded (the Brazilian test) is well illustrated, as is the interaction between particles as the fracturing becomes more intense.

In addition, some idealised mechanisms of deformation at the grain scale are illustrated in Figure 2.44 which shows various possible modes

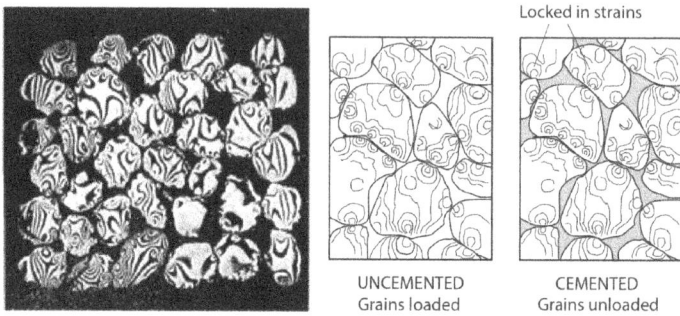

(a)

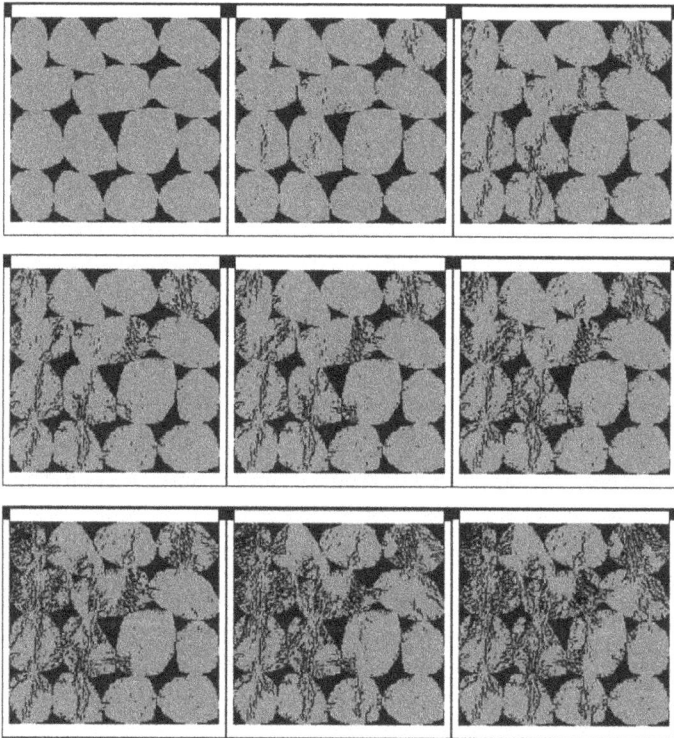

(b)

Figure 2.43. (a) Stress concentrations at 'grain' contacts shown by the photoelastic fringes developed in strained perspex particles subjected to a uniaxial compression, from Price (1966) plus line drawings showing that cement in the pores prevents the elastic strains from dissipating when the uniaxial load is removed. (b) Computer simulation of the progressive fracturing of rock particles during vertical compression (from Tang and Hudson, 2010).

of microscopic deformation which are likely to occur during the compressional (b)–(e) and extensional (f)–(i) deformation of a granular material. These schematic examples of microscopic mechanisms of grain scale deformation, which have been observed to occur in rock during strength tests, include (b) microcrack growth from flaws within mineral grains, (c) wedging of one grain between neighbours with grain boundary sliding, (d) lateral extension of soft grain promotes crack growth in an adjacent stiff grain, and (e) slip of inclined flaw induces wing cracks. These mechanisms which produced local regions of tensile failure all developed in rocks where the applied stress field was compressive.

The sub-figures (f)–(i) show schematic examples of microscopic deformation mechanisms in a damage zone during open fracture propagation, i.e., where the applied stress field was tensile. These mechanisms include (f) microcrack growth within a mineral grain, (g) growth of cracks from flaws, (h) opening of grain boundaries, and (i) shearing of grain boundaries as grains pull apart. These processes are thought to operate in the process zone which develops in front of and around an advancing fracture tip during the formation of shear and extensional fractures. They often occur in combination and an example where small-scale tension fractures (b) combine with sliding crack (e) link to form an incipient macroscopic shear fracture is shown in Figure 2.44(a). This link between extensional and shear failure is discussed further in Section 2.10.4.

## 2.6.4 Displacement on fractures

The magnitude of the displacement and its variation along a fracture, whether it be in shear or extension, is of considerable interest to Earth scientists and engineers — as it has major implications regarding the flow of fluids either along or across the fracture. In the former, the displacements are dominantly parallel to the fracture plane and, in the latter, they are normal. Displacement on fractures has been studied using three main approaches: namely, by the study of field examples, by the use of analogue models, and by theoretical analyses.

The study of naturally occurring shear fractures shows that the displacement gradient often increases gradually from the fracture tips reaching a maximum near the centre of the fracture, Figure 2.45, and it is often

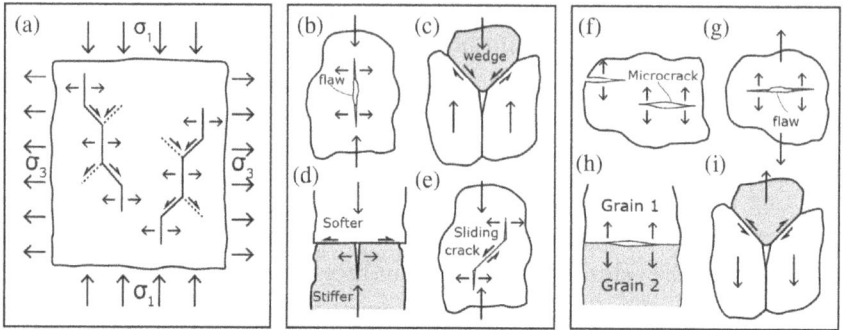

Figure 2.44.  (a) The linking of small scale tension fractures (b) with crack sliding (e) to form an incipient macroscopic shear fracture, (from Mandl, 2000). (b)–(e) Schematic examples of microscopic mechanisms of deformation at the grain scale in rock during strength tests in compression and (f)–(i) in extension, all capable of producing local tensile failure, from Pollard and Fletcher (2005).

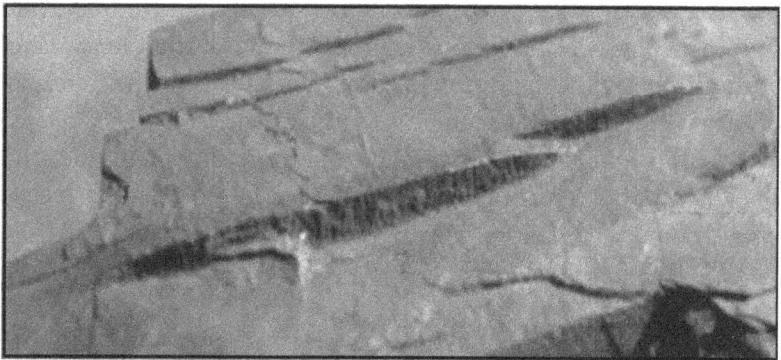

Figure 2.45.  Small-scale shear fractures (normal faults) cutting a limestone bed in the Liassic sediments of Lilstock North Somerset, UK. The nonlinear variation in displacement along the fault can be clearly seen. The displacement reaches a maximum at the mid-point between the fracture tips.

assumed that the displacements are similar to those associated with the flexing of a horizontal strut clamped at both ends and loaded vertically at its centre, Figures 2.46(a) and 2.46(b), the latter showing the displacement profile in the hanging wall of a normal fault adjacent to the fault plane is indicated as a series of circular arcs with an upward (+ve) radius of curvature $R$ at the outer portions of the fault plane, and a downward (−ve)

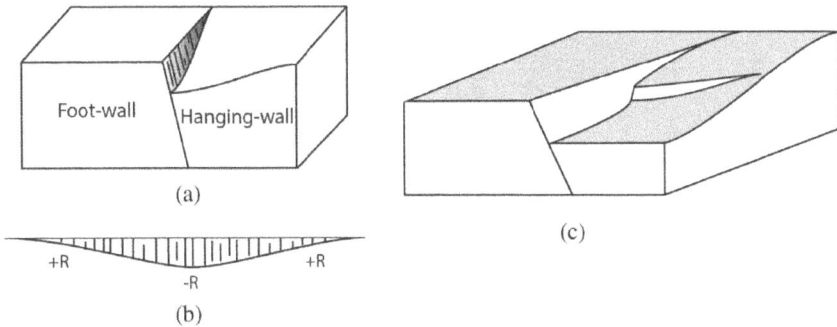

Figure 2.46. (a) Block diagram indicating the displacement variation along the hanging wall of a normal fault. (b) Displacement profile of the hanging wall strata at the fault plane, from Price and Cosgrove (1990). (c) Secondary normal fault branching off a main fault at the location of the greatest extension of the downthrown layer, from Mandl (2000).

radius of curvature in the central regions. Note that the radii of curvature lie in the fault plane and that a similar but mirror image of this displacement profile will occur in the footwall (left-hand block of Figure 2.46(a)).

In theory, the stresses generated in the deflected portion can be retained in the structure as elastic strain energy; however, in nature at least some of these stresses are generally dissipated by either brittle or ductile mechanisms. It can be seen that the length $+R$ to $-R$ to $+R$ in Figure 2.46(b) is greater than the original length, and the stresses associated with this local extension can lead to the formation of secondary fractures. A secondary normal fault branching off a main fault at the point of greatest extension of the downthrown layers is shown in Figure 2.46(c). The displacement profiles shown in Figure 2.46 are only indicative and qualitative. However, attempts have been made to develop a quantitative model for the distribution of displacement on a growing shear fracture such as the normal fault shown in Figure 2.46(a), see Watterson (1986) and Walsh and Watterson (1987). In this model, the fault is assumed to have the form of an elliptical surface completely encased within the rock (it is a 'blind' fault, i.e., one that does not crop out at the Earth's surface or intersect another fault) and the deformation is assumed to be elastic, i.e., strains in the surrounding rock are not dissipated by ductile mechanisms.

Walsh and Watterson propose that the fault grows in a series of slip events after each of which the build-up of stress is released by fault

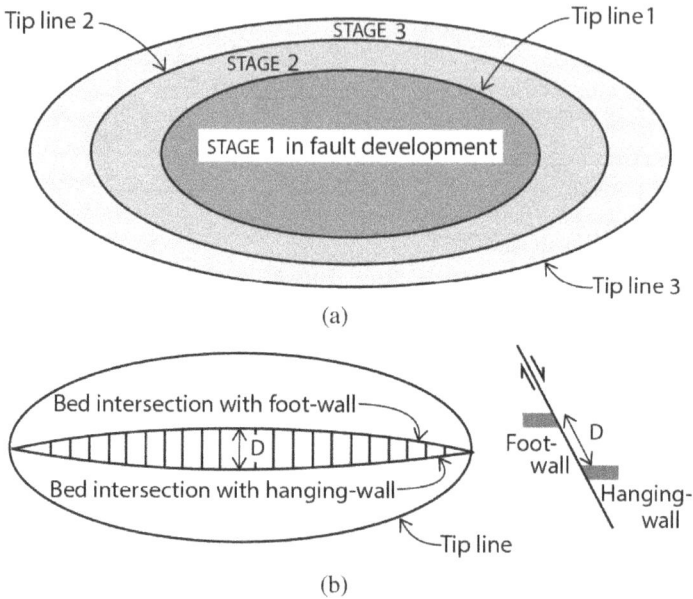

(a)

(b)

Figure 2.47.   (a) Shows the elliptical limit of the normal fault at three stages in its growth, from Mandl, 2000. (b) Plan and profile of an elliptical fault plane and the displacement profile assumed by Watterson (1986) in the analysis of fault growth. Slip on the fault is represented by the striations which vary from zero at the fault edge to a maximum at its mid-point.

movement and deformation. The episodes of fault growth are shown schematically in Figure 2.47(a). At any stage in the fault's evolution, it is assumed to have an elliptical shape.

The displacements range from zero at the fault tips to a maximum at its centre, as seen in the small-scale natural faults shown in Figure 2.45, and it is this displacement profile that is assumed in the analysis of fault evolution proposed by Watterson and Walsh and illustrated in Figure 2.47(b). This profile is observed to occur on all scales, and data from a large-scale fault are presented in Figure 2.48 which shows the slip-rate data for a 200-km slipped portion of the San Andreas Fault in central California. The slip data lie symmetrically around Monarch Peak, the centre of the slipped segment, and despite some 'noise' in the geodolite measurements they show a remarkable correlation to the theoretical slip profile for an elastic crack model.

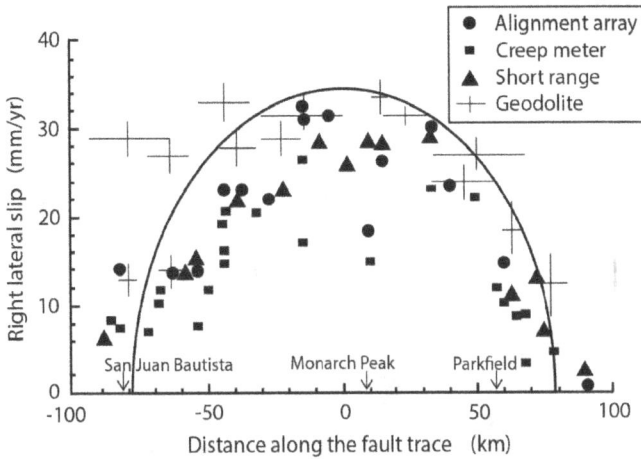

Figure 2.48. Slip rate data from a 200-km trace of the San Andreas fault in central California, after Burford and Harsh (1980), Lisowski and Prescott (1981) and Schulz *et al.* (1982).

## 2.6.5 Growth of a fracture by fracture linkage

In addition to the small-scale processes linked to the growth of fractures by their propagation through a process zone discussed in the previous Section, an alternative mechanism for the growth of faults that involves large-scale processes has been proposed. This involves the propagation, interaction and eventual linkage of fault segments which are parallel and relatively close to each other, but which are offset from each other and initially not overlapping, Trudgill and Cartwright (1994), Cartwright *et al.* (1995, 1996) and Walsh *et al.* (2002). This is in contrast to the model of lateral fault growth described earlier where faults grow as a result of radial propagation and where growth follows a path imposed by a scaling law, Watterson (1986).

Analogue models designed to study the growth of normal faults arrays, such as those commonly found at the margins of rift valleys, show clearly that major faults are formed by the linking of a series of smaller faults, Figure 2.49. The models can be examined at different stages of extension and the originally individual faults are seen to link as deformation proceeds. The displacement profiles of the individual faults

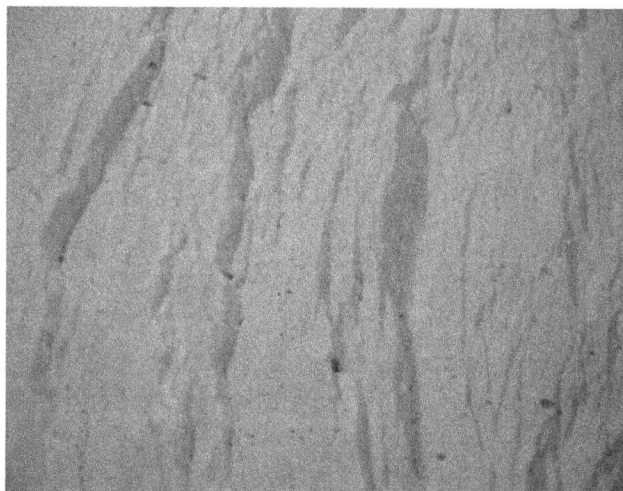

Figure 2.49.   A swarm of small-scale normal faults down-stepping to the right, at various stages in their linkage to form longer faults. Plaster of Paris analogue model, lit from the left-hand side.

approximate to those shown in Figure 2.47; however, when two fault segments meet, this displacement profile is dramatically altered.

The modification of the profile as linkage occurs is illustrated in Figure 2.50. The upper part of each diagram shows the plan view of the faults and the lower part the distribution of displacement. Down-folding occurs in the hanging-wall, (see Figure 2.46) and up-folding occurs in the footwall. At stage (a) (Figure 2.50), the faults do not interact, so the isolated segments will have the simple displacement profile of the type shown in Figure. 2.47. At stage (b) the faults begin to interact, so displacement is transferred between the faults by rotation of the intervening rock block, producing a ramp. As these ramps are features which relay the displacement from one fault to another, they are referred to as 'relay ramps' (Figures 2.51 and 2.52). At stage (c) connecting fractures start to link the overstepping fault segments and at stage (d) the ramp is destroyed, i.e., breached by the connecting fracture(s). The two faults now form a single structure with a marked 'fault bend' in the map view.

As indicated in Figures 2.50(b) and 2.51(a), during the early stages of fracture interaction, the two segments are not linked by a connecting

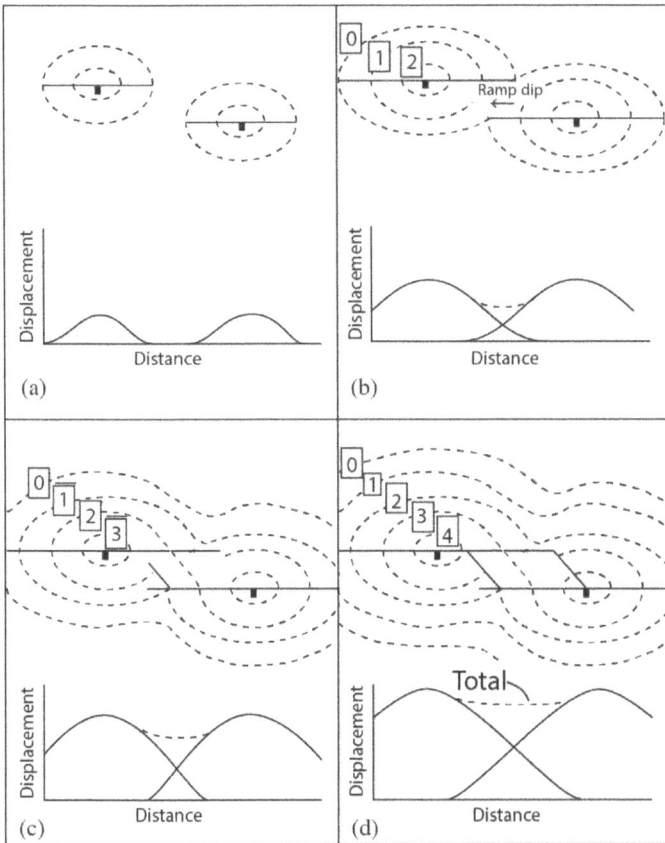

Figure 2.50. Structure contour patterns and distance–displacement graphs at four stages in the relay ramp development between two growing normal faults. The dashed portions of the displacement curves show the 'total' displacement in the zone of overlap between the two faults. The ticks (small black rectangles at the centre of the faults) are on the down-thrown side of the faults. From Peacock and Sanderson (1997).

fracture. Nevertheless the distortion of the stratigraphy in the 'relay zone' during the formation of the relay ramp is likely to result in an increase in fracturing and this fracturing will increase in intensity until the connecting fracture(s) finally link the two segments — which become 'hard linked' as the two originally separate faults form a single structure. The hard linking of originally separate normal faults along the margins of a graben is shown in Figure 2.51(c). The evolution of the relay zone between two

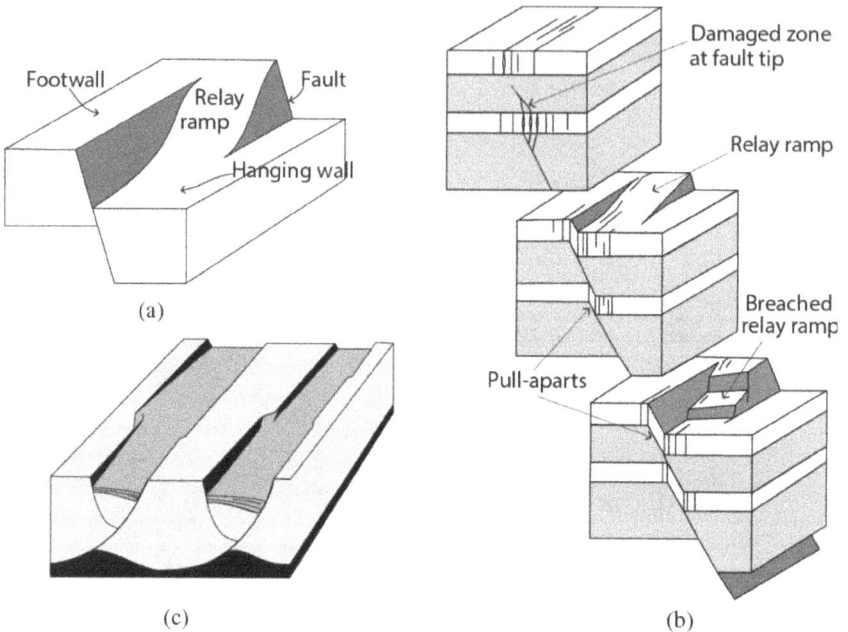

Figure 2.51. (a) Relay ramp formed between two normal faults which over-lap along their strike. (b) The evolution of a relay ramp and dilational jog as displacement on the faults increases. (c) Hard-linked normal faults forming the margins of a major graben. Note that these structures can form on all scales, see Figures 2.52. (a) and (b) after Peacock and Sanderson (1994), (c) after Trudgill and Cartwright (1994).

over-lapping normal faults in both plan and profile is illustrated in Figure 2.51(b). In addition to showing how the overlap along strike of the faults produces the relay ramp across which the two faults eventually link, Figure 2.51(b) also shows the over-lapping of the normal faults in the dip section which results in the formation of pull-apart zones (sometimes referred to as dilational jogs) as movement along the fault opens up the sub-vertical fractures which form to link two overlapping faults. Thus, as normal faults develop by the linking of smaller faults both along strike and down dip, they will be accompanied by the formation of highly fractured relay zones and pull apart features.

The linkage of fractures occurs on all scales and, as intimated in Figure 2.52, relay zones can vary from a few millimetres to several kilometres in size. The large scale example shown in this Figure has the

Figure 2.52.   (a) Small-scale normal faults and relay ramps in a limestone bed. The coin is 18 mm in diameter. (b) Overview from about 500 m height, looking at exposed normal faulting in the Canyon Lands of Utah. Fault throws are typically 10–100 m and their lengths 100–5000 m. The faults are highly segmented and it is apparent that the spatial organisation of the individual normal faults and the formation of the relay ramps are identical in both examples.

dimension of a typical rock engineering site. It follows that, if engineering projects have to be carried out in the vicinity of faults, then, in addition to the need to establish the extent of the damage zone along the fault margins, it is also important to determine whether or not the site is situated within a relay zone between overlapping faults as these areas are likely to regions of particularly high fracturing.

It can be argued that, as the growth of a fault (fracture) by radial propagation is achieved by the linking of microfractures that develop in the process zone of an advancing fracture tip, there is no fundamental difference between this mechanism of fracture growth and that of fracture linkage. The difference is one of scale: in the former mechanism, the growing fracture extends by linking with fractures several orders of magnitude smaller than itself; in contrast, in the process of fracture linkage, the two linking fractures are of approximately the same scale.

### 2.6.6 Fracture size and fracture displacement and the relation between fault width and maximum displacement

Watterson (1986) attempted to quantify the relation between fault width (W) the long axis of the ellipse that defines the fault, Figure 2.47(b) and the maximum displacement ($D$), Figure 2.47(b). He considered data from natural faults of different types and sizes with displacements ranging from 0.4 m to 40 km and widths from 150 m to 630 km and found that there is a near linear relation between these parameters of the form given in Equation (2.32), where '$c$' is a constant of proportionality.

$$D = cW^2. \tag{2.32}$$

The data on which this relation is based are consistent with a simple fault growth model for faults in which the slip in successive events increases by increments of constant size, and which predicts a relation between displacement and width of the form $D = cW^2$. However, it is unlikely that a fault 630 km in width will have formed from a single fracture by radial propagation — and will almost certainly involve both processes of fracture propagation in its evolution. Faults of this size are always 'fracture (fault) zones' and displacement is measured across the zone. Thus, the displacement data shown in Figure 2.48, which fit the theoretical displacement profile of a radially propagating fault, almost certainly involved 'fracture linkage' in its development. The data represented on this graph simply indicate that the gross variation in displacement varies from zero to a maximum and back to zero. The detailed geometry of the displacement profile will however, be more complex.

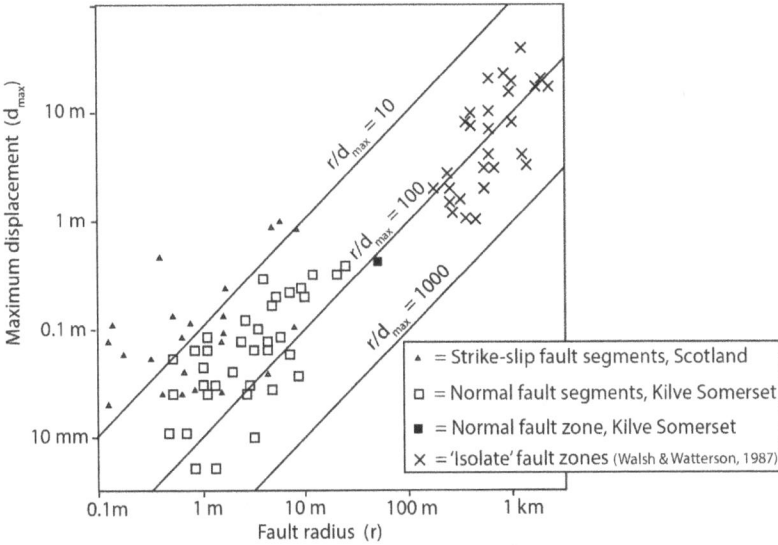

Figure 2.53. Graph of $r$ (distance between the point of maximum displacement and the fault tip) and $d_{max}$ (the maximum displacement) for faults from various sites around the UK, from Peacock and Sanderson (1994).

Peacock and Sanderson (1994, 1997) examined the relation between fracture size and maximum displacement for naturally occurring faults within the United Kingdom and their study included both normal and strike-slip faults, Figure 2.53. It was found that the different types of faults had different size/displacement ratios ($r/d_{max}$), where $r$ is the radius of the fracture (assumed to be circular) and $d_{max}$ is the maximum displacement. The strike-slip fault segments have a mean $r/d_{max}$ ratio of 24.2 (ranging from 1.34 to 109), while the normal fault segments at Kilve in Somerset, UK, have a mean $r/d_{max}$ ratio of 65.3 (ranging from 10 to 307). This is much lower than the mean ratio of 143 found for the isolated British coal-field normal faults and the authors suggest that this is probable because of fault interaction in their study area which would tend to reduce the ratio.

Although Watterson's relation between fault width ($W$) and maximum displacement ($D$), $D = cW^2$, has been tested by studies of natural faults and a statistical width/displacement ratio established for each fault type, it can be seen from the results summarised in Figure 2.53 that the spread of values for these ratios is large, i.e., over two orders of magnitude. This places a limit on the relation's predictive powers. If the value of *c* is not known for a particular fault, then knowing the maximum displacement or width does not enable one to determine the other property with any accuracy. In addition, because of the process of fracture growth by the linking of the smaller fracture, the maximum displacement does not always occur at the mid-point of the fracture, so displacement profiles can be more complex than those predicted by the models shown in Figures 2.46(a) and 2.46(b), and 2.47. An example of this is shown in Figure 2.50.

### 2.6.7 Movement on pre-existing fractures

In the earlier discussion on brittle failure, two rock strengths were identified: the tensile strength and the shear strength (i.e., the cohesion). The cohesive strength of the rock, *c*, relates to the shear strength of the intact rock. However, once shear failure has occurred, there will be a significant reduction in the shear strength of the material. The new failure criterion, which relates to the process of re-shear along the existing fault, rather than to the formation of a new shear fracture, is represented by the lower failure envelope in Figure 2.54(a). In this diagram, the cohesive strength of the shear plane is assumed to be zero and it is also assumed that the angle of sliding friction on the shear plane is the same as the angle of internal friction which defines the slope of the failure envelope for the intact rock.

Thus, in a stressed, fractured, rock mass, the possibility that re-shear will occur on suitably orientated fractures, rather than the creation of a new fracture, needs to be considered. Figure 2.54(a) shows a Mohr circle tangential to the failure envelope for intact rock — thus representing a stress of sufficient magnitude to generate new shear fractures. The state of stress on the fracture plane and its orientation can be determined from the diagram, as discussed earlier. The orientation $\theta$ of the fracture with respect to the maximum principal stress ($\sigma_1$) is obtained by joining the point of contact between the Mohr circle and the failure envelope (point B in Figure 2.54) with the centre of the Mohr circle. This line is inclined at $2\theta$ to the normal stress axis. The state of normal and shear stress on the fracture plane can be read off by projecting point B to the horizontal and vertical axes, respectively. If $\sigma_1$ were vertical, the resulting shear fractures would dip

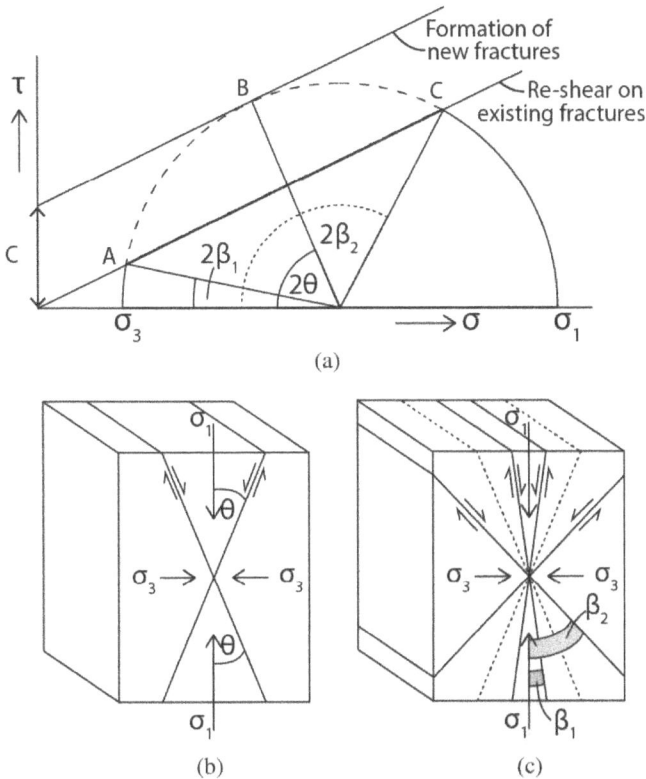

Figure 2.54. (a) Two shear failure envelopes for a particular rock. The upper one relates to fracturing of intact rock and the lower one to 're-shear' on an ideally oriented pre-existing fracture having zero cohesion. The Mohr circle represents a state of stress that satisfies the criterion for the formation of new fractures which will form at an angle of $\theta$ to $\sigma_1$, as shown in (b). It also satisfies the condition for re-shear on pre-existing fractures, see (c), whose angle with respect to $\sigma_1$ falls within the range $\beta_1$ to $\beta_2$, (a) and (c).

at an angle of $\theta$ and strike parallel to $\sigma_2$, Figure 2.54(b). The range of orientations of pre-existing fractures striking parallel to $\sigma_2$ that could be reactivated by this stress state can be determined from the diagram. It can be seen that, under this stress state, the conditions for re-shear on pre-exiting fractures are satisfied for all planes that fall on the dashed portion of the Mohr circle in Figure 2.54, i.e., whose angle with respect to $\sigma_1$ is in the range $\beta_1$ to $\beta_2$, A and C. This occurs at a stress level less than that required to form new fractures, Figures 2.54(a) and 2.54(c).

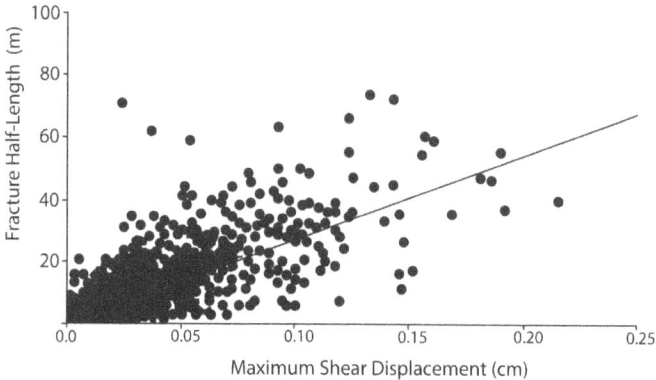

Figure 2.55. A graph showing the induced shear displacements on a population of target fractures of various lengths within a volume of rock 2 km away from a fault generating a Magnitude 6.1 earthquake. Numerical simulation from La Pointe *et al.* (1997).

The link between the maximum slip on a fracture and the fracture length outlined above and summarised in Figure 2.53, although relating to shear fractures, applies equally to re-shear on a pre-existing fracture regardless of its mode of origin. This is illustrated in the results of a numerical simulation, Figure 2.55, which shows the maximum shear displacements induced on a population of fractures of various lengths 2 km away from a fault, movement along which caused an earthquake of Magnitude 6.1. The motivation for this study was a problem relating to safety issues linked to underground radioactive waste disposal, i.e., the requirement to quantify the maximum possible movement on any fracture in the vicinity of a fault caused by the seismicity generated by fault movement. It can be seen that, despite the scatter of data, a relation exists between fracture length and maximum displacement.

There is, therefore, an established link between fracture size and displacement for both newly formed shear fractures and re-shear on pre-existing fractures, regardless of whether they were originally extensional or shear fractures. However, despite these links, an examination of the results in Figures 2.53 and 2.55 shows that these relations are frustratingly limited in their predictive capabilities. Although fracture length can be a useful indicator of possible maximum fracture displacement, on its own it is not an accurate guide.

## 2.7 THE RECOGNITION OF EXTENSIONAL AND SHEAR FRACTURES IN THE FIELD

### 2.7.1 Introduction

In the earlier discussion on brittle failure, it was noted that there are two types of brittle fractures: shear and extensional. These modes of failure are fundamentally different, each having a different failure criterion, a different orientation with respect to the principal stresses causing them, and a different sense of movement. It is therefore not surprising that some significantly different and characteristic features are found associated with these two fracture types which enable them to be recognised and differentiated in the field.

### 2.7.2 Extensional fractures and their recognition in the field

Figure 2.56 shows an example of extensional fractures forming a network of fractures cutting a Liassic limestone bed on the Bristol Channel coast in North Somerset, England. The network is made up of several sets of fractures as a result of separate applied stress regimes: they therefore record the evolution of stress over geological time. The techniques for interpreting the stress history of a region and for determining the detailed geometry of the associated fracture network (which can have a major impact on the geomechanical properties of the fractured rock mass that contains it) are discussed later in the section on the evolution of fracture networks, Section 2.12. In this present Section, we focus on the features that characterise extensional fractures and which enable them to be recognised in the field.

The principal difference between shear fractures and extensional fractures is their mode of displacement. Shear fractures involve slip parallel to the fracture, whereas movement on extensional fractures is normal to the fracture wall. It can be seen than none of the fractures in Figure 2.56 has experienced any shear movement.

Of particular use in differentiating between shear and extensional fractures are their surface characteristics. One of the most striking features of many extensional fractures is the formation of **plumose structures** on their surface, Figures 2.57 and 2.58. These feather-like impressions provide

Figure 2.56. Several sets of extensional fractures making up a fracture network cutting a Liassic limestone bed at Lilstock, North Somerset, UK. The image represents an area 3 m across.

evidence that extensional fractures, like shear fractures, form as elliptical features which grow and maintain this geometry until they reach a free surface or another fracture. Fractures are often initiated at some irregularity in the rock, Figure 2.57(b), and spread from this nucleus. As they do so, a series of small ridges is generated, each of which constitutes an individual barb of the plumose structure. As can be seen from this Figure, they fan out from the fracture nucleus and form parallel to the direction of fracture propagation.

The cause of the plumose striations is thought to be the initial irregularities, i.e., minor steps, on the fracture surface that are apparent on the section through the fracture oriented normal to the fracture and at right angles to the direction of propagation, Figure 2.57(c). These steps consist of two 'runners' parallel to the fracture joined by a 'riser'. Both runners would propagate in the direction of fracture propagation and the resulting riser (plumose line) which separates them would therefore be oriented parallel to this direction. It follows that the lines are always at right angles to the propagating fracture tip and the position of the fracture tip at various times in the evolution of the fracture can be constructed by drawing lines normal to the plumose striations. These 'fracture front lines' are

(a)                                          (b)

(c)

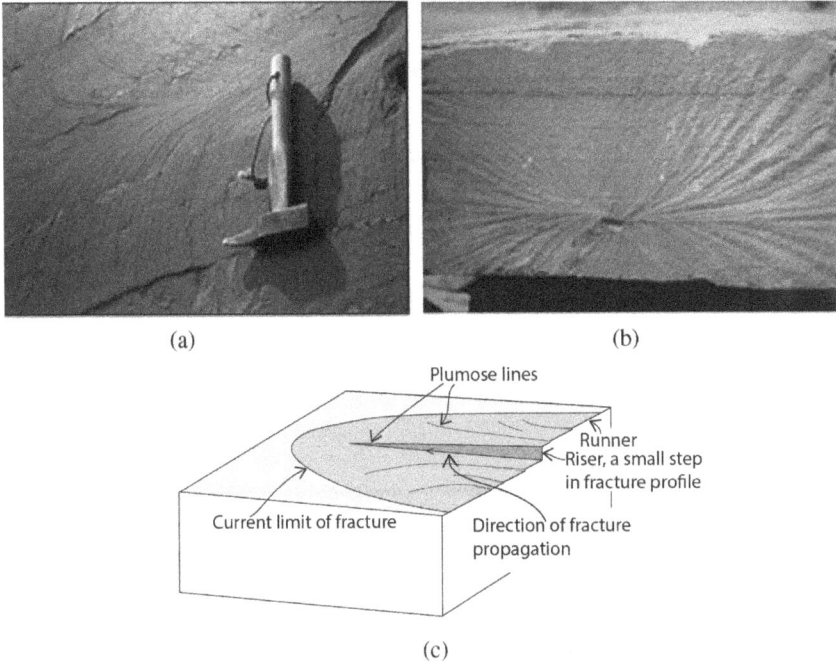

Figure 2.57.   (a) Feather-like plumose structures on a joint surface (courtesy J. Ramsay). (b) Plumose structure radiating from the nucleus where the fracture was initiated (courtesy T. Engelder).  The images represent rock surfaces ~1 m wide. (c) Diagram illustrating the origin of the plumose marks as small steps on the fracture surface. The central step is exaggerated for clarity.

shown in Figure 2.59 and reveal the elliptical shape of the growing fracture at all stages of its evolution. A study of experimentally formed plumose structures confirms that the plume axes (the horizontal line running through the numbers 3, 6 and 10 in Figure 2.59) form parallel to the maximum principal compressive stress $\sigma_1$.

Thus, a study of the surface features on extensional fractures reveals the history of fracture propagation. For example, the plumose structures and arrest lines on the joint face shown in Figure 2.58(c) show that the initiation started in the central circular fracture. Propagation of the fracture to the left occurred without significant breaks. In contrast, propagation to the right was arrested several times as indicated by the formation of a series of arrest lines or ridges.

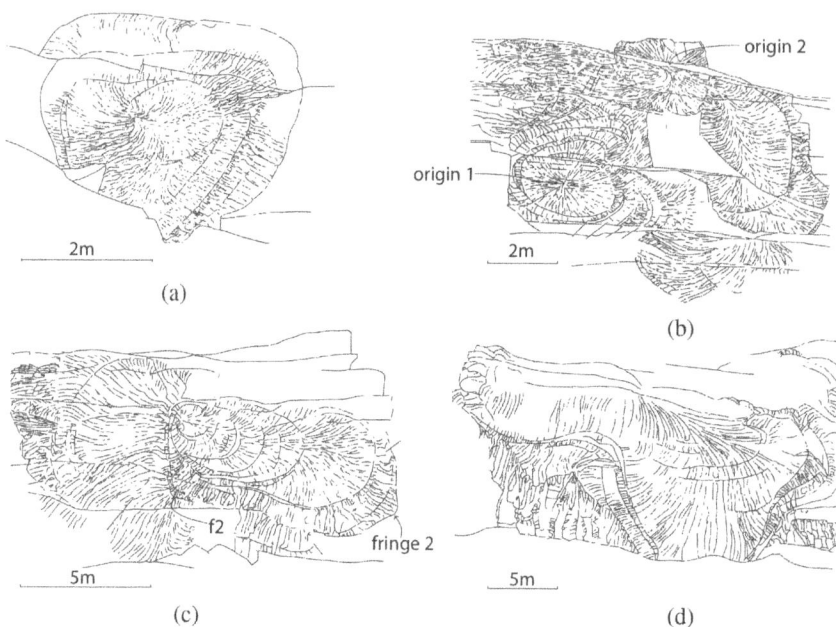

Figure 2.58. Large extensional fractures (2–30 m wide) in the Mrakotin granite, part of the South Bohemian pluton, Czech Republic. Note that the fractures tend to be 'penny shaped', i.e., circular. They are covered with plumose marks and also show well-developed arrest lines (see item 9 in Figure 2.59). In (b) two fracture origins can be seen (Figure 2.59). These two fractures joined to form the joint face. From Bankwitz *et al.* (2004).

It will be recalled from Section 2.3.3, Figure 2.17 and the related discussion, that in the fracture mechanics treatments of extensional failure, the concept of critical crack length indicates that, once a microfracture exceeds a critical length, it will continue to propagate at an accelerating rate. However, as noted above there is clear evidence in the form of surface features found on naturally occurring joint surfaces that this is not necessarily the process by which the fractures propagate in nature. These features indicate that, not only do joints have the same elliptical geometry proposed for shear fractures (Figures 2.59 and 2.60), but that, like shear fractures, they also grow in a series of pulses separated by periods when the fracture tip line is stationary.

The evidence for the periodic growth of these fractures is the formation of a series of arcuate (curved) arrest lines, Figure 2.59. They form

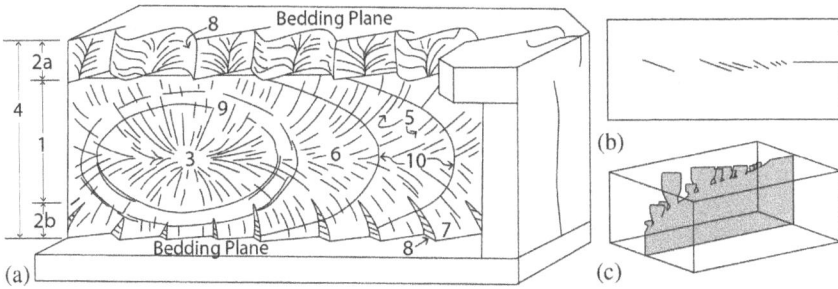

Figure 2.59. (a) Block diagram showing details of surface features observed on extensional fracture planes. 1 — Main joint face, 2a — Abrupt twist hackles, 2b — Gradual twist hackles, 3 — Origin, 4 — Hackle plume, 5 — Inclusion hackle, 6 — Plume axis, 7 — Twist hackle face, 8 — Twist hackle step, 9 — Arrest lines, 10 — Constructed fracture front lines. Note that the twist hackles produce an *en echelon* array of fractures on the bedding plane. Based on Hodgson (1961). (b) and (c) A map and block diagram of idealised *en echelon* cracks in front of a joint trace showing how cracks probably twist and connect to a parent joint at depth Cruikshank *et al.* (1991).

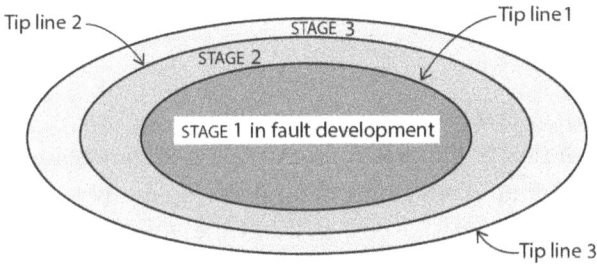

Figure 2.60. Elliptical model for a shear fracture showing its growth in a series of stages. Surface features on extensional fractures indicate that they also have the same elliptical shape and grow in a series of pulses, see Figure 2.47 and the related discussion.

normal to the plumose structures and are parallel to the constructed fracture front lines, feature 10 in Figure 2.59. These lines, as their name implies, are thought to represent the position of the fracture front at a time in the evolution of the fracture when the propagation temporarily ceased. Large scale arrest ridges, termed rib marks, developed in Coal Measures sandstone are shown in Figure 2.61.

Inspection of extensional fractures constrained to a particular bed shows that, as the fracture plane approaches the bed boundary, it often

Figure 2.61. The profile and front view of a joint surface showing large-scale rib structures. They are developed in a block of Coal Measures sandstone used as one of the Gorsedd Stones marking the site of the National Eisteddfod in Aberystwyth, Wales. The face of the joint is ~700 mm wide.

splits into a series of smaller planes with a consistent strike that is different from that of the parent fracture. This is shown schematically in Figure 2.59(b) and 2.59(c) and natural examples are shown in Figure 2.62. Because of the 'twist' in the strike of these features they are termed twist hackles; they form a fringe around the main fracture and, depending upon whether they gradually or abruptly change their strike, they are termed gradual or abrupt twist hackle fringes, Figures 2.59 and 2.62. The reason for the formation of twist hackles is still unclear, but it is thought to reflect a local change in stress state that occurs in the vicinity of a layer boundary or interface. If the twist hackles are Mode 1 fractures, i.e., fractures that form normal to the minimum principal stress, then the implication is that there has been a rotation of this stress as the bed boundary is approached. The formation of twist hackles produces an *en echelon* array of fractures which crop out on the bedding plane, Figure 2.59(a). This linking of *en echelon* fractures to a parent fracture has been observed on several scales (see Cruikshank, 1991) and is illustrated more clearly in Figures 2.59(b) and 2.59(c).

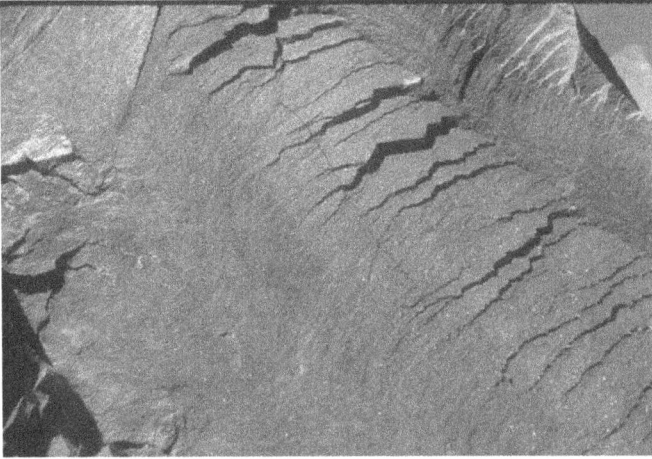

Figure 2.62.   Gradual twist hackles at the boundary of a bed; compare with Figure 2.59.

## 2.7.3 Strata-bound and through-going fractures

As noted earlier, fractures in rock may be restricted to certain beds (i.e., strata-bound fractures) or may propagate as a continuous fracture through a succession of beds (through-going fractures), and a variety of factors are involved in determining which of these develops. The two most important factors are the cohesive strength of the interface between the two beds and their relative stiffnesses, i.e., Young's moduli.

There is a region of enhanced tensile stress ahead of the tip of an advancing extensional fracture and it is this stress which drives propagation. As the fracture approaches the interface between two beds, and if this boundary is one of weak cohesion, the fracture will open upon reaching it — thereby dissipating the driving stress and so preventing further propagation, Figure 2.63(a). If the boundary between the two similar layers is cohesive, the driving stress will be maintained and the fracture will continue into the adjacent layer, Figure 2.63(b).

The other factor controlling fracture propagation from one bed to another is their relative stiffnesses. The critical stress concentration at the fracture tip needed for propagation depends upon the material properties: the stiffer the material, the greater the stress required. Thus, if a fracture in a relatively soft layer approaches an adjacent stiffer layer, it is possible that

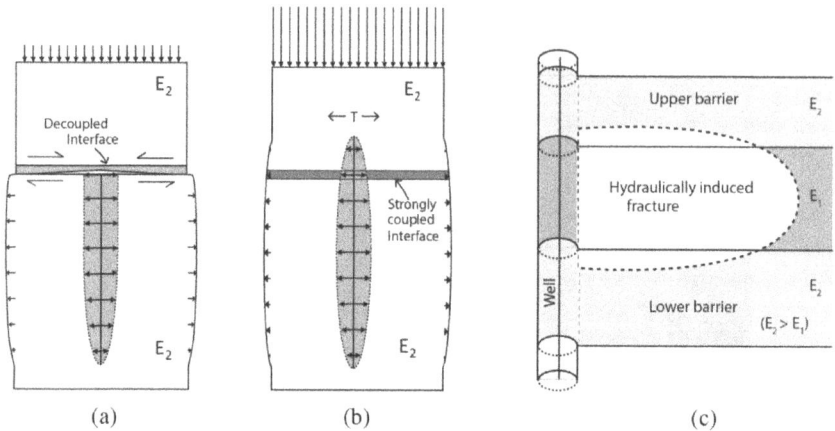

Figure 2.63.  (a) A schematic illustration of the mechanism of 'bed slippage' to impede opening mode fracture propagation. The fracture plane is perpendicular to $\sigma_h$, the least principal compressive stress component, and $E$ is Young's modulus. (b) Two adjacent beds have the same properties and fracture propagation depends on the magnitude of the cohesion along the bedding plane separating them. (c) Hydraulic fracturing from a well into a layered sequence of rocks. Note the fractures are constrained to the beds with the lower Young's modulus. From Casabianca and Cosgrove (2012).

the enhanced stress driving the fracture through the softer rock will not be large enough to propagate the fracture through the stiffer layer. Attempts have been made by Simonson *et al.* (1978), to quantify the magnitude of the moduli contrast between two contiguous layers required to impede fracture propagation across the separating interface. They argue that doubling the moduli across the interface is sufficient to ensure total containment of fractures within the lower moduli layer. The impact of rock strength on the formation and propagation of fractures is illustrated in Figure 2.63(c) which shows a well into which a fluid is being injected. As the fluid pressure increases the weaker horizons in the stratigraphy are the first to fail. For the reason noted above these fractures are unable to propagate into the adjacent, stronger layers and are constrained to the weak horizon.

It is clear from the above discussion that interfaces between two layers, (either similar or dissimilar) and the contrasts in material properties between adjacent layers play an important role in controlling joint

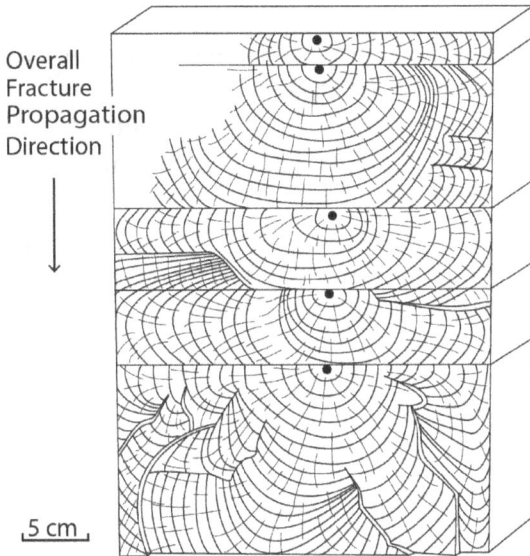

Figure 2.64.    A composite joint in a series of siltstone beds. The concentric lines are fracture front lines constructed from the plumose marks on the joint surfaces, and the black dots represent the initiation point for the five fractures. From Helgeson and Aydin (1991).

propagation in sedimentary rocks, limiting their vertical extent and physical continuity. However, despite the tendency for weak interfaces to impede the propagation of fractures, there is evidence that joints do 'communicate' across interfaces between dissimilar layers to form composite joints. An example of this is shown in Figure 2.64 which is a sketch of a composite joint cutting across several siltstone layers. Each individual layer has its own plumose structure, indicating independent fracturing of each layer. The joints are nevertheless vertically aligned and this in-plane arrangement results in a single composite joint surface. Note the vertical alignment of the 'initiation' points which are all located at the top of each layer and in addition are consistently placed where the above approaching joint first intersects the next layer. This implies an overall symmetrically downward propagation of the composite joint.

In addition to the influence that bedding can have on the propagation of fractures discussed above, it also has some control on fracture geometry. When fractures are constrained to a particular bed they can only propagate along the bedding, resulting in an elliptical fracture as shown diagrammatically in Figure 2.59(a). In contrast, when the fractures form in more massive rocks, the fractures tend to be more circular. This is well illustrated by the fractures in the massive granite shown in Figure 2.58. Construction of the fracture front lines (see Figure 2.59) in Figure 2.58 shows that the fractures were originally circular and maintained this geometry as they propagated. For example, inspection of the plumose structures and arrest lines on the fracture surface shown in Figure 2.58(c) shows that it nucleated in the centre of the face as a circular fracture and then propagated to the right maintaining this circular geometry.

## 2.7.4 The kinematic implications of crystal fibres

A number of diagnostic features have been described that can be used to demonstrate that a particular fracture is an extensional fracture. They are all 'surface features' and are summarised in the block diagram shown in Figure 2.59. However, there are other features that can be used to determine the type of fracture in a rock. One of the most convincing is the formation of fibrous minerals along the fracture. Minerals often grow in a fibrous habit when being precipitated into a developing fracture and it has been established that the fibres grow in the direction of fracture opening, (see e.g., Ramsay and Graham, 1970). Consequently fibres form normal to extensional fractures and sub-parallel to shear fractures, Figures 2.65(a) and 2.65(b), respectively.

A convincing demonstration that the mineral fibres form parallel to the direction of opening of the fractures is provided by the satin spar veins (a fibrous form of gypsum) shown in Figure 2.66. During the formation of this vein, the overburden stress was the minimum principal stress and the ideal orientation of the extension fracture is therefore horizontal. However, the existence of the irregular, sub-horizontal bedding provides a weak fabric which, although not perfectly oriented, can be used in preference to the formation of new perfectly horizontal fracture. The fracture follows the irregularities of the bedding and opens parallel to the least principal stress which is acting vertically. As can be seen, regardless of the local dip of the bedding, all the fibres form parallel to this direction.

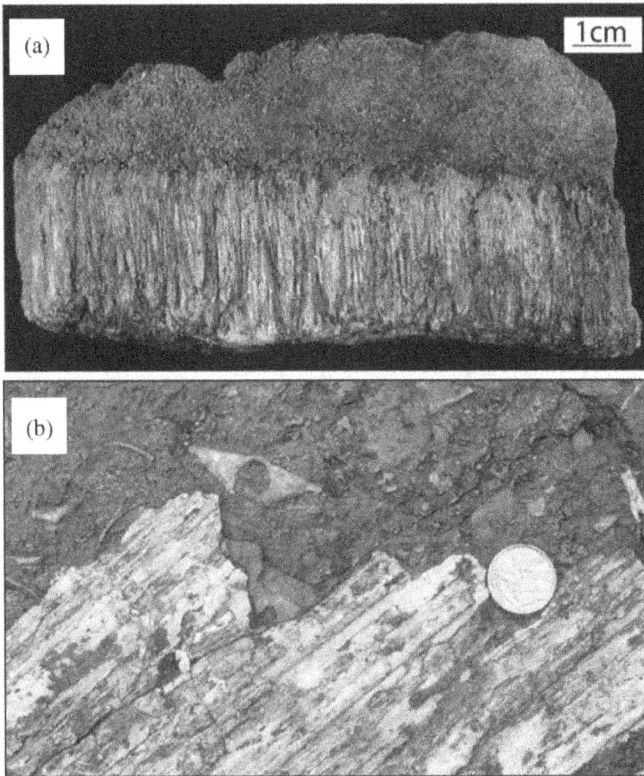

Figure 2.65. Two calcite infilled veins that formed parallel to bedding. (a) A vein of fibrous calcite in which the fibres formed at right angles to the fracture wall. The fibres formed parallel to the direction of fracture opening. (b) Calcite fibres that grew sub-parallel to the fracture wall (bedding) indicating that the relative motion of the 'fracture' walls was dominantly parallel to the bedding. This is clearly a shear fracture.

## 2.7.5 Shear fractures and kinematic indicators

Because of the fracture-parallel movement associated with shear fractures their surfaces are unlikely to be decorated with delicate features, such as the plumose structures that characterise extensional fractures. Instead, they are found associated with a variety of structures and features linked to the fracture-parallel displacement. These structures not only indicate the direction of movement but often reveal the sense of movement and are therefore termed kinematic indicators.

Figure 2.66. Satin spar vein formed parallel to an irregular bedding plane. Note the consistent, vertical orientation of the fibres which form parallel to the opening direction which is determined by the least principal stress (in this example the overburden load). Mercia mudstone, Watchet, North Somerset, UK. The photograph represents an area 100 mm × 35 mm.

Figure 2.67. Slickenside on a normal fault surface cutting a succession of Liassic limestone beds. The lineations are not scratch marks left by the two sides of the fault grinding past each other but are defined by long crystals of calcite lying sub-parallel to the fault.

The most well-known kinematic indicators are slickensides — which are smoothly polished rock surfaces that are the result of frictional movement associated with slip on a fault, Figure 2.67. As one side of the fault moves past the other, the fault plane becomes grooved and striated in the direction of movement. However, as in the example shown in Figures 2.65(b), the fault plane may be coated by mineral fibres that grew

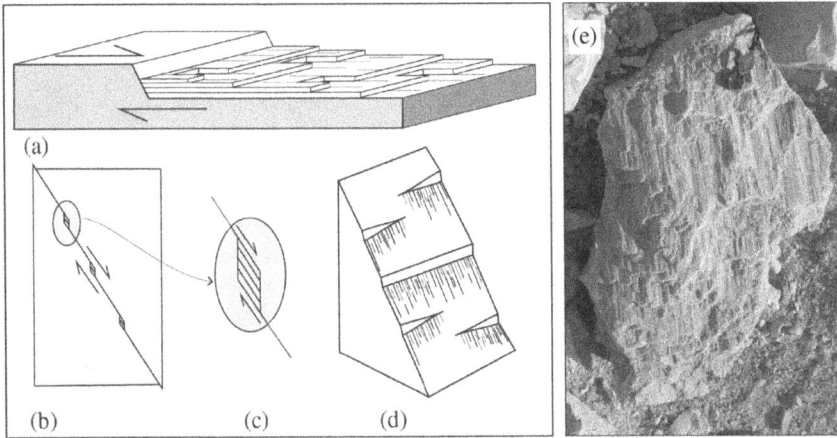

Figure 2.68.  (a)–(d) Diagrams illustrating the mechanism by which crystal fibres (slickenfibres) form sub-parallel to a fault surface. The fibres grow in the 'shadow' region of the small steps on the fault plane. The stepped geometry of the plane provides a mechanism for determining, not only the direction, but also the sense of shear across the fault (in this example, dextral, i.e., top to the right). (e) An example of calcite slickenfibres in the Chalk from Lulworth Cove, Dorset, UK.

during the fault movement. These are known as 'slickenfibres', and also show the direction of displacement. They are thought to be formed as a result of irregularities in the fault plane, as indicated in Figure 2.68.

Irregularities in the fault surface produce steps that either ascend or descend in the direction of fault movement. Small descending steps are shown on the dextral fault plane in Figures 2.68(a)–2.68(d). As the fault moves, gaps are generated at the steps and these small-scale dilational jogs can become infilled with crystal fibres, thus generating the 'slickenfibres'. Steps which ascend are areas of high frictional resistance to shear. They become rapidly eroded and are sites where true slickensides are likely to develop. By running a hand over the surface in the direction of the fibres, it is often possible to determine the sense of movement across the fault. In one direction, the surface feels rough and, in the other, smooth. The hand thus represents the missing side of the fault which moved in the direction which feels smooth.

The shear deformation associated with faulting can be approximated by simple shear. This deformation is illustrated in Figure 2.69 and can be

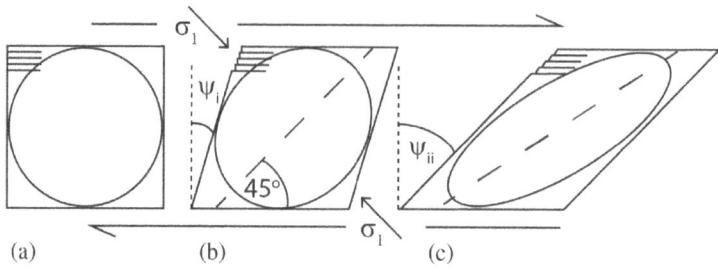

Figure 2.69. Theory of simple shear. (a) Undeformed state with a circular marker. (b) Deformation after the first increment of shear. Note that the eccentricity of the strain ellipse and the magnitude of the shear strain, $\psi$, has been exaggerated for clarity. (c) The incremental strain ellipse has its major axis inclined at 45° to the shear direction and this angle decreases as the deformation proceeds.

thought of as the deformation that occurs when a pack of cards moves by the movement of one card past another as indicated in the Figure. The magnitude of the strain associated with this type of deformation can be determined by stamping a circle on the undeformed state, (a) in Figure 2.69, and tracking the change in geometry and orientation of the deformed circle as deformation proceeds. An intriguing property of simple shear is that the first increment of shear deforms the circle into an ellipse whose maximum and minimum axes are inclined at 45° to the shear direction, (b) in Figure 2.69. For the first increment of deformation, the axes of stress and the resulting strain are co-axial and this implies that the maximum principal compression associated with simple shear is at 45° to the shearing direction, Figure 2.69(b). Based on this premise, it is possible to consider what second order brittle and ductile structures might form in response to such a stress regime and which therefore might be found in association with faults and shear zones.

If it is assumed that the conditions under which the shear deformation is developing are appropriate for brittle deformation, then the small-scale structures that develop can be either extensional fractures or shear fractures — depending on the magnitude of the local differential stress $(\sigma_1 - \sigma_3)$, see Section 2.3.4. If the differential stress is less than four times the tensile strength of the rock, then extensional fractures will form and these will be oriented at a right angle to the minimum principal stress, $\sigma_3$, i.e., parallel to $\sigma_1$, Figure 2.70(a)(i) and (ii). Thus, the fault or shear zone may

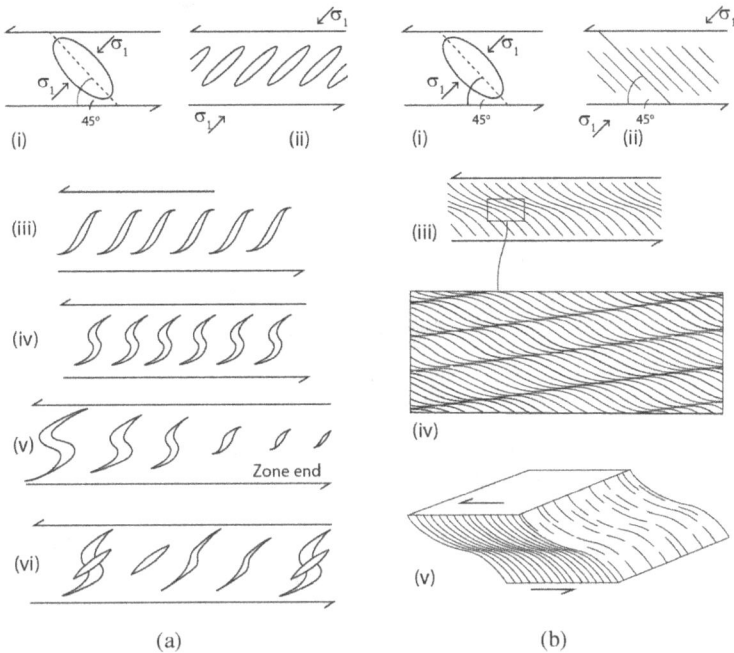

Figure 2.70. (a) The formation of *en echelon* tension gashes and their relevance as a kinematic indicator. The orientation of the local maximum principal compressive stress with respect to the fault or shear zone (represented by the half arrows) is shown. (b) The formation of a local cleavage along the fault or shear zone. From Price and Cosgrove (1990).

be characterised by an *en echelon* array of extensional fractures. As shearing continues, these tension gashes rotate and continue to grow. They grow parallel to $\sigma_1$, i.e., at 45° to the fault zone and, as a result, the veins take on a sigmoidal geometry. A natural example of this is shown in Figure 2.71. The asymmetry of these veins with respect to the shear zone or fault enables them to be used to determine the sense of shear along these major shears. Such structures are termed **kinematic indicators** and are extremely useful when attempting to trace a feature from one side of a fault to the other.

If the local differential stress is greater than four times the tensile strength of the rock, then second-order shear fractures can form, Figure 2.72. These form at approximately 30° to $\sigma_1$, see Figure 2.13, i.e., at either 15° or 75° to the major fault and, in the example illustrated in Figure

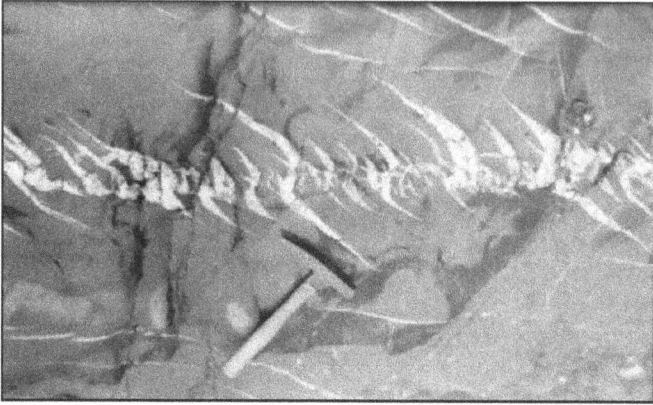

Figure 2.71. Arrays of *en echelon* quartz tension gashes. They indicate a dextral sense of shear, i.e., top to the right, see Figure 2.70(a).

2.72(b), the sense of shear on the more gently inclined of these two fractures is sinistral and on the more steeply inclined one is dextral. Because the orientation of the gently inclined shear fractures is close to that of the first-order fault and because the sense of shear along them is the same, this low angle shear tends to be developed in preference to its conjugate partner, Figure 2.72(c). Thus, faults are often characterised by an array of second-order shear fractures developed at a low angle to the fault. These are referred to as Riedel shears after the German experimenter who first studied their development in clay slabs caused by horizontal slip between two underlying wooden planks. As illustrated in Figure 2.72(c), the systematic displacement of a marker layer cut by these low angle shears provides another kinematic indicator that can be used to determine the sense of movement on a fault.

In addition to the second-order brittle structures shown in Figures 2.70(a), 2.71 and 2.72 that can form in response to the local stress regime associated with a fault or shear zone (Figure 2.69) and which can be used as kinematic indicators, the rock may also respond locally in a ductile manner and form a locally developed cleavage, Figure 2.70(b). This flattening fabric forms at right angles to the maximum principal stress and, like the *en echelon* tension gashes, will rotate as shear on the fault continues. As deformation continues, the zone of cleavage widens and the newly formed cleavage at the edge of the zone

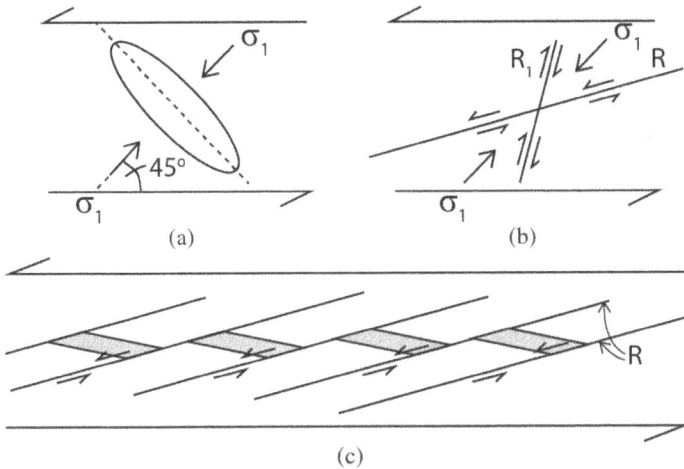

Figure 2.72. The formation of Riedel shears (R and R$_1$) within a shear zone. (a) The stress state within the shear zone. (b) The orientation and sense of shear of the predicted shear fractures. (c) The preferred formation of one set of the Riedel shears producing a consistent off-set of a marker band that can be used as a kinematic indicator to determine the sense of shear on the shear zone.

forms normal to $\sigma_1$. Thus, the geometry of the cleavage takes on a sigmoidal form as shown in Figure 2.70(b)(iii) which can also be used as an indicator of the sense of movement on the fault. This induced cleavage is often referred to as an 'S' fabric. As movement on the major fault continues, the intensity of the locally induced cleavage increases and it can develop to the point where it becomes unstable with respect to the stress generating it. The result is the formation of planar zones of localised shear, Figure 2.70(b)(iv). These are referred to as 'C' bands, or normal kink-bands, and their consistent asymmetry means that they too can be used as kinematic indicators to reveal the sense of shear on the major fault. These features are discussed in more detail in Section 2.13.6.3 and are shown in Figures 2.150 and 2.151.

The *en echelon* tension gashes and the local sigmoidal cleavage (Figure 2.70) represent pure brittle and pure ductile deformation, respectively. However, the two types of deformation can occur at the same time in rocks and this is clearly illustrated in Figure 2.73 which shows the synchronous formation of an array of *en echelon* tension gashes and a

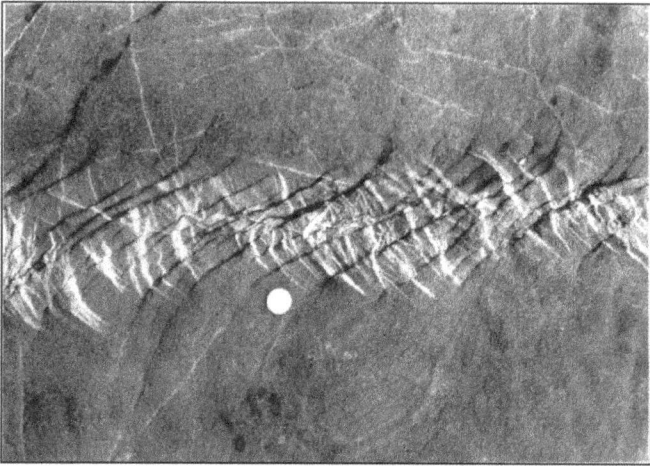

Figure 2.73. The synchronous formation of brittle structures (*en echelon* tension gashes) and ductile structures (locally developed sigmoidal cleavage) in a shear zone. Both structures indicate a dextral shear, i.e., top to the right. Marloes Sands, Pembrokeshire, South Wales.

localised cleavage with a sigmoidal geometry. We will return to the transition between brittle and ductile deformation in Section 2.10.

A variety of structures that are found in faults and shear zones and that can be used as kinematic indicators is shown in Figure 2.74. These include: (1) the rotation of a pre-existing rock fabric or one generated as a result of movement on the fault or shear zone; (2) the rotation and stretching of layers and other markers; (3) the formation of asymmetric intrafolial folds; (4) local 'C' bands in either the marginal; or (5) central fabric of the shear zone; (6) sheared porphyroclasts, i.e., fragments of resistant rocks or minerals); (7) the rotation of fragments of rock or mineral generated by either shear fractures; or (8) tensile fractures; (9) asymmetric trails growing around rotating clasts; and (10) non-rotating clasts; (11) the asymmetry of elongate, recrystallized quartz grains; (12) dragged out mica porphyroclasts; and (13) quartz '*c*' axis fabrics.

The recognition of these and the other kinematic indicators mentioned above enables geologists to determine the direction and sense of movement on faults and thus determine the orientation of the ancient stress fields responsible for their formation. Mining geologists frequently encounter faults when excavating mineral veins and coal seams and need

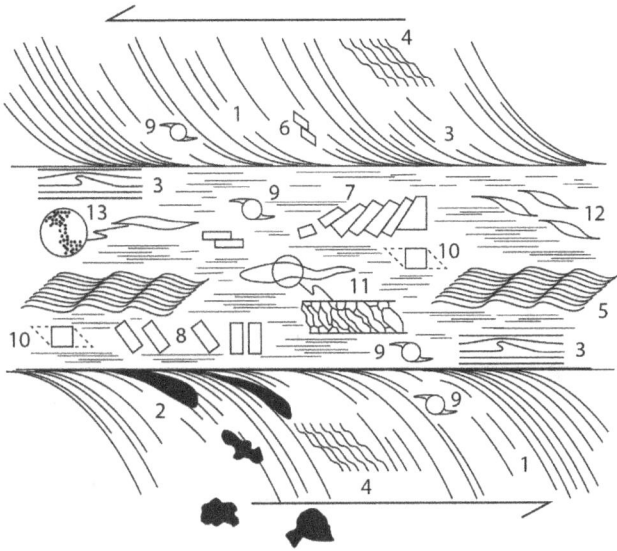

Figure 2.74.   A variety of structures that can be used as kinematic indicators to determine the sense of movement on a shear zone. From White *et al.* (1986).

to determine the position of these seams on the other side. Kinematic indicators are invaluable aids in this task. Similarly, at any engineering site, faults and shear zones are likely to displace other shear zones and fracture corridors, i.e., features that have an adverse effect on the bulk rock properties. Kinematic indicators can enable the likely position of the displaced 'damaged zones' to be determined even in areas of poor outcrop at the surface or underground where outcrops are limited to boreholes, tunnels and underground caverns. In addition, many fracture zones are too large to be recognised easily when examined on the scale of an engineering site. However the recognition of kinematic indicators will indicates the presence of these major shear features at or in the immediate proximity of the site.

## 2.8 FRACTURE SPACING

### 2.8.1 Fracture spacing in thin beds

Observations of fractures in layered rocks show that they often occur as fracture sets, i.e., arrays of fractures with similar orientations, and that a

Figure 2.75. Two major sets of extensional fractures cutting a limestone bed exposed on the wave-cut platform at Lilstock, North Somerset, UK.

fairly regular spacing exists between the fractures. This can be seen in Figure 2.75 which shows a limestone bed cut by two well-developed sets of extensional fractures. The study of fracture spacing and of the parameters that control spacing has involved a variety of techniques including field observations, theoretical studies, and analogue and numerical modelling (see e.g., Bogdanov, 1947; Price, 1966; Hobbs, 1967; Rives *et al.*, 1992; Olson, 1993; Renshaw and Pollard, 1994; Wu and Pollard, 1995; Bai and Pollard, 2000a, 2000b).

The regularity of fracture spacing, as for example shown in Figure 2.75, has long been recognised and the results of early work on the link between fracture spacing and layer thickness is shown in Figure 2.76. Several attempts have been made to account for this relation. Price (1966) suggested that joint frequency can be related to the frictional forces that exist between adjacent beds. The proposed model is illustrated in Figure 2.77 which shows the build-up of tensile stress in a horizontal competent layer by the ductile flow of adjacent, less competent beds. Such a scenario would develop during burial and compaction of sediments as a result of the gradual increase in overburden stress.

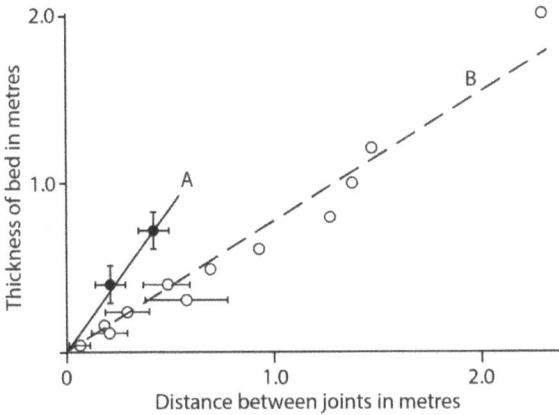

Figure 2.76. The relation between bed thickness and joint spacing for two rock types, A and B. From Bogdonov (1947).

At some point in the deformation history, the tensile stress in the layer will reach the tensile strength of the rock and the rock will fracture. Price considered the stress state in the competent layer after the formation of this initial fracture A–A, Figure 2.77(b), and pointed out that, if the competent bed were completely free to move, the joint would continue to open, relieving the stress over a wide region. However, even if there were no cohesion between the layers, frictional shearing stresses $\tau$ ($\tau = \mu \sigma_n$, where $\mu$ is the coefficient of sliding friction and $\sigma_n$ is the normal stress acting across the interface between the competent layer and the incompetent matrix) would develop along the interface to oppose the opening. Price noted that, since the joint forms a free surface, the horizontal stress in the competent bed at the joint plane is reduced to zero but argues that, because of the friction along the bedding plane, it gradually increases in intensity away from the joint until at some distance, $L$, it attains its original magnitude. This represents the minimum distance at which a second tensile fracture can develop.

The equilibrium conditions for this portion of the competent layer, length $L$, represented in Figure 2.77(c), can be easily determined by balancing the horizontal forces attempting to extend the layer ($\sigma_T z$, where $z$ is the layer thickness) against the forces resisting the extension, i.e., the shear force ($\tau_{AV}$) acting along the top and bottom of the bed along the distance $L$ ($2L\tau_{AV}$), where $\tau_{AV}$ is the mean shear stress, Equations (2.33) and (2.34).

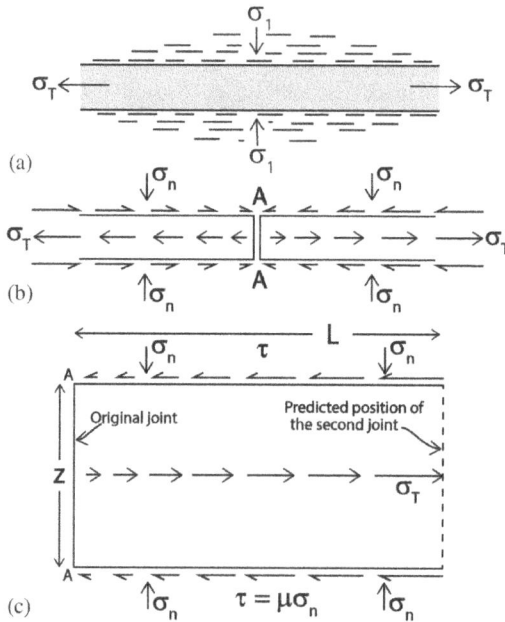

Figure 2.77. (a) Uniform tensile stress $\sigma_T$ acting in a single, competent unit caused by the lateral flow of the adjacent, less competent beds in response to a layer-normal load caused by the overburden. (b) Indicates the reduction in tensile stress due to the formation of a single joint, coupled with the development of a shear stress along the bedding planes which prevent excessive opening of the joint. (c) Details of stress intensity in a section of competent bed length, $L$, and thickness, $z$, adjacent to the initial joint. After Price (1966).

$$\sigma_T z = 2L\tau_{AV} \tag{2.33}$$

$$L = \sigma_T z / 2\tau_{AV} \tag{2.34}$$

This relation predicts that there is a direct connection between the spacing of joints and the thickness of the bed in which they form. This is in agreement with the field observations summarised in Figure 2.76. In addition to the geometric control the bed has on determining fracture spacing, it can be seen from Equation (2.34) that spacing is also dependent on an intrinsic property of the bed i.e., its tensile strength ($\sigma_T$).

This model assumes no cohesion along the bedding plane and that resistance to slip relates only to friction. An analysis where the layers are assumed to be welded together by strong cohesive bonds which exclude any interfacial slippage has been considered by Hobbs (1967). His model is made up of a relatively strong, elastic layer (Young's modulus, $E_1$) sandwiched between two weaker layers (Young's modulus $E_2$, where $E_1 > E_2$) subjected to a uniform layer parallel extension. As the strain increases, the higher tensile stress in the stronger layer eventually reaches the tensile strength and a first joint is formed.

The mechanical models presented by Price (1966) which assumes no cohesion along the bedding and the occurrence of slip, and Hobbs (1967) which assumes total cohesion and no slip, are the two end member possibilities. A number of other models have been developed to explain the regular spacing of extensional joints in competent layers and these are compared and discussed in Chapter 4 of Mandl (2005). All these models predict a linear relation between layer thickness and fracture spacing (Figure 2.78) and show that, in addition to layer thickness, the spacing is also influenced by the intrinsic properties of the rock, such as Young's modulus, Poisson's ratio and tensile strength.

## 2.8.2 Fracture spacing in thick beds

The relation between bed thickness and fracture spacing illustrated in Figures 2.76 and 2.78 does not hold when the layer thickness

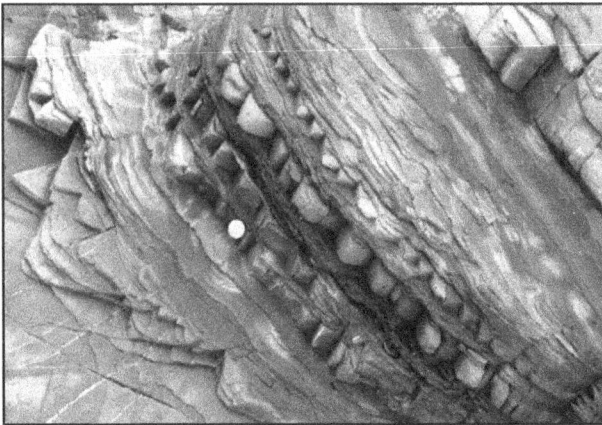

Figure 2.78.   Competent limestone beds sandwiched between weaker shales. A direct relation between layer thickness and fracture spacing is apparent.

Figure 2.79. Closely spaced fractures developed in thick sandstone beds in Monument Valley, USA. In such thick units, the linear relation between bed thickness and fracture spacing that characterises thinner beds breaks down.

becomes large. This is apparent when the spacing of the fractures that give rise to the Mesa and Butte topography that characterises the south western states of the USA is examined, Figure 2.79.

Another example of closely spaced fractures in thick sedimentary layers is provided by the cliffs exposed behind Deir el-Bahari, Queen Hatshepsut's funeral temple at Luxor, Figure 2.80. Here the Theban limestone is cut by closely spaced fractures which facilitate toppling, thus jeopardising the temple; a topic discussed more fully in Section 9.1.

A study of fracture spacing in Carboniferous turbidites and Jurassic limestones by Ladeira and Price (1980), provide an indication of the magnitude of the layer thickness when the linear relation between fracture spacing and layer thickness begins to break down. The results of these and other studies are shown graphically in Figure 2.81. The layer thickness/ fracture spacing plots illustrated in this Figure show an initial linear dependence between these two parameters but, as the layer thickness increases beyond a certain limit which seems to be different for different rock types, this relation breaks down and fracture spacing remains constant regardless of the layer thickness.

Figure 2.80.   Closely spaced fractures in the thick marly limestone that makes up Unit 1 of the Lower Tertiary Thebes limestone formation, exposed in the cliffs above the Temple of Hatshepsut, Luxor Egypt. These limestones rest on a weak shale, the Esna shale, the erosion of which constantly undermines them — thereby contributing to their instability. (See also Chapter 9.)

It has been suggested by Ladeira and Price that the data presented in Figure 2.81 can be represented by two straight line relations (Figure 2.81(c)), which relate to two different mechanisms of fracture formation. The first, line 0A, shows a linear relation between fracture spacing and bed thickness which has been discussed above. The second, which operates when the thickness exceeds a certain value, shows that the fracture spacing for a particular rock is independent of layer thickness, Line BC.

The mechanism proposed for the formation of fractures in thick beds is that of hydraulic fracturing. Fluids will induce extensional fractures when the differential stress is less than four times the tensile strength and when the fluid pressure equals or exceeds the least principal stress and the tensile strength of the rock, Section 2.5, Equation (2.28) (Equation (2.35)):

$$p > \sigma_3 + T. \qquad (2.35)$$

When this condition is satisfied in the rock, an extensional fracture will form. This will produce a local conduit along which fluids can flow

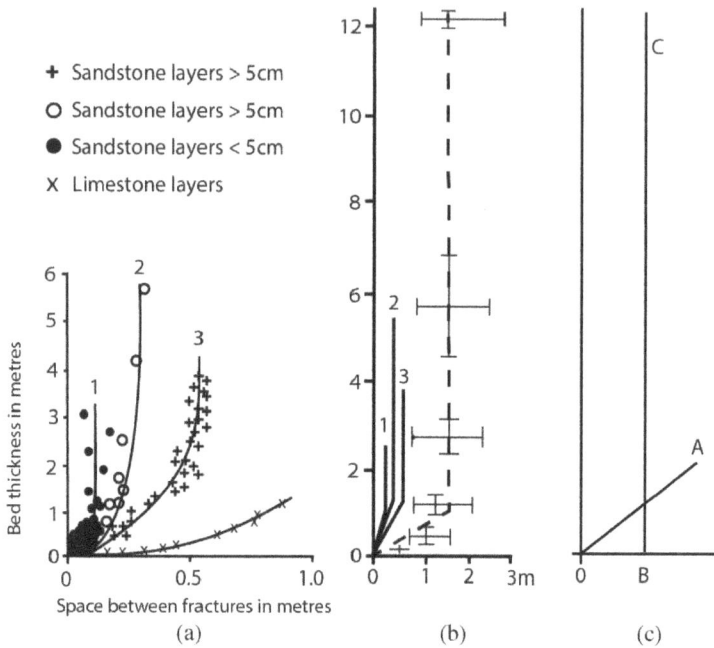

Figure 2.81.   (a) Plot of field data relating to bed thickness and fracture spacing from four different localities. (b) A representation of the curved distribution of data in (a) as two straight segments. (c) The generalised form of the 'curves' shown in (b). The two lines represent two mechanisms for the controlling of fracture spacing. From Price and Cosgrove (1990), after Ladeira and Price (1980).

and there will therefore be a draw-down of fluid pressure around the fracture as indicated schematically in Figure 2.82. The reduction in fluid pressure immediately adjacent to the fracture will inhibit the formation of other fractures within this zone. In this way, a minimum fracture spacing will be imposed on the rock. The width of this zone ($L$) will be determined by the permeability of the rock: the greater the permeability the wider the zone. Fracture spacing is independent of layer thickness, a geometric property and is determined by an intrinsic property of the rock.

### 2.8.3   The concept of fracture saturation

Although as discussed above, fracture sets often show a somewhat regular spacing, Figure 2.75, work by Priest and Hudson (1976), Hudson and

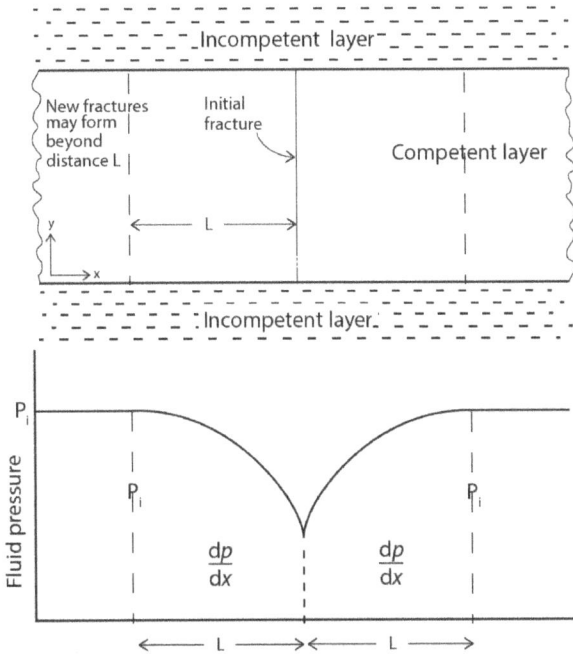

Figure 2.82.  Mechanism for the control of the spacing of hydraulic fractures by the permeability of the rock. Upper sketch: a thick bed under a high fluid pressure *Pi* which has resulted in the formation of an initial fracture. Lower sketch: the drawdown of fluid pressure around the initial fracture as a result of release of pressure by fluid expulsion along the fracture; the horizontal direction is the *x* direction. After Ladeira and Price (1980).

Priest (1983) and La Pointe and Hudson (1985) has shown that different patterns of fracture spacing can occur in natural rocks. These include negative exponential, power law, log normal and normal. These authors noted that, when several fracture sets are superimposed, each represented by a different statistical distribution, the overall result is a negative exponential distribution, i.e., as found along a scan line across a rock mass surface containing several sets.

Interestingly, Rives *et al.* (1992) also found these different patterns of fracture spacing even when they constrained their observations to a single fracture set. They account for these different distributions by suggesting that they represent different stages in the **evolution** of a fracture set. They argue that the fractures are initiated at randomly distributed irregularities

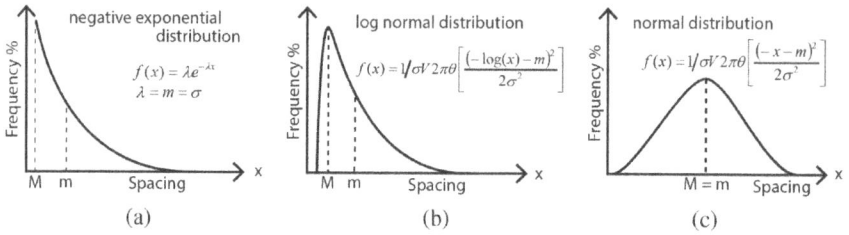

Figure 2.83.   Three types of spacing distribution found in natural fractures: (a) negative exponential distribution, (b) log-normal distribution, and (c) normal distribution. They are thought to represent three stages in the evolution of a regularly spaced set of fractures in an originally unfractured bed. $M$ = mode, $m$ = mean and $\sigma$ = standard deviation. Note that the mode of the negative exponential distribution in (a) is where the density distribution intersects the frequency axis, i.e., for a spacing of zero. From Rives *et al.* (1992).

within the material and that, as a consequence, initially the histogram of fracture spacing plots is a negative exponential curve, reflecting the random spacing of the flaws, Figure 2.83(a). It is argued that the formation of a fracture causes a local reduction in the extensional stress around the fracture and that this inhibits the formation of another fracture in the immediate vicinity. Thus, as the fracture set develops and the number of fractures increases, they begin to 'sense each other's presence', i.e., they are close enough for the zones of reduced stress which surround them to begin to overlap, and consequently begin to interact. This interaction leads to a self-organisation which influences the fracture spacing and the histogram changes from negative exponential to 'log normal', Figure 2.83(b). As the fracture set continues to develop, the histogram begins to exhibit a 'normal' distribution, Figure 2.83(c). However, as noted above, the superimposition of a series of different fracture spacing distributions (i.e., associated with different sets) can also result in an overall negative exponential distribution along a borehole or scanline on a rock outcrop.

The gradual change from a negative exponential to a normal distribution as a fracture set evolves in a given set of circumstances can be observed in analogue models. The results of such a model are illustrated in Figure 2.84 which shows the spacing distributions at three stages in its evolution, i.e., after the formation of 65, 109 and 140 fractures.

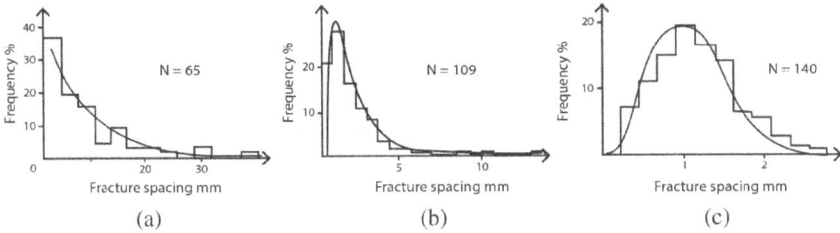

Figure 2.84.   Histograms and fitted distribution curves of the spacing observed at different stages during the generation of fractures in an analogue model, i.e., (a) after a few fractures ($N = 65$), (b) at an intermediate fracture density ($N = 109$) and (c) at a high fracture density ($N = 140$). From Rives *et al.* (1992).

As noted above, there are zones around natural fractures where the regional stress field is modified, Figures 2.77 and 2.82. Fractures that form outside these zones are not affected by the existence of the fractures and form randomly. However, if the density of fractures increases to the point where the zones of influence overlap, the fractures begin to 'self-organise' and develop a regular spacing. The differences between the histograms linked to a random process and those linked to a process where feedback from pre-existing fractures impacts on the generation of new fractures is shown in Figure 2.85.

The work of Rives *et al.*, discussed above and illustrated in Figures 2.83–2.85, has introduced some fundamental concepts relating to the different distribution laws of fracture spacing and their implication regarding the process of fracture development. To summarise, they point out that, if the process of fracture initiation remains a random process during the propagation of all the fractures during the evolution of the complete fracture set, i.e., if at all stages in the evolution of the fracture set the system remains uninfluenced by the generation of the fractures, the fracture spacing distribution will always plot as a negative exponential distribution, Figure 2.85(a). However, they also note that this does not generally occur and that, at some point in the evolution of the fracture set, the influence of earlier formed fractures is 'felt' by the new fractures and consequently begins to influence the location of fracture initiation. This point in the evolution of the fracture set is marked by a change in the spacing distribution from negative exponential to log-normal and eventually

RANDOM PROCESS                    INTERACTION PROCESS

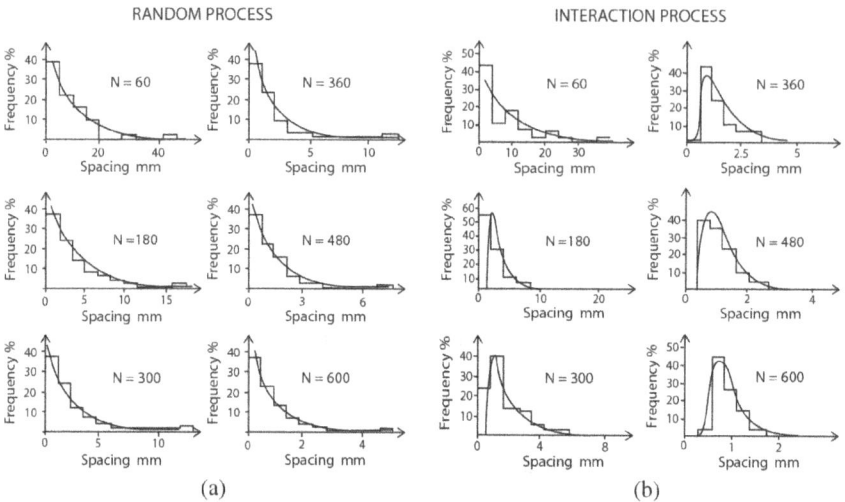

(a)                                      (b)

Figure 2.85.    (a) Histograms plus fitted distribution laws at different stages in the generation of a fracture set in a model with a *random* process. With increasing fracture density
(increasing *N*) the type of distribution (negative exponential) stays the same.
(b) Histograms plus fitted distributions at different stages in the generation of a fracture
set in a model with an *interaction* process. With increasing fracture density (increasing *N*)
the type of distribution varies from negative exponential, to log normal to almost normal
after 600 fractures have formed. From Rives *et al.* (1992).

normal. Note, however and as mentioned earlier, it is emphasised that this
discussion relates to a single fracture set. Moreover and as also pointed out
earlier when several fracture sets are superimposed, the fracture spacing
along a scanline on a rock exposure or along a borehole core will be of
negative exponential form — because the superposition of a variety of
probability density distributions (of whatever type) will converge to a
negative exponential distribution.

Rives *et al.* (1992) introduce the concept of *saturation* to describe the
evolution of the fracture set from an initial random distribution of a few
early fractures (a state of under-saturation) to a point where the material
is saturated with fractures and the formation of new fractures ceases.
An example of limestone beds undersaturated with fractures is shown in
Figure 2.86. One of the most convincing explanations for the linear relation between layer thickness and fracture spacing first recorded by

Figure 2.86. An example of a limestone bed containing a few, randomly spaced, through-going fractures. The bed is 'unsaturated' with fractures and, had deformation continued and more fractures developed, they would have developed into a regular spacing.

Bogdanov (1947), working on coal seams in the USSR (see Figure 2.76) is provided by Bai and Pollard (2000b). They examined the stress distribution between two adjacent opening mode fractures as a function of the fracture spacing to layer thickness ratio. The results show that, when the fracture spacing to layer thickness changes from greater than to less than a 'critical value' (approximately 1.0), the normal stress acting perpendicular to the fractures changes from tensile to compressive. This stress state transition precludes further infilling of fractures unless there are existing flaws and/or the fractures are driven by an internal fluid pressure or other mechanism. Hence, for fractures driven by tectonic extension, the critical fracture spacing to layer thickness ratio defines a lower limit, which also defines the condition of *fracture saturation*.

Bai and Pollard (2000b) found that the critical value of fracture spacing to layer thickness increases with increasing ratio of Young's modulus of the fractured layer compared to the neighbouring layers and with increasing overburden stress (depth), but it decreases with increasing Poisson's ratio of the adjacent layers. They note that, for geologically realistic values of the elastic constants and depths, the critical fracture spacing to layer thickness ratio varies between 0.8 and 1.2, a range which encompasses the often cited ratio of 1.0 in the literature for well-developed fracture sets.

Despite the numerous examples of a direct relation between fracture spacing and bed thickness described in the literature and the various explanations for this phenomenon, examples occur in nature where the spacing of fractures is significantly lower than that normally observed in saturated fracture sets. Theoretically, as noted above, saturation is marked by a range of spacing/thickness ratios ($S/T$) between 0.8 and 1.2, and values greater than 1.2 are explained by arguing that the fracturing process has not yet reached saturation level. In an attempt to explain ratios of <0.8, Bai and Pollard (2000a) studied the possibility of 'further fracture infilling' by considering flaw distributions between adjacent fractures loaded by extension of the layer. It was found that, depending on their position with respect to the existing fractures, some of these flaws could grow into fractures cutting across the layer. For example, if the elastic constants are the same for the fractured and neighbouring layers, then, if the new fracture has an $S/T$ ratio of greater than 0.55, it can grow and completely cut the layer.

### 2.8.4 Clustering versus uniform distribution of fractures

A fracture zone is a general term used by geologists to describe a planar or sub-planar concentration of fractures in a rock. They can be either shear fractures or extensional fractures. Shear fracture zones have the same spatial relation to the stress field that generated them as do discrete shear fractures and the process of development of a shear fracture is one involving progressive deformation of the wall rock. The shear deformation becomes distributed and forms a shear zone, the result of continued movement on the shear fracture and the development of more and more second-order fractures. In addition, the transition from brittle to ductile deformation is accompanied by progressive reduction in strain localisation. Shear deformation then ceases to be localised along a discrete fracture; the shear deformation becomes distributed and forms a shear zone, see Section 2.10 and specifically Section 2.10.6.

Figure 2.87. (a) Fracture corridor in the Lower Cretaceous, moderately layered, marly limestones of the Calvisson quarries, Gard, South France. The left side of the fracture corridor is partly brecciated. (b) Fracture corridor in the Oxfordian limestones of the Coulazou Valley, Languedoc. Photographs courtesy of J.-P. Petit.

Extensional fracture zones also occur in rocks and one of the characteristic features of many extensional fracture sets observed in nature, and modelled either via analogue or numerical models, is that of 'fracture clustering', Figure 2.87. Instead of uniform fracture spacing, clusters of closely spaced fractures (sometimes referred to as 'fracture corridors') develop, separated by areas of low fracture density. These fracture corridors, which can have significant hydrogeological ramifications, may be laterally continuous or of limited lateral extent, Figures 2.88(a) and 2.88(b) respectively, and can have a major impact on the properties of a rock mass. Attempts have been made to understand why this clustering of fractures occurs. Two numerical model studies of fracture distribution (Renshaw and Pollard, 1994; Olson, 2004) illustrate the phenomenon of fracture clustering well and offer an explanation as to why such a distribution might develop. The numerical modelling used in the study simulates the mechanical interaction between fractures and the results compare well with experimentally generated fracture sets in analogue models. The numerical models are seeded with initial flaws from which the fractures grow and the

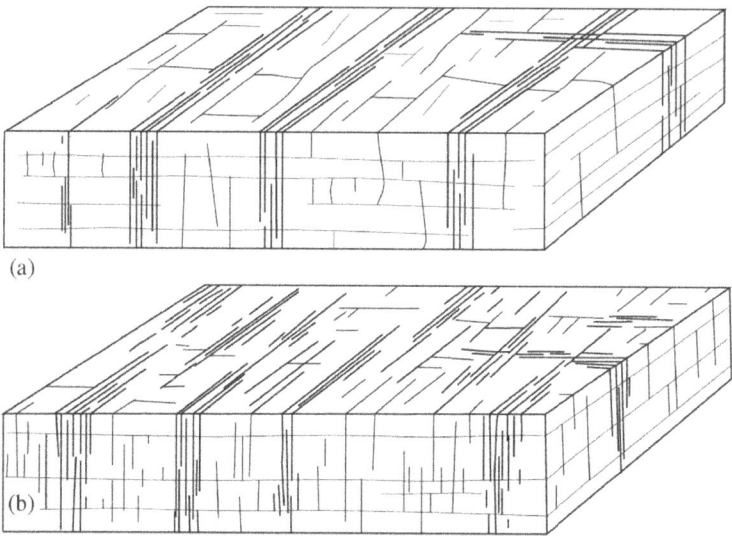

Figure 2.88. Two types of systematic fracture corridor: (a) Longitudinally persistent type. (b) Longitudinally impersistent type. More than one set of fracture corridors can be present in a rock mass and these may be either orthogonal or oblique. Diagram courtesy of J.-P. Petit.

deformation is achieved by applying successive increments of extension to the model. Renshaw and Pollard found that, once the flaw geometry is specified, only one parameter, the velocity exponent, $\alpha$ (which relates the fracture propagation velocity, $v$, to the stress intensity factor at the fracture tip, $K_I$, Section 2.3.3 and Equation (2.19)) controls the geometric evolution of the fracture set.

Sensitivity analysis shows that this governing parameter also controls the extent to which fracture growth is concentrated within zones or clusters. The velocity exponent, a fracture mechanics parameter, controls the amount each fracture tip is advanced ($l_{adv}$) for each increment of extension. This is given by Equation (2.36).

$$l_{adv} = 2c \left( \frac{G}{G_{max}} \right)^{\alpha}, \tag{2.36}$$

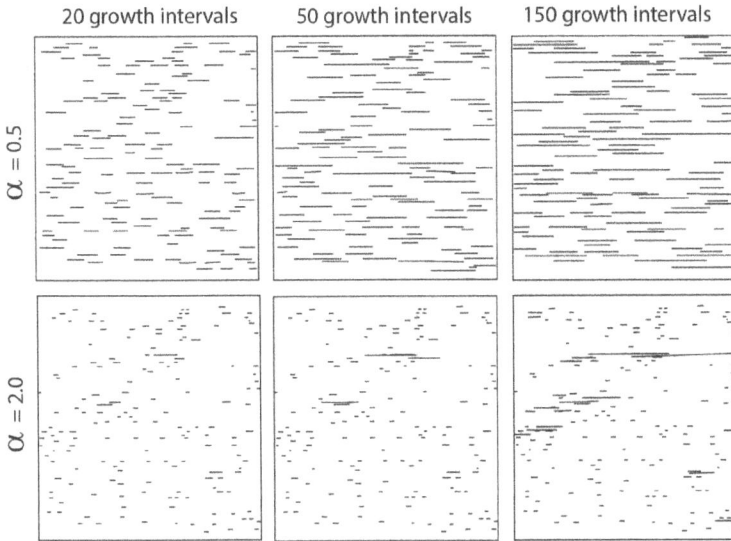

Figure 2.89. The evolution of typical fracture sets under two different velocity exponents α, indicated after 20, 50 and 100 growth iterations. The higher the velocity exponent, the greater the tendency of the fractures to form in clusters, rather than to be uniformly spaced. From Renshaw and Pollard (1994).

where $2c$ is the flaw length, $G$ is the elastic energy release rate at the fracture tip, $G_{max}$ is the maximum energy release rate in the fracture set at the current growth iteration, and $\alpha$ the velocity exponent.

Figure 2.89 shows the development of typical fracture sets in two models which only differ in the values of their velocity exponents. The initial distribution of flaws in both experiments is the same. As $\alpha$ is decreased, the growth rate of the fractures becomes less dependent on the stress intensity factor (i.e., the stress concentration at the fracture tips) distribution across the network, Equation (2.36), and all the fractures tend to grow at a similar rate, regardless of the presence or absence of nearby fractures, $\alpha = 0.5$, Figure 2.89. Thus, any clustering present in the network tends to reflect the original clustering of the flaws — rather than the selective growth of the clustered fractures. In contrast, as $\alpha$ is increased, only those fractures with the largest stress intensity factors will grow. Growth will be concentrated in those areas where the geometry of the fractures is such that the mechanical interaction of the flaws and the fractures leads to amplification of the energy release rates rather, than to 'shielding'. The restriction of growth to fewer and fewer fractures as $\alpha$ is increased, leads to the development of fracture zones which are separated by zones of little or no fracture growth, $\alpha = 2.0$, Figure 2.89.

Similar results are presented by Olson (1993). He also investigated the sub-critical crack growth of a fracture set using numerical modelling techniques. The model calculates the mechanical interaction between simultaneously propagating fractures and uses the sub-critical crack propagation rule where propagation velocity during stable crack growth scales with the crack tip stress intensity factor $K_I$. The propagating velocity, $v$, is related to the opening-mode (extensional fracture) stress intensity factor $K_I$, with an empirically quantifiable, power-law relation, Equation (2.37):

$$v = A \left( \frac{K_I}{K_{Ic}} \right)^n ,$$

(2.37)

see Equations (2.19) and (2.20), where $K_{Ic}$ is the critical fracture toughness, $n$ is the sub-critical index, and $A$ is a proportionality constant. It is found that $n$ is dependent on rock type and environmental conditions (such as wet versus dry). Reported values vary from 20 or less for sandstones submerged in water (Atkinson, 1984) to values greater than 250 under dry conditions in carbonates (Olson *et al.*, 2002). The evolution of fracture sets in two models, identical in all respects except for the value of the sub-critical indices, is shown in Figure 2.90. It can be seen that the tendency for the fractures to cluster, rather than form in a uniform distribution, is directly related to the index value.

### 2.8.5 Strata-bound and through-going fractures

The discussion so far in this Section has related to the fracturing of layers and the relation between the layer thickness and fracture spacing. Field observations indicate that fractures are not always constrained to a single layer but can cut through many layers. The reasons for fractures remaining 'strata-bound' or becoming 'through-going' will now be discussed.

Fractures in layered rocks can be grouped into two categories: those that are restricted to a single layer or rock unit, Figure 2.91; and those that

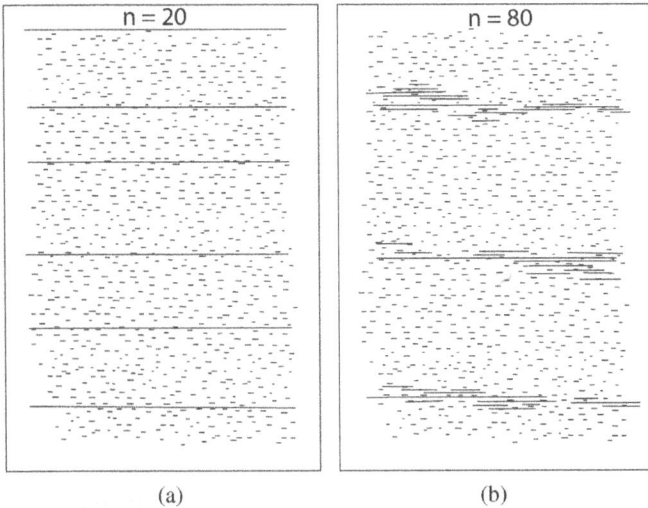

Figure 2.90. Examples of sub-critical fracture growth for models with different sub-critical indices, (*n*). (a) *n* = 20 and (b) *n* = 80. All simulations start with the same randomly located parallel flaws. An extensional strain was applied in the vertical direction at a constant strain rate. *n* is determined by rock type and environmental conditions and, as can be seen from the experimental results, has a dramatic influence of fracture distribution. From Olson (2004).

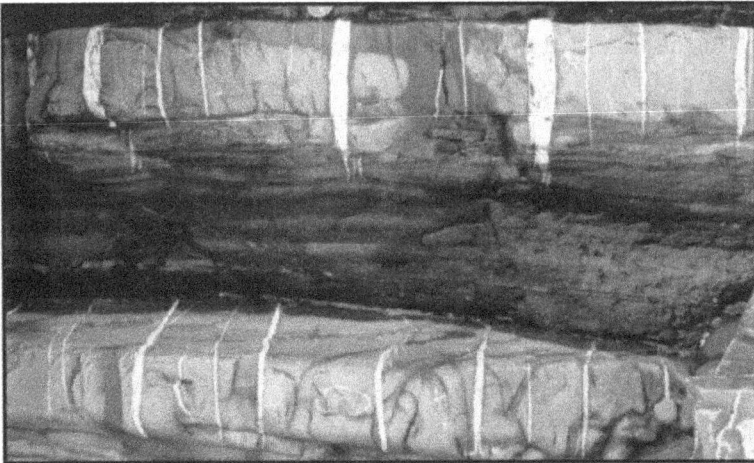

Figure 2.91. Regularly spaced, strata-bound fractures infilled with quartz, developed in sandstone beds of the Carboniferous turbidites of Millook, North Cornwall, UK.

Figure 2.92.   Sub-vertical fractures cutting through several Carboniferous limestone beds. Tytherington Quarry, Somerset, UK (see Chapter 5 for a description of the Tytherington Quarry and the evolution of the fracture network).

cut through many layers, Figure 2.92. A variety of factors determines which of these two types of fracturing will occur. These include the cohesion or lack of cohesion along the bedding planes between adjacent beds, the relative strength of the two adjacent layers (i.e., the relative magnitudes of their Young's moduli) and the magnitude of the normal stress on the bedding. These effects are illustrated in Figure 2.93.

As an extensional fracture propagates, the region just in front of the fracture tip is experiencing a high tensile stress, Figure 2.93(b). As this fracture reaches a bedding plane along which there is only low cohesion, this stress is dissipated by slip on the bedding plane, as illustrated in Figure 2.93(a) and the fracture is therefore unable to propagate. Such low cohesion bedding planes therefore favour the formation of strata-bound fractures. Conversely, if there is a high cohesion across the bedding plane and the layers are strongly coupled, then the stress at the tip of the fracture is not dissipated when it reaches the bedding plane and the fracture can propagate into the adjacent layer generating a through-going fracture,

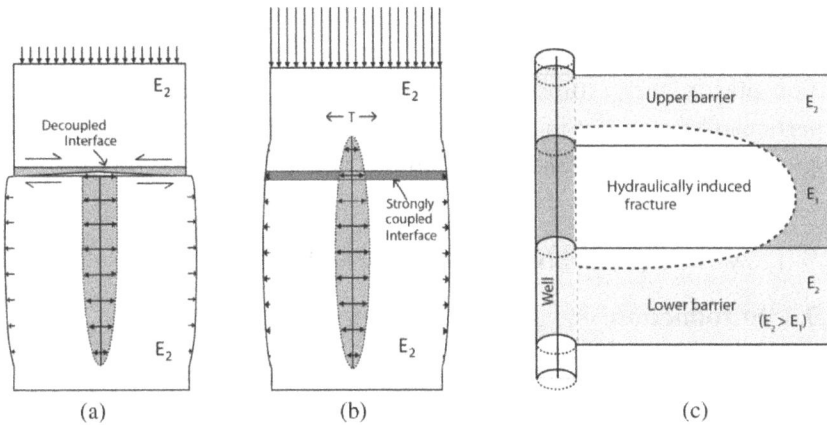

Figure 2.93. Schematic illustration of three mechanisms that condition fracture propagation or arrest. (a) Low cohesion between layers permits bed slippage and inhibits fracture propagation from one layer to another. (b) Strong coupling on the interface between layers allows fractures to propagate across the interface. (c) Changes in the elastic properties of the layers can inhibit fracture propagation. $E$ is Young's modulus and $E_2 > E_1$. From Casabianca and Cosgrove (2012).

Figure 2.93(b). In addition to low cohesion along bedding planes, a difference in the mechanical properties of two adjacent beds can also impede fracture propagation and result in fractures being strata-bound. It is found that fracture propagation is impeded if a fracture attempts to propagate from a layer with a relatively low Young's modulus to one with a higher value, Figure 2.93(c). It has been shown (Simonson *et al.*, 1978) that doubling the modulus across an interface is sufficient to completely prevent fracture propagation.

\*\*\*\*\*

In conclusion, we note that understanding the distribution of fractures within a particular fracture set is essential if one wishes to predict the likely location of fractures within that set. A study of fracture distributions in natural rocks, analogue experiments and numerical models has shown that the fracture distribution is influenced by a variety of parameters, depending primarily on the degree of development of the fracture set and upon the geometric and rock mechanics properties. An understanding of this dependency and the mechanisms involved in fracture propagation is important when establishing the likely distribution of fractures at an excavation site, or when estimating the likely position of adjacent fractures of a particular fracture set in connection with the requirements of a specific engineering project, e.g., for underground radioactive waste disposal.

## 2.9 FRACTURE ORIENTATION AND REGULARITY

### 2.9.1 Introduction

The link between stress and brittle failure has been discussed in Section 2.3 and is summarised in Figure 2.19, reproduced below for convenience as Figure 2.94, which shows the intimate relation between the orientation of both extensional and shear fractures and their causative stress fields.

It will be recalled that extensional fractures form normal to the least principal compression and contain the $\sigma_1 - \sigma_2$ plane (Figure 2.94(b)) and that shear fractures form as a conjugate set symmetrically arranged about the maximum principal compression and containing the $\sigma_2$ stress axis (Figure 2.94(c)). It can be seen from Figure 2.94 that a low differential

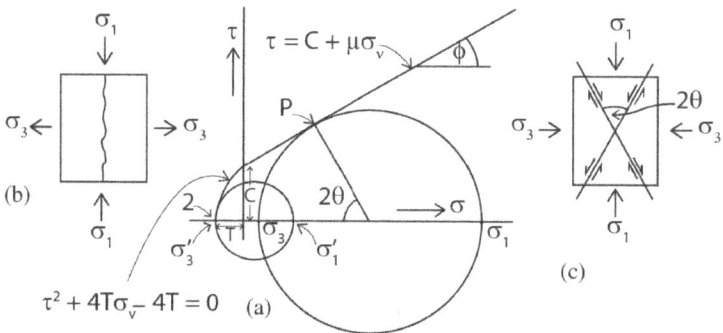

Figure 2.94. (a) The combined brittle failure criterion derived by linking the linear Navier–Coulomb shear failure criterion with the parabolic extensional failure of Griffith, previously discussed in Section 2.3. The stress conditions required for the two modes of failure are apparent. In order to satisfy the extensional failure criterion, a small differential stress $(\sigma_1 - \sigma_3)$ is required; whereas, shear failure requires a larger differential stress. (b) and (c) show the two modes of fracture and the stress fields causing them.

stress is required for the formation of extensional failure $((\sigma_1 - \sigma_3) < 4T)$ and a high differential stress is required for shear failure $((\sigma_1 - \sigma_3) > 4T$, Section 2.3.4). Thus, the orientation of fractures is controlled by the orientation of the stress that causes them, and the type of fracture is controlled by the differential stress. As is discussed in the following Section, the differential stress also has a major impact on the regularity of fractures.

## 2.9.2 The effect of the differential stress on the regularity of fractures

In Section 2.5, where the influence of fluid pressure on brittle failure was discussed, it was shown how the regularity (straightness) of fractures is determined by the differential stress, Figure 2.33 reproduced below as Figure 2.95. The four Mohr circles shown in this Figure represent four different stress fields, all with a differential stress of less than $4T$, and therefore all of which will give rise to extensional failure. In this example, the stress

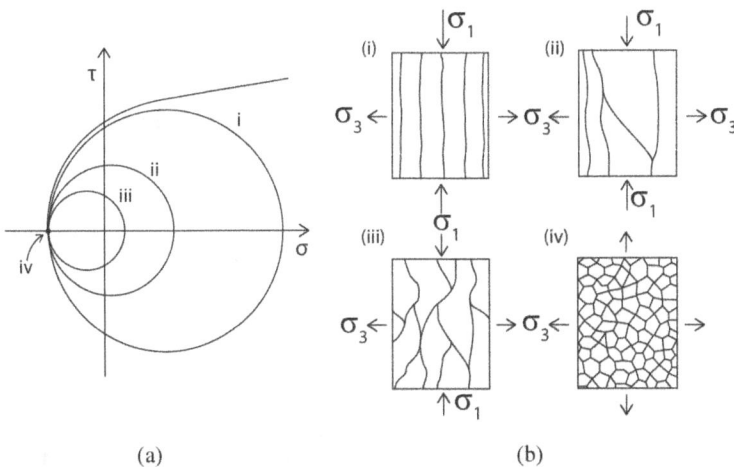

Figure 2.95. (a) Four stress states represented by their Mohr stress circles, each of which will produce extensional failure of the rock. (b) The expression of these fractures for each of the stress conditions shown in (a). As the differential stress increases so does the regularity and alignment of the resulting fractures. From Cosgrove (1997).

fields are considered to be associated with an overburden stress, i.e., $\sigma_1$ is vertical, and it is assumed that no tectonic stresses are acting. The predicted fractures for each stress field are shown in Figure 2.95(b) which illustrates their expression on the vertical plane containing $\sigma_1$ and $\sigma_3$.

When the differential stress is relatively large (but still less than $4T$), there is a definite direction of easy opening for the fractures (i.e., perpendicular to the least principal stress, $\sigma_3$, according to the least work principle) and consequently the resulting fractures tend to be straight, well organised and parallel to the $\sigma_1 - \sigma_2$ plane, Figure 2.95(b)(i). However, as the differential stress drops and approaches zero (i.e., a hydrostatic stress), it is equally easy for extensional fractures to open in all directions. There is no direction of relatively easy opening and therefore no tendency for the fractures to form in any particular direction. The result is the generation of randomly oriented fractures which would result in a polygonal array as shown in Figure 2.95(b)(iv). Thus, as the differential stress drops from just below $4T$ to zero there is a progressive decrease in fracture regularity, Figure 2.95(b).

### 2.9.3 The impact of the local fracture tip stress on fracture regularity

As noted above, the orientation and degree of regularity of the fractures shown in Figure 2.95(b) is controlled by the regional stress which in this example is the overburden. However, in addition to the regional stress field, there are also local stress fields that develop around the fracture tips. If two adjacent fractures of the same set are propagating towards each other and are closely enough spaced, then, as they approach each other and begin to overlap, the local stress fields will interact and cause irregularities in the orientations of the two fractures.

This interference has been examined by Olson and Pollard (1988, 1989) whose results are summarised in Figure 2.96 The local stress field around the tip of an extensional fracture, A, is shown in Figure 2.96(a), together with the position of the tip of a second, parallel, but offset, extensional fracture, B. The dashes indicate planes perpendicular to the local maximum tensile stress, i.e., the orientation that fracture B would follow if it were to propagate into this region. As can be seen, this stress field will

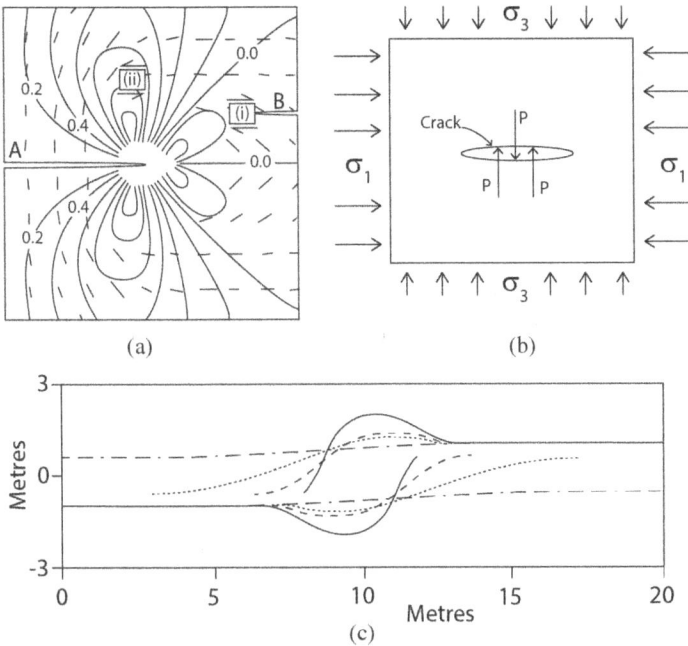

Figure 2.96.   (a) The local stress state induced at the tip of an opening mode crack, centre at A. The contours are of shear stress magnitude acting on planes parallel to crack A. Dashes indicate planes perpendicular to the local maximum tensile stress and show the possible path of crack B. (b) An idealised crack in an elastic solid with an internal fluid pressure, P, and horizontal remote principal stresses $\sigma_1$ and $\sigma_3$. The remote stress components act perpendicular and parallel to the crack plane. (c) Graphs of the theoretical crack paths under four different differential remote stress fields. Solid line, –0.5 MPa; dashed, 0 MPa; dotted, 1 MPa; and dot–dash, 5 MPa. From Olson and Pollard (1989).

cause fracture B to rotate from being parallel to A into an orientation at a high angle to it — causing it to curve and link producing a hook relation, Figure 2.97.

Figures 2.96 and 2.97 show the effect that the local stress field around the tip of a propagating fracture can have on the regularity of adjacent, parallel fractures which form at the same time and which are part of the same fracture set. We will return to this phenomenon in Section 2.12, where the build-up of fracture networks by the successive superposition of different fracture sets is considered. In this scenario, the propagating

Figure 2.97. Two parallel fractures, (a) and (b), in the process of linking as they overlap during fracture propagation. Fracture (b) has curved towards and linked with (a); the white dashed line shows the expected path of propagation of fracture (a) had the propagation continued. From the Liassic limestones of the Bristol Channel, UK.

fractures interact with fractures which are not parallel to them and which belong to an earlier fracture set.

The deflection and linking of the two fractures discussed above and illustrated in Figure 2.97, occurs in response to the local stress fields around the fracture tips. However, if a regional stress field is acting on the system, it is also necessary to consider the effect of this on the propagation of the fractures. Olson and Pollard (1989) examined this by considering the effect of a regional stress field appropriately oriented for fracture propagation, i.e., with $\sigma_1$ parallel to and $\sigma_3$ normal to the fractures, Figure 2.96(b). There is a differential stress, $\sigma_1 - \sigma_3$, related to this regional stress field which is compressional and aligned with the crack. By varying the magnitude of this, it is possible to determine the effect of the regional stress field on fracture propagation.

Olson and Pollard (1989) point out that, if crack B in Figure 2.96(a) begins to curve away from A as the two fractures begin to overlap (as it would do in response to the local stress trajectories, dashes, in the region of the shear box (i)), a left lateral shear is resolved across its tip by the remote stresses. If, on the other hand, B were to curve towards A, as it would do in the region of shear box (ii), a right lateral shear would be resolved. In both cases, the remotely induced shear stress works against the locally induced shear stress, as indicated by the shear boxes, thus promoting a straighter crack path. As the differential stress of the remote stress field increases, so does its influence on the propagation of the fractures. This has been quantified and the results summarised in Figure 2.96(c) which shows the theoretical crack propagation paths under four different differential remote stress fields: solid line, −0.5 MPa; dashed, 0 MPa; dotted, 1 MPa; and dot–dash, 5 MPa. As the differential stress of the regional stress field increases, the impact of the local stress becomes progressively suppressed and the overlapping fractures show less and less tendency to deflect and link.

This spectrum of interaction between the two fractures (which ranges from marked deflection and linking at a high angle to the fractures propagating past each other with very little interference as the differential stress of the regional stress field progressively overrides the effect of the local fracture tip stress field) also occurs as the separation between the two interfering fractures increases. Here the decrease in influence of the fracture tip stresses is simply related to an increase in the separation of the fractures.

Olson and Pollard (1988), suggested that this relation between fracture straightness and the differential stress at the time of fracturing could be used in palaeo-stress analysis. They introduce a method to infer the remote differential stress magnitude from the curvature of overlapping fracture traces. Curving paths imply the predominance of local crack-induced stresses over the remote stresses during propagation and indicate a low regional differential stress.

Renshaw and Pollard (1994) studied the impact of heterogeneities on the propagation pattern of fractures and confirm that the dominant factor controlling fracture straightness is the magnitude of the far field differential stress. They observe that, as the differential stress increases, so the effect of local fractures on the fracture propagation path becomes progressively weaker and the fractures become straighter, Figure 2.98. They also note that the 'driving force' available to propagate a fracture, which controls the propagation rate and therefore the straightness of the fractures, is determined by the magnitude of the stress ratio, $R$, which is related directly to the far field differential stress $(\sigma_1 - \sigma_3)$ as shown in Equation (2.38).

$$R = (\sigma_1 - \sigma_3)/(\sigma_3 + P), \tag{2.38}$$

where $P$ is the fluid pressure in the fracture.

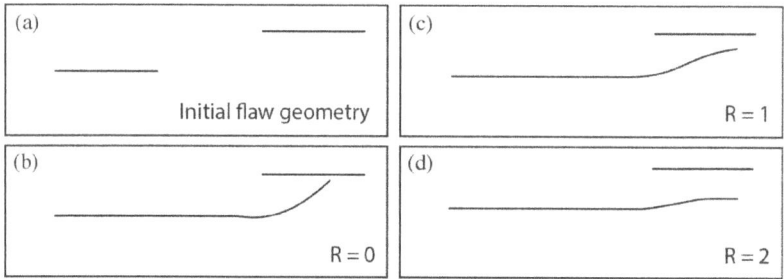

Figure 2.98. Illustrating the sensitivity of fracture paths to the differential stress (stress ratio $R = (\sigma_1 - \sigma_3)/(\sigma_3 + P)$ where $P$ is the fluid pressure in the fracture, $\sigma_1$ horizontal, $\sigma_3$ vertical). (a) Shows the two initial flaws, with only the left-hand flaw being allowed to propagate. As the differential stress ($R$) of the regional stress field increases, so the influence of the adjacent fracture on fracture propagation decreases. From Renshaw and Pollard (1994).

### 2.9.4 Fracture regularity in the $\sigma_2-\sigma_3$ plane

We have considered the regularity of fractures in the $\sigma_1-\sigma_3$ plane, Figure 2.95(b), which is a vertical plane in an overburden dominated stress regime. We also need to consider its regularity in the $\sigma_2-\sigma_3$ plane which is the horizontal plane in the overburden stress regime — and fracture regularity in this plane will also be controlled by the differential stress in the plane, i.e., $(\sigma_2-\sigma_3)$.

This has been discussed in Section 2.4.2 (see Figure 2.24) and is illustrated in Figure 2.99 which shows a stress regime (a) in which $\sigma_2$ is greater than $\sigma_3$, and (b) in which they are the same. In the former case, there is a definite direction of easy opening of the fractures (normal to $\sigma_3$) and the fractures generated will therefore be aligned and have a clearly defined strike direction. In contrast, when $\sigma_2 = \sigma_3$, there is no preferred direction of fracture opening and the fractures will form with random orientations in this plane. This leads to the formation of a polygonal array, Figure 2.99(b). Thus, in a regime where the overburden stress dominates, the regularity of the strike of a fracture set is determined by the value of $(\sigma_2-\sigma_3)$.

A fracture network in an outcrop from south-eastern Utah containing two fracture sets is shown in Figure 2.100(a). The two sets are shown individually in Figures 2.100(b) and 2.100(c) and it can be seen that, although the two sets are produced in the same bed, the trace patterns are fundamentally different. The E–W traces are regularly spaced and uniformly abundant throughout the outcrop. They exhibit an anastomosing (branching network)

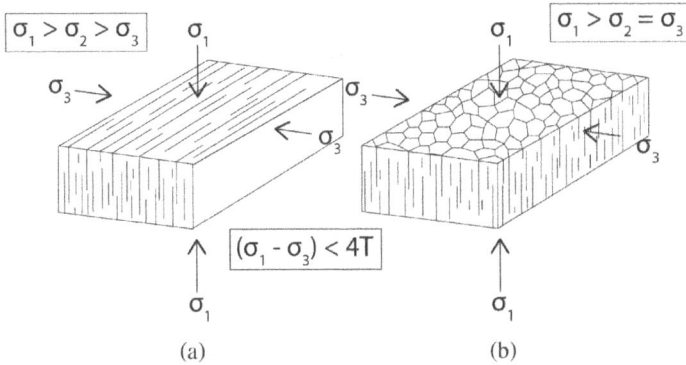

Figure 2.99. Extensional fractures caused by an overburden stress. The regularity of fractures in the $\sigma_1-\sigma_3$ plane, i.e., the vertical front face of the block diagrams, is controlled by the differential stress $(\sigma_1-\sigma_3)$. This is less than $4T$ and consequently extensional fractures form. The regularity in the $\sigma_2-\sigma_3$ plane, i.e., the horizontal plane, is controlled by the differential stress $(\sigma_2-\sigma_3)$.

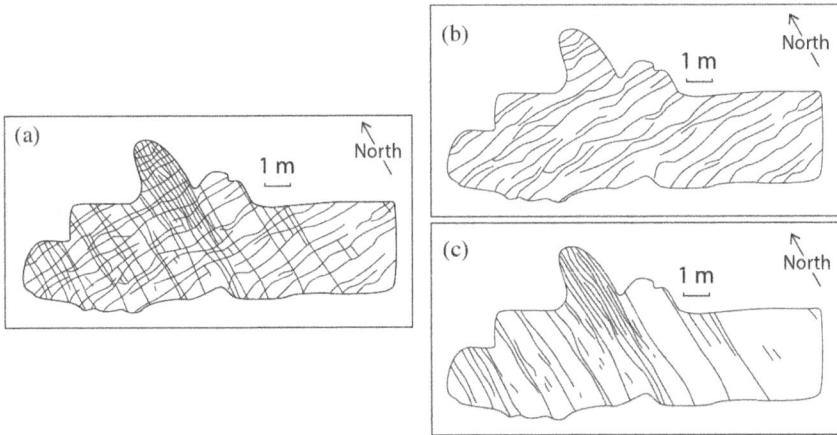

Figure 2.100. (a) Map of joint traces exposed on a bedding plane surface. (b) The older joint set formed under a near isotropic remote stress. (c) The younger set formed under strongly deviatoric remote stress. From Olson and Pollard (1988).

geometry with numerous curving overlaps and crack intersections. This is taken to indicate that, at the time of their formation, the remote differential stress was very low and exceeded by the local stress field around the fracture tip determined by the fracture propagating driving pressure (assumed to be provided by the water pressure inside the fractures, see Figure 2.98). In contrast, the N–S trending joints occur in clusters and display little curvature. The linearity of this set is taken to indicate a larger differential stress was operating during their formations.

In addition to the change in fracture regularity between the two fracture sets which, as noted, implies fracture formation under different differential stress states, the fractures also display a difference in distribution, the early set being evenly spaced, the later set more clustered. The tendency to cluster is determined by the velocity exponent $\alpha$ (which relates the fracture propagation velocity, $v$, and the stress intensity factor $K_1$, Section 2.8.4), i.e., an intrinsic property of the rock. Thus, a study of the fracture sets allows one to track not only the change in the orientation of stress and magnitude of the differential stress with time but also the change in rock properties as the processes of burial and diagenesis (changes undergone by a sediment after initial deposition) progressively lithify the rock.

In the above discussion on the regularity of fractures, it has been assumed that the differential stress was less than $4T$ and therefore that extensional fractures formed. However, it will be recalled from Section 2.4 (Figure 2.23) that the stress generated in the crust loaded only by an overburden is one in which the differential stress increases with depth, Figure 2.101. Thus, Figure 2.99 can be expanded to include fractures that form deeper in the crust where the differential stress $(\sigma_1 - \sigma_3)$ is greater than $4T$, with the result that shear fractures form, Figure 2.101(b) and 2.101(d). As with the extensional fractures, Figure 2.99, the regularity of the strike of these shear fractures will be controlled by the magnitude of $(\sigma_2 - \sigma_3)$.

### 2.9.5 The effect of the scale-independent process of fracture growth on fracture regularity

Another factor that influences the regularity of fractures is the process by which large fractures form. There is considerable evidence that the process of fracture growth observed on a small scale also occurs on a large-scale. The large-scale single fracture shown in Figure 2.102(a), often shown on geological maps and seismic sections, is in fact generally made up of numerous smaller fractures which develop and link with each other as the large fault zone evolves, Figure 2.102(b). This process of linking has been discussed in detail in Section 2.6.5 and it is only briefly mentioned here in the context of its impact on fracture regularity.

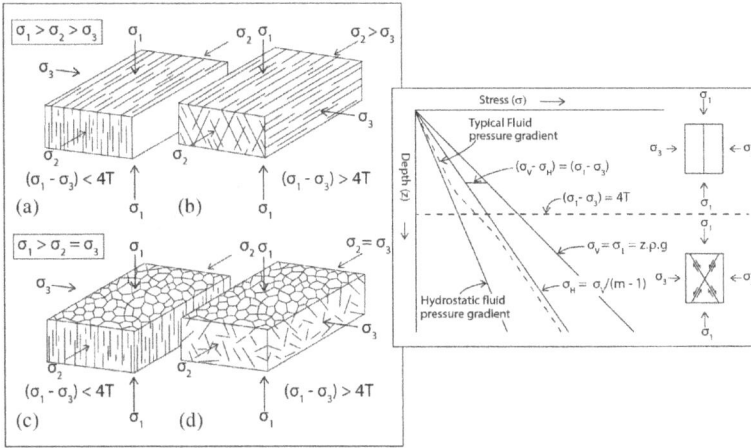

Figure 2.101. The impact of the magnitude of the differential stress ($\sigma_2 - \sigma_3$) on the regularity of the strike of normal and shear fracture sets. The inset shows the increase in the differential stress ($\sigma_1 - \sigma_3$) with depth and the accompanying change in mode of failure from extensional to shear.

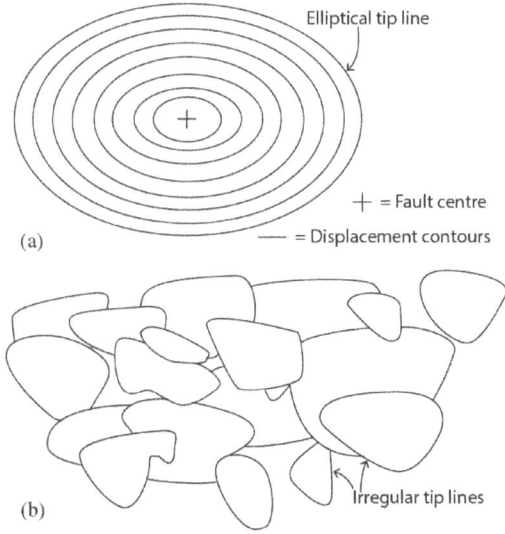

Figure 2.102. (a) Model for an isolated normal fault. The tip line is elliptical, with contours of equal displacement being around a central displacement maximum. (b) Model for the three-dimensional geometry of a fault zone. Segmentation occurs in both cross-section and map view with interaction between segments causing deviation from the model of the elliptical tip line. From Peacock and Sanderson (1997).

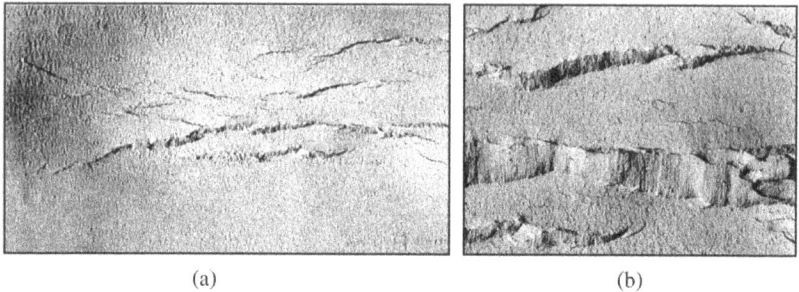

(a)                                                    (b)

Figure 2.103.   Stages in the evolution of a fracture zone in an analogue model. (a) Shows
an early stage clearly displaying the individual fracture. (b) Shows a more evolved stage
where the smaller normal faults have linked. Note the lineations (slickensides) on the fault
plane which reveal the direction of fault movement. A schematic representation of this
process of fault linkage is shown in Figure 2.104.

The process of fracture linking is illustrated in Figure 2.103 which
shows two stages in the evolution of a fracture zone generated in an analogue model made up of a ductile layer of plaster of Paris, prior to its
setting, which is caused to flex by the lowering part of its underlying
support. The early stage shows a series of small normal faults and the later
stage when these have linked to form a single structure. It is marked by an
irregular trace showing sharp bends at relay zones. A schematic representation of the linking of fractures in a vertical (dip) section during the
formation of a normal fault is shown in Figure 2.104 which also shows the
evolution of the fault traces of the two overlapping faults during the development of the relay ramp, stages (a)–(e).

The tick marks depict the down-thrown block of the faults. At stage
(a) in Figure 2.104, the faults do not interact. By stage (b) the faults have
started to interact and a relay ramp has developed to transfer the displacement among the fault segments. Stage (c) shows that the accumulated
strain in the relay ramp has resulted in the initiation of fracturing. At stage
(d), the relay ramp is broken by a breaching fault which links the two faults
to form a single fault zone with strike irregularity. At stage (e), the part of
the fault in the upper bench is abandoned (see the map view) and the two
segments joined through breaching of the lower ramp. The breaching fault
forms an along-strike bend in the course of the main fault.

*****

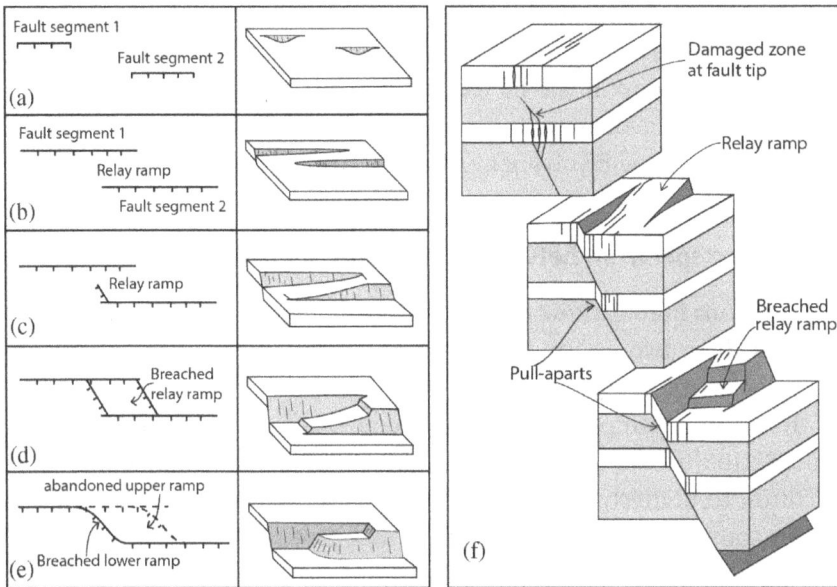

Figure 2.104. Diagram showing the impact of fracture linkage in both the strike (a)–(e) and dip (f) directions on fracture regularity. (a)–e) after Ciftci and Bozkurt (2007), and (f) after Peacock and Sanderson (1994).

In summary, the factors controlling the regularity (straightness) of fractures in a fracture set are as follows.

— The differential stresses linked to the regional stress field, $(\sigma_1 - \sigma_3)$ and $(\sigma_2 - \sigma_3)$. In an overburden stress regime, $(\sigma_1 - \sigma_3)$ controls regularity in the vertical $(\sigma_1/\sigma_3)$ plane and $(\sigma_2 - \sigma_3)$ controls regularity in the horizontal plane, i.e., controls, the strike of the fractures.
— The fracture tip stresses.
— The separation of the individual fractures.
— The amount of fracture linkage.

## 2.10 THE BRITTLE-DUCTILE TRANSITION

### 2.10.1 Introduction

So far, we have only considered purely brittle deformation of rocks. It is now appropriate to develop the discussion to consider ductile deformation and to examine the spectrum of structures that can form as conditions in the Earth's crust change from those favouring brittle to those favouring ductile deformation. It will be recalled that brittle deformation is defined as deformation involving a 'loss of continuity' in the material and 'ductile deformation' as one involving no loss of continuity.

## 2.10.2 The spectrum between brittle and ductile structures

Observations in rock show that there is a complete spectrum of structures between these two end members which show a progressive decrease in strain localisation, Figure 2.105.

Extreme examples of brittle deformation are illustrated in Figure 2.105(b) and 2.105(c) which show discrete extensional and shear fractures, respectively. The fractures represent planes of highly localised deformation which separates blocks of rock that remain essentially undeformed. Natural examples of these structures are shown in Figures 2.106(a) and 2.106(b). At the other end of the deformation spectrum is homogeneous flattening, Figure 2.105(f). In contrast to brittle failure, which is characterised by localised deformation, homogeneous flattening results in a perfectly uniform distribution of deformation. Natural examples of such deformation occur during the formation of rock cleavage and schistosity, Figure 2.106(f). The properties of the structures (i.e., the shear zones) that develop as conditions change from those conducive for brittle deformation

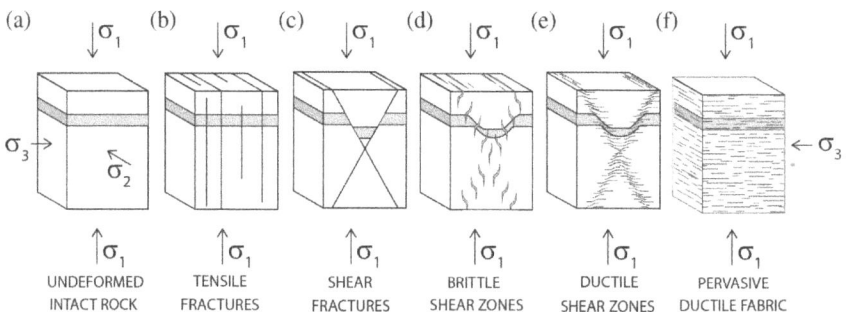

Figure 2.105.   The spectrum of structures between pure brittle (b) and (c) and pure ductile (f) deformation. Many parameters determine whether a rock behaves as a brittle or ductile material including, lithology, pressure, temperature and strain rate.

Figure 2.106.   A range of natural deformation structures between purely brittle (a) and (b) and ductile (f). (a) Quartz infilled, strata-bound joints in turbidite sandstones. (b) Conjugate shear fractures defining a small graben; (a) and (b) Carboniferous sandstones from Devon, UK. (c) Conjugate *en echelon* extensional veins defining conjugate shear zones. (d) Conjugate shear fabrics defining conjugate shear zones. (e) Conjugate shear zones defined by the formation of both the brittle extensional veins and the ductile shear fabric; (d) and (e) from Marloes Sands Pembrokeshire, Wales. (f) Fold with an axial plane cleavage, Cambrian slate, Sardinia.

to those favourable for ductile deformation, particularly their strength and permeability, are generally significantly different from the host rock. Such structures can form on all scales and can present major problems if encountered at rock engineering sites.

It can be seen from Figures 2.105(d) and 2.105(e) and 2.106(c)–2.106(e) that both types of shear zone are more diffuse examples of the conjugate shear fractures shown in Figures 2.105(c) and 2.106(b) and the deformation that occurs within these zones can be understood in terms of the local stress field linked to the generation of zones of 'simple shear' within the rock, as discussed in Section 2.7.5. It will be recalled (see Figure 2.69) that the maximum principal compression $\sigma_1$ is inclined at 45° to the zone as shown in Figure 2.107(a)(i).

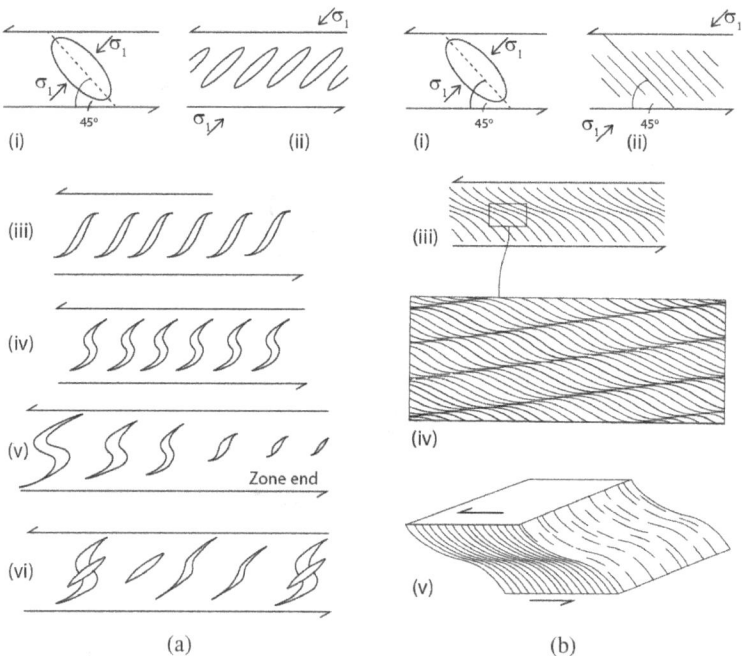

Figure 2.107.   (a) The formation of *en echelon* tension gashes and their relevance as a kinematic indicator. (a) (i) Shows the orientation of the local maximum principal compressive stress with respect to a fault or shear zone represented by the half arrows. (b) The formation of a local cleavage along the fault or shear zone. From Price and Cosgrove (1990).

The deformation within the shear zones illustrated in Figures 2.105(d) and 2.106(c) is brittle, as indicated by the extensional fractures that formed parallel to the local $\sigma_1$ and normal to $\sigma_3$. The result is the formation of an *en echelon* array of extensional fractures that define the shear zone and that can be used as kinematic indicator to reveal the sense of shear. In the example shown in Figure 2.107(a) this is sinistral, i.e., when one looks across the shear zone, displacement is to the left. In contrast, the deformation within the shear zones illustrated in Figures 2.105(e) and 2.106(d) is ductile and characterised by a flattening fabric which forms normal to the maximum principal compression, Figures 2.107(b)(ii) and (iii). As noted in Section 2.7, as the shear zone develops, the early formed fabric rotates towards parallelism with the shear zone but the newly formed fabric at the margins of the shear zone continues to form at 45° to the zone. This results in the characteristic S shaped geometry of the fabric, Figure 2.107(b)(iii) which like the *en echelon* extension fractures, can also be used as a kinematic indicator.

These two types of shear zone represent the transition from brittle to ductile deformation. There is a variety of parameters that determine whether a rock behaves as a brittle or ductile material. These include the lithology, and temperature, pressure and strain rate at which the deformation occurs.

### 2.10.3 Rheology

In the discussion of brittle failure, Section 2.3, it has been assumed that, prior to a rock fracturing as a result of the applied stress reaching the rock strength, the rock behaves as an elastic material. The stress–strain plot for an ideal linear elastic material is shown in Figure 2.108(a) and, before discussing the brittle–ductile transition and its expression in rocks in more detail, it is useful to recall the basic rheological models that describe elastic, viscous and plastic deformation. These three types of behaviour can be represented graphically and by simple mechanical models as shown in Figure 2.108.

The deformational behaviour of an ideal linear elastic material, Figure 2.108(a), when plotted on a stress–strain graph appears as a straight line which passes through the origin. Its slope, which represents the resistance to deformation is defined by $E$, the Young's modulus. In addition, the

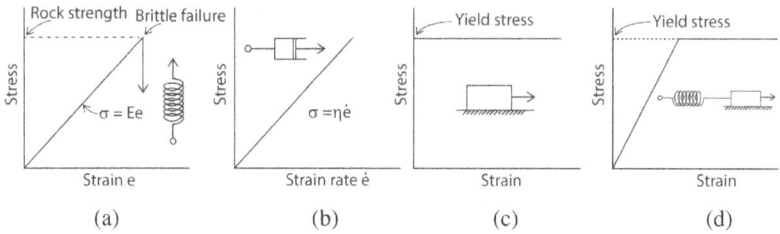

Figure 2.108. (a) Stress–strain plot for a linearly elastic material. The relation between the stress and strain, i.e., the slope of the line, is $E$, Young's modulus. The spring is the analogue model that represents this material. (b) The stress–strain rate plot for a linear viscous fluid, the slope of the line, $\eta$, is the viscosity. The dashpot is the analogue model representing this material. (c) The stress–strain relation for a plastic material. The analogue model representing this behaviour is a sliding block. (d) The stress–strain plot for an elastic-plastic material (by combining (a) and (c)).

deformation is reversible and, if the load is removed, the material returns to its unstressed state without any loss of energy. Such a material can be represented by a spring. The strength of the material is defined as the maximum stress it can sustain before it fails; at this stress, the elastic limit is reached and the material deforms permanently by the formation of either extensional or shear fractures. In an ideal elastic material, this failure is represented by an abrupt drop in the stress–strain curve, Figure 2.108(a).

In contrast, a viscous material has no strength and cannot support a shear stress. The application of a load, no matter how small, would cause the material to deform (flow) and it would continue to do so until the stress was removed. If this deformation behaviour were plotted on a stress–strain graph, it would simply plot along the strain axis. A more useful representation of this behaviour is obtained by plotting it on a stress–strain rate graph, Figure 2.108(b). An ideal, linear viscous material (a Newtonian fluid) would plot as a straight line which passes through the origin and its slope, which defines the resistance to flow, is defined by $\eta$, the coefficient of viscosity. This deformation is irreversible and can be represented by a dashpot, Figure 2.108(b).

An ideal plastic material is similar to a viscous material but has a yield strength below which deformation will not occur. Its stress–strain behaviour is shown in Figure 2.108(c). As the applied stress is increased in this idealised material, no deformation occurs until the yield stress is reached.

At this point, deformation is initiated and continues as long as the stress level is maintained. The deformation is irreversible and can be represented by a sliding block, Figure 2.108(c), where the yield stress is determined by the frictional resistance to sliding which needs to be overcome before the block can move. This resistance ($\tau$) is quantified by Amonton's law of sliding friction which can be written as $\tau = \mu\sigma$ where $\mu$ is the coefficient of sliding friction and $\sigma$ the normal stress acting between the block and the surface on which it rests. A more realistic model of a plastic material is one in which elastic, i.e., reversible, deformation occurs during the build-up of the load to the yield strength after which plastic deformation begins. Such a material is termed an elastic-plastic material and its stress–strain curve is shown in Figure 2.108(d).

It can be seen that these various rheological models imply specific stress–strain or stress–strain rate relations. This is in contrast to the generic terms 'brittle' and 'ductile', which simply imply that the material deforms by losing its continuity, i.e., by fracturing or by maintaining the continuity. When geological deformation is sub-divided into brittle and ductile, it is implicit that the deformation of the rock occurs by different mechanisms. For example, when a sandstone, which is made up primarily of quartz grains, deforms in a brittle manner, microfractures form either along grain boundaries or within the grains. As discussed in Section 2.6, as deformation proceeds these microfractures eventually link to form larger fractures and the rock loses its continuity. However, when it deforms in a ductile manner, the rock maintains its continuity and the grains deform either by processes of recrystallization or by the movement of dislocations. These two types of rheological behaviour result in different stress–strain histories which are characterised by different stress–strain plots, e.g., Figures 2.108(a) and 2.108(d).

However, experimental studies of rock deformation have shown that the bulk stress–strain behaviour of a sample during deformation is not always a reliable indicator of the micromechanism by which the rock is deforming. This is well illustrated by the experimental data summarised in the graphs of Figure 2.109. These results relate to the triaxial loading of cylindrical rock specimens (e.g., Figure 2.110) and the graphs show the plot of differential stress against strain under different confining stresses for a marble and a sandstone. The experiments were carried out at the

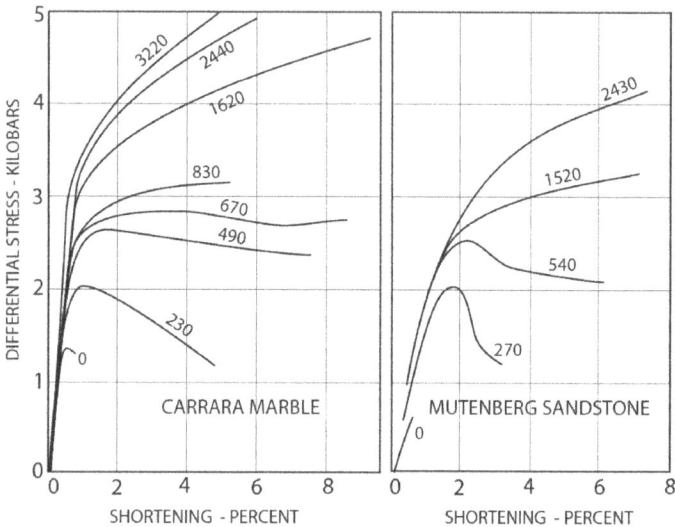

Figure 2.109. Stress–strain curves for Carrara marble and Mutenberg sandstone, showing the dependence of the bulk behaviour on the confining stress which is shown in bars on the individual deformation curves. After von Karman (1911).

same strain rate and at room temperature, conditions under which it would be expected that brittle deformation would occur. Under low confining stress, 100 bars in the marble and 250 bars in the sandstone (noting that 1 bar ≡ 100 kPa ≡ 0.1 MPa ≡14.5 psi), the stress–strain plots of the rocks approximate well to that of a linear elastic material which shows an abrupt drop in stress once the strength of the material is reached and brittle failure occurs. However, as the confining stress increases, the bulk stress–strain behaviour changes and becomes more characteristic of an elastic–plastic material, Figure 2.108(d), i.e., ductile deformation involving no loss of continuity, as shown by the experiments carried out under confining stresses of between 500 and 800 bars. However, this does not imply that the micromechanism of deformation has changed, but only that the deformation has become less localised. This process can be clearly seen in Figure 2.110 which shows rock samples forming a single extensional fracture, a single shear fracture, and a progressively widening zone of diffuse shear deformation as the confining pressure increases.

Deformation in all samples is still primarily by the formation of microfractures. However, as the confining stress increases, the tendency for

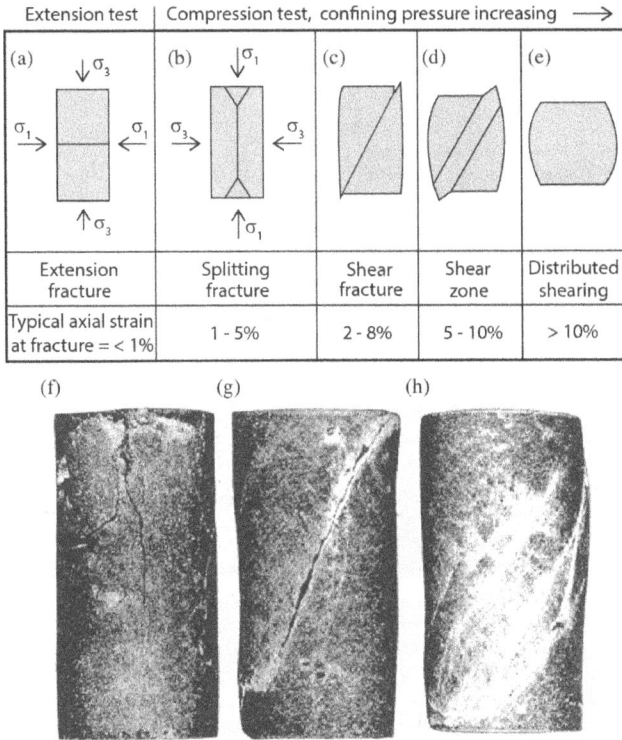

| Extension test | Compression test, confining pressure increasing $\longrightarrow$ | | | |
|---|---|---|---|---|
| (a) | (b) | (c) | (d) | (e) |
| Extension fracture | Splitting fracture | Shear fracture | Shear zone | Distributed shearing |
| Typical axial strain at fracture = < 1% | 1 - 5% | 2 - 8% | 5 - 10% | > 10% |

Figure 2.110.  Top diagrams show schematic representations of brittle failure styles in triaxial tests. (a) Extension test. (b)–(e) Compression tests with confining pressure increasing to the right (from Griggs and Handin, 1960). Bottom diagrams show examples of brittle failure in triaxial test specimens of Ottawa basalt. (f) 0.1 MPa confining pressure, (g) 49 MPa confining pressure, (h) 98 MPa confining pressure. From Hoshino *et al.* (1972).

the microfractures to link and form a single, discrete shear plane (Figure 2.110(c) and 2.110(g)) is reduced and more diffused zones of microfractures form (Figure 2.110(d), 2.110(e) and 2.110(h)) characterised by uncemented grains and small rock fragments. The process of progressive fracturing and comminution of the rock is known as cataclasis and the resulting rock as a cataclasite. It is the result of brittle failure on a microscale. The effect of cataclasis on the expression of shear failure in a rock can be seen in Figures 2.111(b) and 2.111(c) which show a network of conjugate 'granulation seams'. These are in marked contrast to the discrete shear fractures shown in Figure 2.111(a).

(a)                                    (b)

(c)

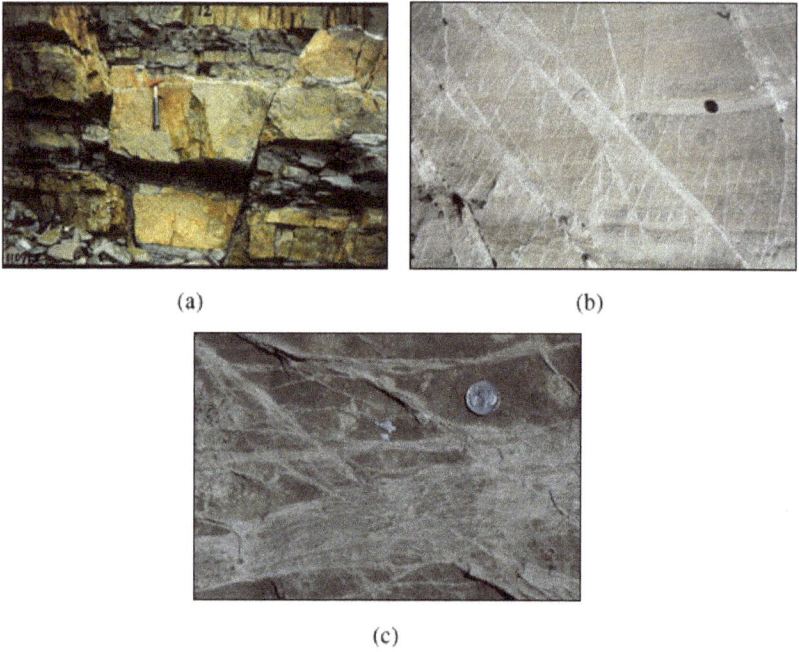

Figure 2.111.    Examples of brittle shear failure in which the failure planes become increasingly more and more diffuse. (a) Shear failure is defined by single localised fracture planes. (b) Shear failure defined by granulation seams, a less localised form of failure. (c) Broader granulation seams. Note that (b) and (c) were formed in a porous sandstone.

## 2.10.4  Brittle failure revisited

The major differences between the two principal modes of brittle failure were discussed in detail in Section 2.3, are summarised in Figure 2.112, and are listed below.

— The two modes have *different* failure criteria.
— The resulting fractures have a *different* orientation with respect to the causative stress field.
— They form under *different* stress conditions: ($(\sigma_1 - \sigma_3) > 4T$ for shear failure, $(\sigma_1 - \sigma_3) < 4T$ for extension).
— The sense of movement on the fractures is *different*. Movement on shear fractures is parallel to the fracture walls and on extensional fractures is normal to the walls.

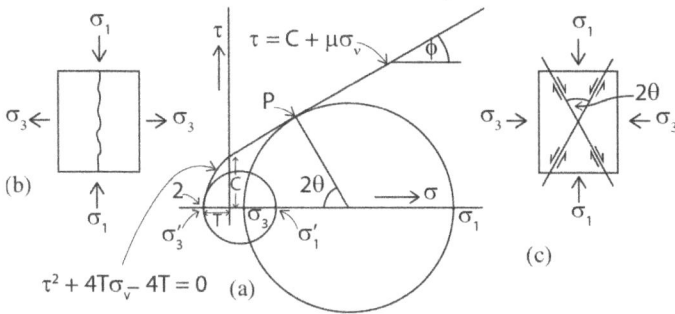

Figure 2.112. (a) The combined brittle failure criterion derived by linking the linear Navier–Colomb shear failure criterion with the parabolic extensional failure criterion of Griffith. The stress conditions required for the two modes of failure are apparent. In order to satisfy the extensional failure criterion, a small differential stress is required, whereas shear failure requires a larger differential stress. (b) and (c) show the relation between the two modes of fracture and the causative stress fields.

However, despite these important differences, it is clear from Figures 2.105(d) and 2.106(c) and 2.106(e) that the two mechanisms of failure do in fact appear to be more closely related than these observations would suggest. For example, if the spectrum of structures associated with the brittle ductile transition are considered, Figure 2.105, it can be seen that, as the transition occurs, the highly localised deformation associated with the formation of a shear fracture gives way to a less localised deformation in which the single shear plane, Figure 2.105(c), is replaced by a set of *en echelon* extensional fractures, Figure 2.105(d). It is clear from this Figure and the naturally occurring vein sets shown in Figure 2.113(a) that large-scale shear failure can be initiated by the formation and linking of smaller-scale extensional fractures. These field observations indicate that, prior to the formation of either extensional or shear fractures in a rock, the formation of small-scale extensional cracks oriented normal to the least principal compression, $\sigma_3$, form randomly through the rock, Figure 2.113(c). If the stress conditions for extensional failure are satisfied (i.e., when the differential stress is less than $4T$, where $T$ is the tensile strength of the rock, see Section 2.3.4), with continued deformation some of these then extend and link to form macroscopic extension fractures. Conversely, when the differential stress is greater than $4T$, the extensional cracks organise themselves into a systematic *en echelon* array, Figures 2.113(a)–2.113(d).

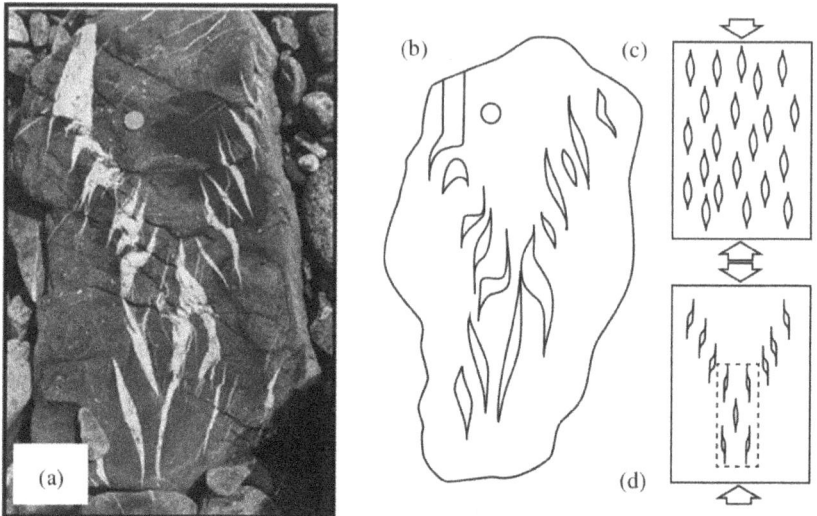

Figure 2.113. (a) Conjugate sets of *en echelon* tension gashes defining zones of shear failure in the rock. (b) line drawing of (a). (c) The uniform distribution of tension gashes in a rock representing a 'bulk' extensional failure. (d) The organisation of small-scale extensional fractures to produce macroscopic shear failure. From Kidan and Cosgrove, (1996).

It can be seen from Figure 2.113(d) that within a conjugate array of *en echelon* fractures, i.e., within a domain undergoing shear deformation, there will be areas of tension fractures within this array (e.g., the dotted rectangle in this Figure) where the distribution of the fractures will be identical to the distribution in Figure 2.113(c). Clearly the scale of observation is important (see Section 2.13.4, Figure 2.142 and Principle 7 in Chapter 3) and we shall return to the impact of the 'scale' of geological structure and the scale of the engineering site in Chapters 4–8 when presenting practical engineering cases.

The role of extensional fractures in the generation of faults is particularly clear in the example illustrated in Figure 2.114 which is based on observations made of faults cutting the well-bedded Liassic limestones and shales outcropping on the Bristol Channel coast of North Somerset, UK. The linking of shear fractures (faults) along *strike* by the formation and deformation of relay ramps has been discussed in Section 2.6.5. Figure 2.114

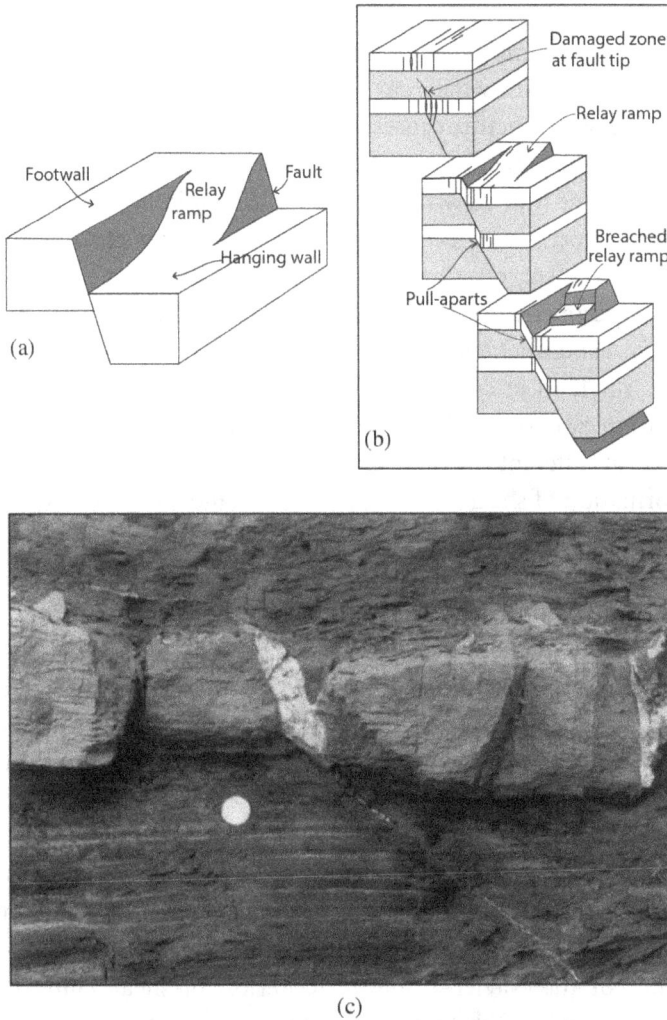

(a)

(b)

(c)

Figure 2.114. (a) The linking of small shear fractures along strike by the formation of relay ramps (see Section 2.6.5). (b) The role of extensional fractures in the formation of a macroscopic shear fracture. The extensional fractures in the limestones link along the dip direction by the formation of cross-fractures in the shales. Movement on the relatively low angle cross-fractures results in a dilation of the sub-vertical extensional fractures. (c) Calcite deposited within a dilated extensional fracture in a limestone bed. Note the cross-fractures in the shale. Lilstock, North Somerset, UK.

illustrates the linking of faults in the *dip* direction. Bedding-normal extensional fractures form in the competent limestones and these are linked by inclined cross fractures which cut through the shales. Movement on the resulting large-scale fracture causes the extensional fractures to open forming rhomb-shaped gaps known as pull-aparts or dilational jogs. These frequently become infilled with minerals as shown in Figure 2.114(c).

A similar process of shear failure in rocks is observed to occur at the microscopic level, i.e., on a grain scale. Mandl (2000) notes that there is abundant experimental evidence that the macroscopic shear fractures developed in experimentally deformed rock samples do not form by the growth of a single shear crack in its own plane, but that shear failure is preceded by complex pervasive brittle damage of the rock. By arresting triaxial experiments at different stages of failure, it is possible to examine the progressive formation of shear fractures in rocks. Mandl describes experiments by Hallbauer *et al.* (1973), who axially loaded cylinders of quartzite encased in a strong confining jacket which caused the confining stress to rise during loading. At some critical load, arrays of microcracks begin to form throughout the rock. This is the onset of microscopic brittle failure which corresponds to the onset of microseismic activity within the sample. They are mainly inter-granular and tend to be parallel to $\sigma_1$. Their density increases as the differential load increases and they eventually localise in a shear band just after the differential load has passed its peak value, i.e., at the point where the shear strength of the material is reached and the sample fails macroscopically. This process is shown diagrammatically in Figure 2.115.

One of the reasons why the link between the two modes of failure was not realised earlier becomes apparent when it is recognised that, despite the elegance of the Navier–Coulomb criterion of shear failure (unlike Griffith who considered the mechanism that gave rise to the tensile failure of a material and the role of microcracks in that process), Navier and Coulomb did not consider the detailed mechanism involved in shear failure. Their failure criterion is phenomenological, i.e., a theory that expresses mathematically the results of observed phenomena without paying detailed attention to the fundamental failure mechanism involved.

More recently the link between small-scale extensional cracks and shear fractures has also been illustrated by numerical studies (see e.g., Tang and Hudson, 2010). Figure 2.116(a) shows the results of a series of

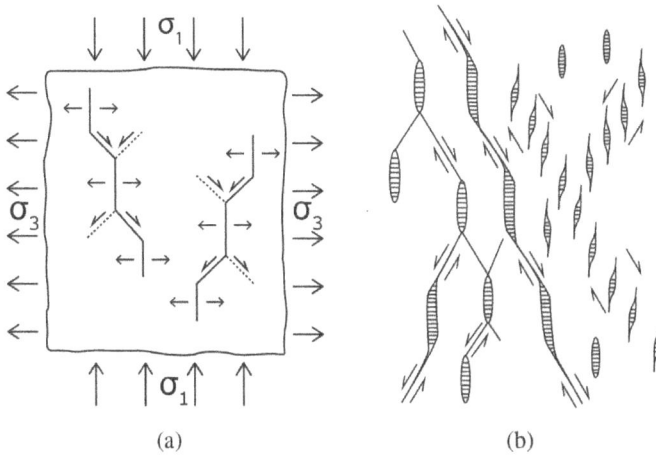

Figure 2.115. Idealised model showing the formation of through-going shear fractures by the linking of individual, isolated, extensional fractures. (a) From Mandl (2000); (b) from Sibson (2004).

numerical models performed to determine the impact of the dimensions (height:width ratio) of a slab sample on the distribution of deformation when it is loaded axially to failure. It can be seen that that the longer samples tend to show vertical splitting (i.e., extensional) failure, whereas the shorter specimens show a more complicated failure mode, one in which splitting occurs first followed by the formation of internally inclined faults (shear fractures) which dominate the overall fracture process. Later in the deformation process, these show a marked tendency to link with other microfractures to form macroscopic zones of shear deformation. A similar process can be seen in the numerical simulation of crack propagation shown in Figure 2.116(b). In this experiment, an inclined fracture is incorporated into the numerical sample which is then loaded axially. Initially, extensional wing fractures form at the end of the crack and, as deformation proceeds, more extensional microfractures form which eventually link to form macroscopic shear fractures.

## 2.10.5 A modification to the brittle failure envelope

Earlier in this Chapter, the effect of confining stress on the expression of brittle deformation in a rock was noted (see Figure 2.109 and the related

(a)

(b)

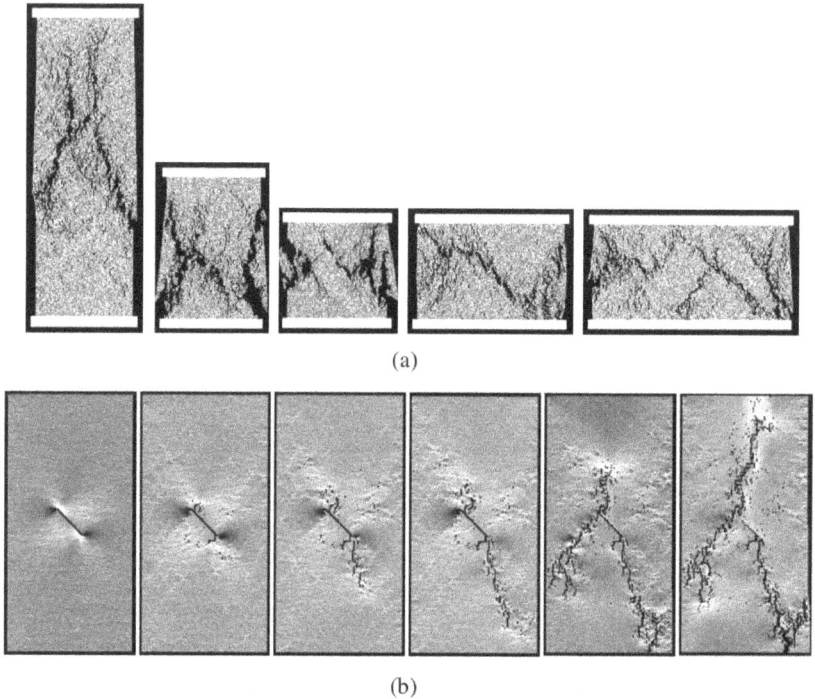

Figure 2.116. (a) Numerical models showing the evolution of brittle failure in axially loaded (top to bottom of the page) elastic samples of different shapes. Although the samples undergo macroscopic shear failure, it is apparent that the evolution of shear failure requires the formation of numerous, small, tensile fractures. (b) Six stages in the deformation of an axially loaded (top to bottom of the page) elastic slab in which an inclined crack is initially present. The deformation is dominated by the formation of extensional microfractures which coalesce to form macroscopic shear fractures. From Tang and Hudson (2010).

discussion). It was observed that, with increasing confining stress, the 'bulk' behaviour of the rock changed from that of a brittle material towards that of a ductile material. It has been argued (Ingram and Urai, 1999) that this transition will have an impact on the brittle failure envelope (see Section 2.3). The proposed modification is shown in Figure 2.117.

Thin section studies of the textures linked to shear failure discussed above show that two types of shear fracture can develop in rocks. The first are those whose origin involves the linking of extensional fractures by cross fractures (Figure 2.115) and whose formation initially involves an

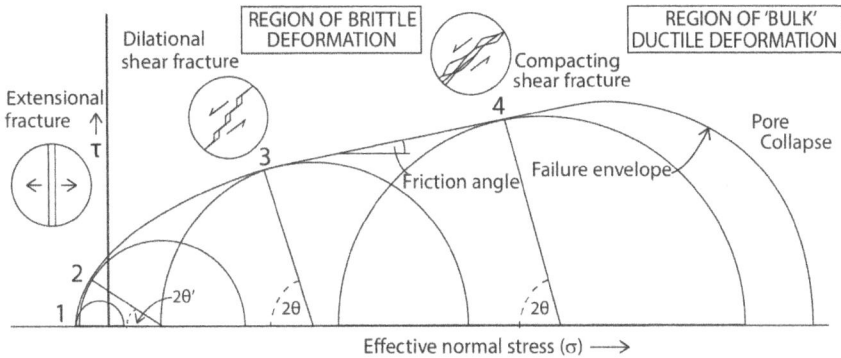

Figure 2.117. Modified brittle failure envelope which takes into account the tendency for rocks to display a bulk, ductile rheology as the confining stress increases (see Figure 2.109 and the related discussion). The deformation field above the failure envelope is divided into a region where both the microscopic and macroscopic deformation is brittle (labelled the 'region of brittle deformation') and one where the microscopic deformation is brittle but the macroscopic deformation is ductile (labelled the 'region of 'bulk' ductile deformation'). After Ingram and Urai (1999).

element of dilation (Figures 2.114–2.116). The second type of shear fracture leads to the formation of a granulation seam, Figures 2.110(d), 2.110(e) and 2.110(h) and Figures 2.111(b) and 2.111(c), and this process does not appear to involve dilation. The experimental work illustrated in Figure 2.110 indicates that this second mode of shear failure characterises rocks deformed under relatively high confining stress. It was noted that, as the confining stress increases, so the localisation of the brittle deformation decreases to the point where it can totally pervade the rock. Although the rock is behaving as a brittle material on a grain scale where microfractures and brecciation lead to cataclastic deformation, on a bulk scale the rock behaves as a 'ductile' material. It is suggested that rocks that deform without dilation and without the formation of discrete fracture planes, can be classified as ductile materials. The tendency to dilate is a function of several parameters including the mechanical properties of the rock, the effective stress and the shear fracture geometry. At a given confining pressure, a stronger rock is more likely to dilate than a weaker one. The stress regimes linked to the formation of discrete brittle fractures and pervasive brittle deformation (i.e., bulk ductile behaviour) are shown in. Figure 2.109

and it can be seen that the transition between the two is linked to the gradual increase in confining stress.

Wood (1990) has determined the stress conditions that mark this transition. He demonstrates that the key factor that determines the onset of dilatancy and therefore of brittle behaviour is the ratio of the compressive strength of the rock (defined by the effective stress state, $\sigma_1' - \sigma_3'$, that causes failure) and the mean effective stress, Equation (2.39).

$$(\sigma_1' - \sigma_3')/(\sigma_1' + \sigma_2' + \sigma_3'), \tag{2.39}$$

where $\sigma_1'$, $\sigma_2'$ and $\sigma_3'$ are the effective principal stresses; $(\sigma_1 - p)$, $(\sigma_2 - p)$ and $(\sigma_3 - p)$ and $p$ is the fluid pressure.

Extensional fractures form when the least effective principal stress reaches the tensile strength of the rock ($T$) and when the differential stress is less than four times the tensile strength, Mohr circle 1 in Figure 2.117. Hybrid, shear fractures occur in rocks during deformation in low confining stress regimes when the least effective principal stress is tensile and where the differential stress is in the range $4T$–$5.66T$, Mohr circle 2 in Figure 2.117, see Section 2.3.4. This type of failure has also been referred to as 'extensional shear failure' to distinguish it from 'compressional shear failure' which occurs when the differential stress is greater than $5.66T$, e.g., Mohr circle 3, i.e., the value above which the normal stress on the fracture plane becomes compressive, Figure 2.22. At relatively low confining stresses, this shear failure is a dilatant shear failure, Figure 2.117. As noted above, as the confining stress increases so dilation decreases, Mohr circle 4, and the deformation becomes less localised.

## 2.10.6 The generation of rock textures linked to shear deformation

The progressive comminution of a rock during the initiation and evolution of a shear fracture can be divided into two phases. These relate to the break-up of the rock prior to the formation of a through-going fracture (pre-fault damage) and the formation of fault products generated by subsequent movement on the through-going fracture or fracture zone (post-fault damage and the production of fault breccias, gouges and mylonites).

### 2.10.6.1 *Pre-fault formation damage*

The process described earlier in this Section and illustrated in Figures 2.115 and 2.116, in which extensional microfractures build up in the rock prior to the formation of a macroscopic, through-going shear, is based on numerical models and a study of thin sections of experimentally deformed rock samples arrested at various stages in the evolution of shear fractures. These microfractures are approximately grain-sized and occur on the scale of a few millimetres. However, as noted earlier, this process occurs **on all scales**. Figures 2.113 and 2.114 show the process on a scale of a few centimetres and examples of it occurring on a larger scale are illustrated in Figure 2.118. In Figure 2.118(a), the formation of *en echelon* tension fractures, described in Section 2.7.5 is shown. In (b) and (c), two fracture zones are shown representing different stages in the formation of a through-going shear failure.

In these examples, the initial *en echelon* organisation of the extension fractures can be clearly seen trending approximately WNW–ESE as in a. They indicate the orientation of the maximum principal compression and act as kinematic indicators showing that the sense of shear along these embryonic shear zones was dextral, i.e., top to the right. In Figure 2.118(b), the linking of the individual *en echelon* extension fractures by cross joints is apparent. In Figure 2.118(c), the process of shear failure is more advanced and the linking of the individual *en echelon* extension fractures by cross joints is more developed and complex. The result is the local formation of breccia zones between some of the extensional fractures. As the process continues, so the array of extensional fractures is progressively destroyed and replaced by a breccia zone which then represents a shear zone separating the two blocks of rock each side of it — which are therefore free to move with respect to each other. A through-going fracture zone has been formed.

### 2.10.6.2 *Post fault formation damage*

The various stages in the evolution of a through-going fracture from an incipient shear are shown in Figure 2.119. These are based on field observations in dolomite and the evolution of a shear fracture from early formed

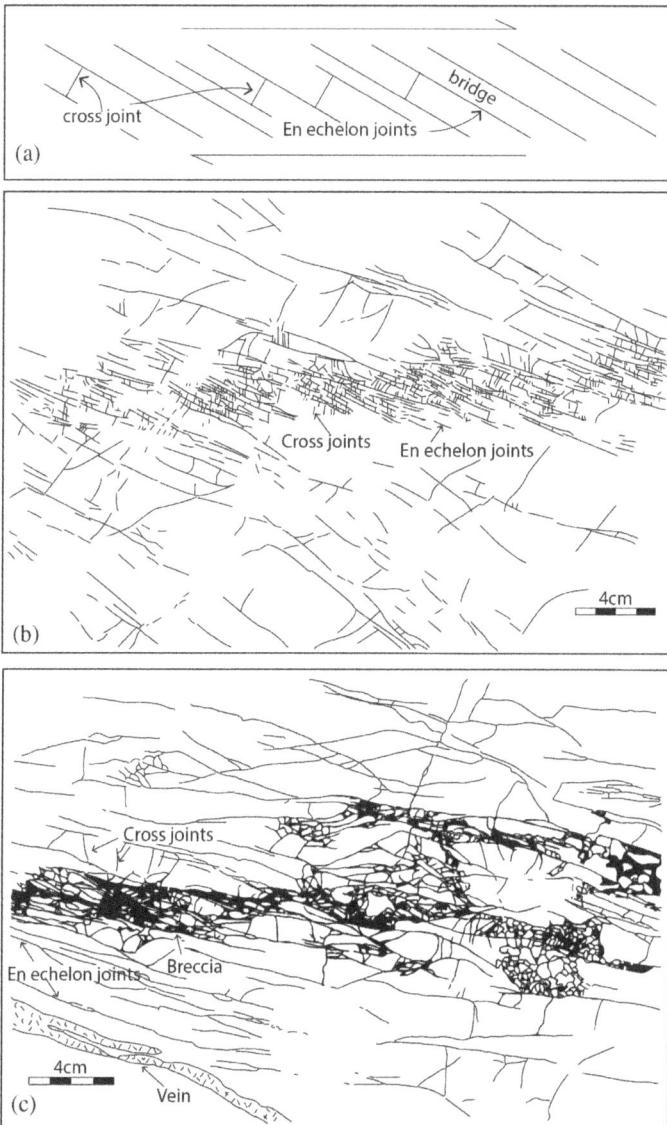

Figure 2.118.  (a) *En echelon* extensional fractures generated in a dextral shear zone, see Figure 2.107(a). These fractures become linked by cross fractures which cut across the bridges of rock which separate them, eventually forming a through-going fault. (b) and (c) Maps of zones of *en echelon* and cross joints accommodating dextral shear in a dolomite in the Sella Group, Northern Italy, showing two stages in the evolution of a through-going fracture. After Mollema and Antonellini (1999).

extensional fractures is apparent. The first stage is the formation of exten-sional fractures orientated normal to the least principal compression and which form by the linking of microcracks Figure 2.119(a). The localisation of the joints into *en echelon* arrays and their linking by the generation of cross joints is shown in Figures 2.119(b)–2.119(d). Further movement along these incipient shears generates pockets of breccia as a through-going fracture is formed, Figures 2.119(e)–2.119(h). Subsequent movement along the fracture is likely to generate a variety of deformation products in addi-tion to the fault breccia shown in Figure 2.119(h). These fault products vary, depending on a variety of parameters including rock type, pressure and temperature condition, the strain rate and the amount of slip.

The fault products include angular fault breccias, Figure 2.120, rounded breccias, fault gouges and mylonites, Figure 2.121. Although the first three of these can be considered as progressive stages in the evolution of the fault products as movement on the fault increases, as indicated, they can also form at the same time as each other at different sites on the fault as a result of irregularities in the geometry of the fault plane. Fault breccias can be initiated during the early stages of fault development, Figure 2.118(b). As the fault evolves and a through-going shear plane or shear zone devel-ops, these breccias become crushed and the fragments rounded. Eventually the breccia becomes ground down into a fine-grained, clay-like material known as a fault gouge. Both fault breccias and fault gouges are the prod-uct of brittle deformation and represent different stages in the comminu-tion (grinding down or milling) of rock fragments as the two sides of a fault move past one another.

When faulting occurs deeper in the Earth's crust, the fault products may be the result of ductile deformation. Then the fault product is termed a mylonite which is a ductilely deformed, extremely fine-grained rock generated by the accumulation of large shear strains by movement along ductile fault zones. The reduction in grain size is the result of crystal-plastic deformation which results in a rock with a strong foliation. They are the deep crustal counterparts of cataclastic brittle faults that generate fault breccias and gouges. It will be recalled from earlier in this Section that there are a number of factors that determine whether a mate-rial behaves in a brittle or ductile manner. These include the pressure, temperature and strain rate at which the deformation is occurring and

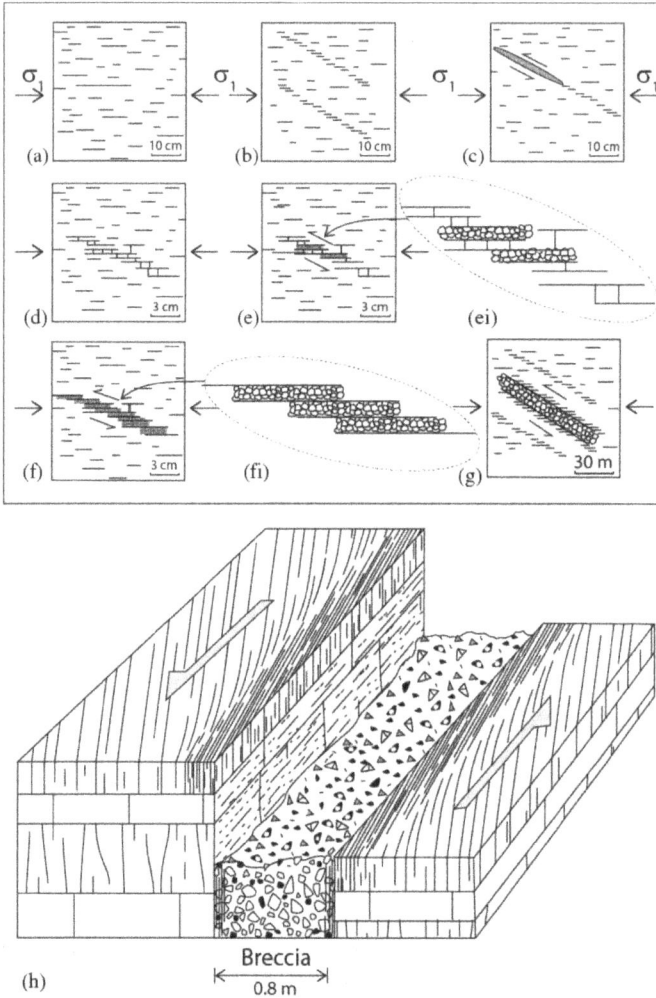

Figure 2.119. Model for fault development based on observations in dolomite. (a) microcracking and jointing parallel to the regional maximum principal compression. (b) Localisation of joints in *en echelon* arrays. (c) Localisation of joints at the tip of a sheared vein. (d) The formation of cross joints. (e) The formation of arrays of breccia pockets; (ei) a detail of (e). (f) The formation of a continuous breccia zone parallel to the *en echelon* array of joints; (fi) a detail of (f). (g) Large, wide, cataclastic fault zone, with a high joint density next to the fault. (h) A schematic diagram of a fault showing a central zone of fault generated breccia and marginal zones which shows *en echelon* fractures (b) from which the fault zone evolved. From Mollema and Antonellini (1999).

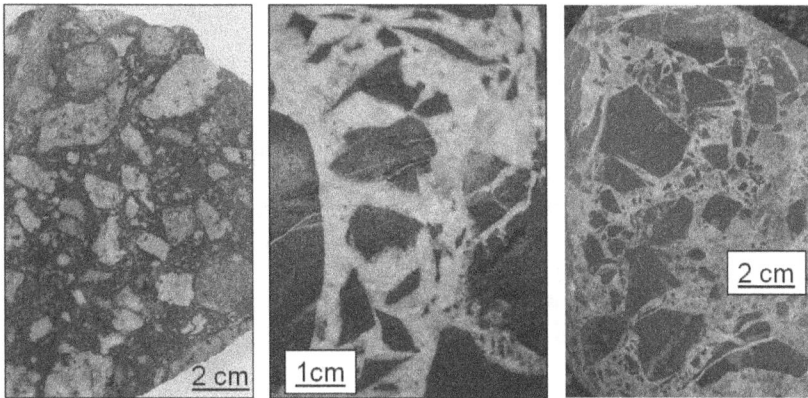

Figure 2.120. Three examples of fault breccias generated as a result of movement along a through-going fracture zone. Left-hand example from the Bodmin Granite showing feldspar clasts (now weathered to kaolin) in a finer grained, breccia matrix, Cornwall, UK. Middle and right-hand examples from the Carboniferous turbiditic sandstones of Bude, Cornwall, UK. The sandstone clasts are suspended in a quartz matrix.

also the lithology (mineralogy). Because rocks are often made up of several different minerals, it is not surprising to find that, when fault products are examined in the field, they often display characteristics of both modes of deformation. This is illustrated in Figure 2.121 which shows a mylonite taken from a large shear zone cutting a granite in the Pre-Cambrian basement of Tanzania. The rock consists primarily of quartz and feldspar and, under the conditions prevailing during the formation of the shear zone, the quartz behaved as a ductile material producing a fine-grained foliated mylonite (the dark matrix). In contrast the feldspar behaved in a brittle manner and forms isolated breccia clasts floating in the mylonite. These clasts sometimes display an asymmetry which enables them to be used as kinematic indicators revealing the sense of shear along the fault, see Figure 2.74, feature 9.

Other features that can be used to determine the sense of shear have been discussed earlier and include *en echelon* extension fractures, Riedel shears and shear zone fabrics. These indicators pre-date the formation of a through-going fracture or fracture zone. Others, including slickensides and slicken-fibres, which have been discussed and illustrated in Section 2.7.5, form after their formation.

Figure 2.121.   Light coloured feldspar clasts 'floating' in a dark quartz mylonite. During faulting, the two minerals behaved in different ways, the feldspar in a brittle manner and the quartz in a ductile manner. From a shear zone in the Pre-Cambrian basement of Tanzania. The specimen is 120 mm wide.

It was suggested above that fault breccias and gouges represent progressive stages in the evolution of fault products as the movement on the fault increased. However, it is possible for these two rock types to be formed synchronously if the fault plane contains local bends or irregularities. For example, when the small irregularities on the fault plane that cause the generation of the slicken-fibres (see Figure 2.68) occur on a larger scale, they produce local zones of dilation and compression along the fault, Figure 2.122.

The jogs or bends in the fault are termed dilational jogs or compressional jogs and represent areas of anomalously low or high compression respectively. If the rocks are under a high fluid pressure at the time of faulting, then it is likely that an implosion breccia will form in the vicinity of the dilational jogs as a result of the drop in pressure. An example of an implosion breccia is shown in Figure 2.123(b). In contrast, at the compressional jogs, normal stresses across the fault will be high and there will be a rapid grinding down of any fault product to produce a gouge.

*****

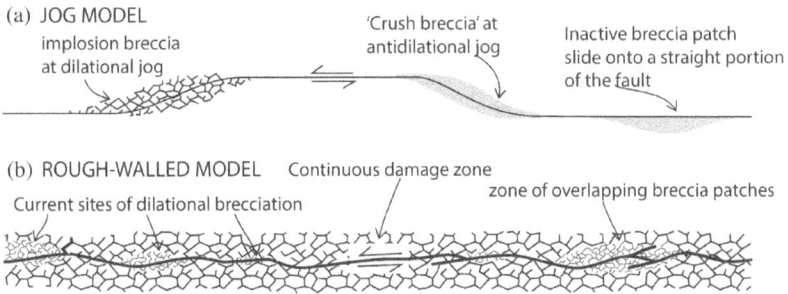

(a) JOG MODEL
implosion breccia
at dilational jog

'Crush breccia' at
antidilational jog

Inactive breccia patch
slide onto a straight portion
of the fault

(b) ROUGH-WALLED MODEL    Continuous damage zone

Current sites of dilational brecciation

zone of overlapping breccia patches

Figure 2.122.    (a) The formation of a dilational and compressional jog on a fault plane as a result of fault movement. (b) Collateral damage along a fault generated as a result of irregularities on the fault plane.

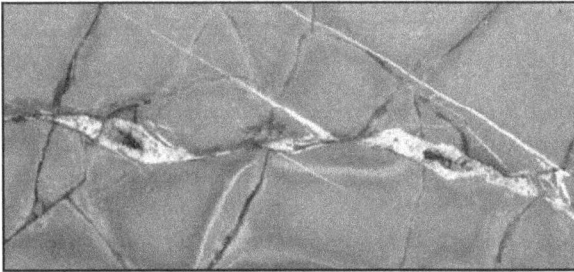

(a)

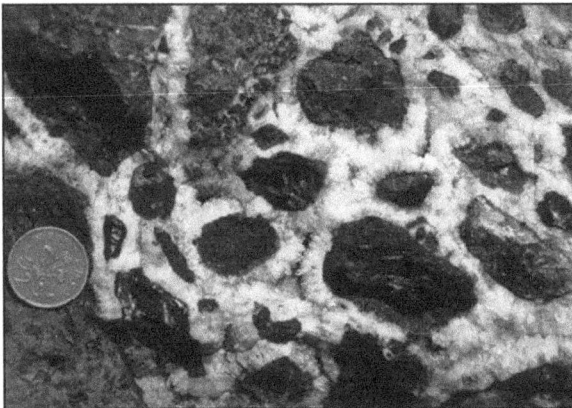

(b)

Figure 2.123.    (a) The formation of dilational jogs at steps in the fault plane. The sense of movement on the fault is dextral (top to the right) as indicated by the *en echelon* tension gashes. (b) An implosion breccia formed in a dilational jog along a fault in Rusey, Cornwall, UK. The angular slate clasts are floating in a quartz matrix.

|  | Rock mechanics significance of foliation increasing from left to right | | |
|---|---|---|---|
| Rock mechanics significance of foliation increasing from top to bottom | **G1** gneissic/low **App. 1A(b)** | **G2** gneissic/intermediate **App. 1A(c)** | **G3** gneissic/high **App. 1A(d)** |
|  | **B1** banded/low **App. 1B(b)** | **B2** banded/intermediate **App. 1B(c)** | **B3** banded/high **App. 1B(d)** |
|  | **S1** schistose/low **App. 1C(b)** | **S2** schistose/intermediate **App. 1C(c)** | **S3** schistose/high **App. 1C(b)** |

Legend

|  | |
|---|---|
|  | Foliation of low rock mechanics significance (RMF 1) |
|  | Foliation of intermediate rock mechanics significance (RMF 2) |
|  | Foliation of high rock mechanics significance (RMF 3) |

Figure 2.124. The two variables, type of foliation and intensity of foliation, are combined in the matrix which is constructed to reflect the rock mechanics significance of the nine classes. The foliation type classes (G, B, S) are arranged vertically, with the mechanics significance increasing from top to bottom, while the foliation intensity classes (1, 2, 3) are arranged horizontally, with the mechanics significance increasing from left to right. This means that the combined significance increases diagonally, from top left to bottom right, as indicated by the shading in the matrix. Milnes *et al.* (2006).

Sometimes on rock engineering projects, it is necessary to quantify, or at least codify, the geological variation that is occurring — for example, in site investigation borehole core or the sidewall of a tunnel. As can be deduced from the text and photographs in this and the other Chapter 2 sections so far, such engineering codification is not an easy task. However, empirical systems of rock mass characterisation have generally proved to be most successful in rock characterisation and engineering design. An example of this in relation to foliation and the link with the design of an underground radioactive waste repository is shown in Figure 2.124 from Posiva's site investigation work in Finland. A well developed, pervasive foliation is a characteristic feature of the migmatites and gneisses at the Olkiluoto western Finland bedrock site, and could have a significant influence on the underground construction, the design and layout and the groundwater flow regime of the deep spent nuclear fuel repository.

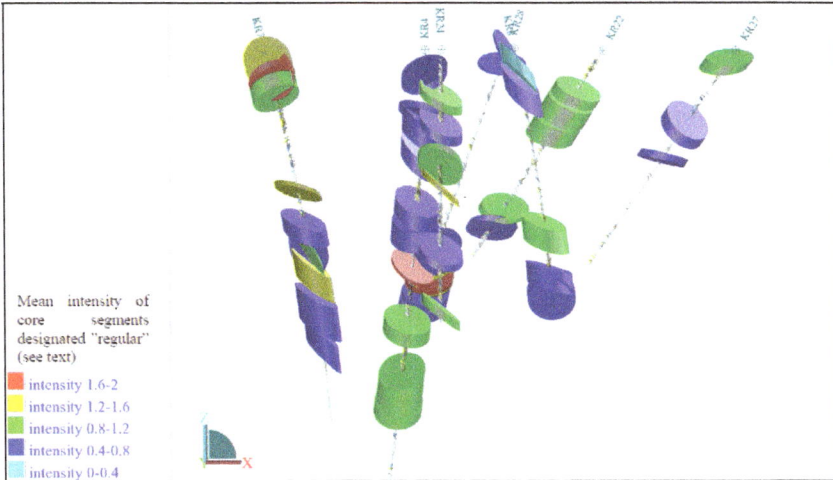

Figure 2.125. Diagram showing the mean orientation of regularly foliated core segments in the investigated drillholes, using Surpac Vision (viewed from the South). The intensity values given in the legend are the mean intensities of foliation determined from the 1 m core lengths identified along each regular segment of borehole core (from Milnes *et al.*, 2006).

A methodology for the systematic acquisition of foliation data in cored boreholes and in tunnels at the Olkiluoto site was developed in order to provide the necessary basis for future geological, rock mechanics and hydrogeological modelling, Milnes *et al.* (2006).

At the suggested reference scales (1 m length of core, 10 m$^2$ area of tunnel wall), the most representative foliation type and intensity were assessed using a standard set of core photographs, providing a systematic description in terms of the nine descriptive types (G1, G2, G3, B1, B2, B3, S1, S2, S3). As a further step, the rock mechanics significance of these types is assessed and a rock mechanics foliation (RMF) number is assigned (RMF 0 = no significance, RMF 1, RMF 2 and RMF 3 = low, intermediate, and high significance, respectively).

Experience at the time of the Milnes *et al.* (2006) report indicated that RMF 2 is the most common designation and that rocks with this number or below are not expected to require special measures during underground excavation. The interest of rock engineers is likely to be focussed on RMF 3 rocks, which at Olkiluoto seem to be relatively rare and to occur only in thin zones.

## 2.11 COMPLEX FRACTURING EVEN IN THE SIMPLEST OF STRESS REGIMES

### 2.11.1 Introduction

The fractures that form in the crust reflect the state of stress and, as noted earlier, the two main causes of stress in the crust are the overburden and plate tectonics. The weight of the overburden causes both a vertical stress component and a resultant horizontal component. The forces associated with plate tectonics cause horizontal stresses and resultant vertical stresses. In our earlier discussion, Section 2.4, the simplest stress regime was considered, i.e., where no tectonic stresses are acting (a tectonically relaxed region) and where the only stresses in the crust were those caused by the overburden.

### 2.11.2 Stress state induced in the crust by the overburden

We now explore this simple stress regime further and consider the fractures that can form in response to it. Recall that in this scenario the vertical stress ($\sigma_v$) is the maximum principal stress ($\sigma_1$) which can be determined using Newton's definition of force (force = mass $\times$ acceleration) and is given by Equation (2.40):

$$\sigma_1 = z\rho g, \tag{2.40}$$

where $\rho$ is the mean density of the overburden, $z$ the depth, and g the acceleration due to gravity. It was also noted earlier that, if the rock at depths is prevented from expanding laterally by the confining effects of the surrounding rocks, the horizontal stress induced to prevent this expansion is determined by Poisson's ratio ($v$), a measure of the compressibility of the rock as in Equation (2.41):

$$\sigma_h = [\sigma_1/(m-1)], \tag{2.41}$$

where $m$ is Poisson's number, the reciprocal of Poisson's ratio. Both the vertical and horizontal stresses increase with depth and, if it is assumed that the density remains constant, they will increase linearly as shown in

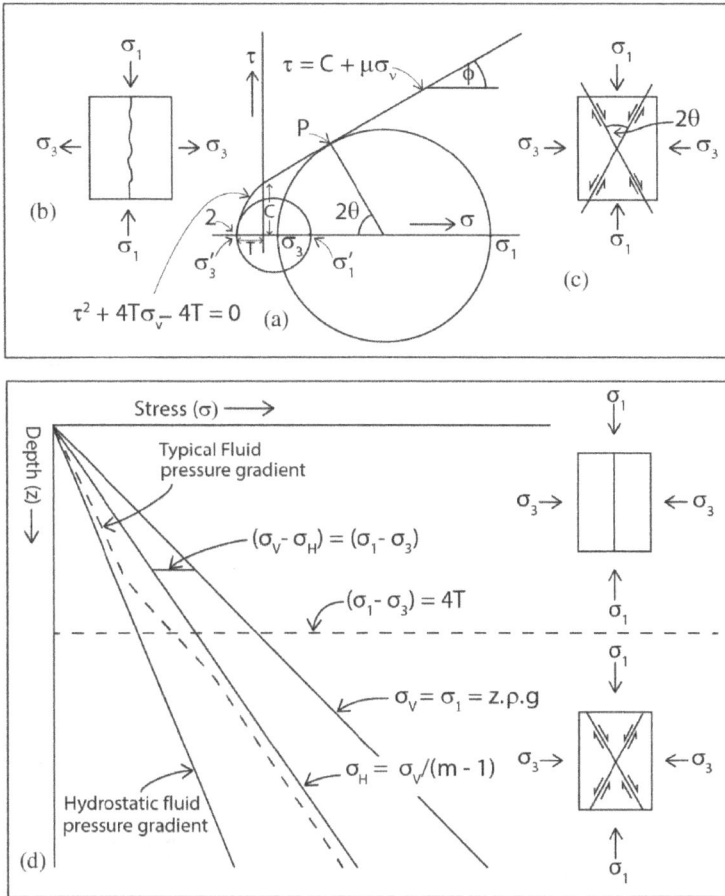

Figure 2.126. The upper diagram is the graphical expression of the brittle failure criteria already discussed in Section 2.3 (see Figure 2.19), and shows the relation between the two types of failure and the stress field generating them. The lower diagram shows the stress profile in the crust generated by the overburden. The assumption of a uniform density with depth results in a linear increase in vertical and horizontal stress with depth.

Figure 2.126(d). It follows that the differential stress ($\sigma_v - \sigma_h = \sigma_1 - \sigma_3$) increases with depth. The differential stress controls the type of brittle failure, i.e., extensional or shear, that forms, see the upper diagram in Figure 2.126, and the transition between these two failure modes occurs when the differential stress reaches four times the tensile strength of the rock. Thus, the upper portion of the crust, which has a relatively low differential stress, is characterised by extensional fractures and the lower part by shear fractures, Figure 2.126(d).

Typically, sedimentary rocks are formed by the accumulation of sediments in various types of basins and their subsequent burial and diagenesis (i.e., the name given the various processes that occur during burial that convert the sediments into sedimentary rocks and which include compaction and cementation). During diagenesis, the rocks undergo geometric and thermal changes both of which generate stresses which can be quantified (Price, 1974).

### 2.11.3 Stresses induced as a result of basin geometry

There are many types of sedimentary basin and they fall into two main categories, namely those that are bounded by faults, e.g., a rift valley or graben and those that are unfaulted. A typical unfaulted basin with an elliptical plan is represented in Figure 2.127. A vertical section AB along its

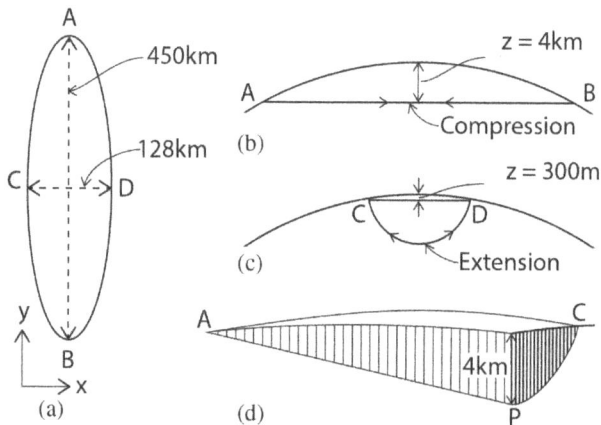

Figure 2.127.   Schematic plan view (a) and cross sections (b) and (c) through an unfaulted basin. For large basins, the curvature of Earth must be taken into account. The process of burial of a layer in the basin leads to strains both along and across the basin and these generate stresses that can be quantified. After Price (1974).

long axis is also shown in, Figure 2.127(b). When considering the geometry of the profile, it can be seen that for large basins the curvature of the Earth has to be taken into account. The sediments are deposited at the water–sediment interface which coincides with the arc AB. From the geometry of the cross section, it follows that as the sediments become buried there is a shortening along the long axis of the basin and that this will continue until a depth $z$ is reached when the length of the sediment has been reduced to that of the straight line AB. For a basin with the dimensions shown on the Figure, this depth is 4 km. The strain associated with this shortening ($e_y$) can be calculated. It is the change in length (arc length AB minus the line length AB) divided by the original length (the arc length AB). Burial beyond this depth will result in the onset of extension along the $y$-axis.

The vertical section along the short axis of the basin CD is shown in Figure 2.127(c). It can be seen from the geometry of this section that the depth at which the shortening along the axis changes to extension is much less, i.e., 300 m. Thus, the state of strain induced by the geometry of the basin at a depth of 4 km in the centre of the basin, point $P$ in Figure 2.127(d), will be one of compression along the long axis and extension along the short axis (Equations (2.42) and (2.43)).

$$\sigma_y = [\sigma_1/(m-1)] + Ee_y, \tag{2.42}$$
$$\sigma_x = [\sigma_1/(m-1)] - Ee_x. \tag{2.43}$$

As noted, these strains, $e_y$ and $e_x$, respectively, can be readily calculated and depend only on the geometry of the basin. By assuming the lithified sediments at depth $P$ behave as linear elastic materials, the corresponding stress can be determined from Hooke's law ($\sigma = Ee$ where $E$ is the Young's modulus). Thus, the stresses in the $x$ and $y$ directions are $-Ee_x$ (extension) and $+Ee_y$ (compression). These can be added to Equation (2.41) to give the horizontal stresses generated by the overburden stress and the basin geometry: Equations (2.42) and (2.43).

## 2.11.4 Stresses induced by the geothermal heat

A geothermal gradient exists in the Earth's crust with a value typically of between 25 and 30 °C/km. This increase in temperature with depth means that, in addition to compaction and cementation that accompany the processes of burial, the rocks are also heated. They attempt to expand but lateral expansion is inhibited by the confining effect of the surrounding rocks. As a result, additional horizontal stresses are generated to prevent this expansion. The amount of expansion (strain) is determined by the coefficient of thermal expansion $\alpha$ and the increase in temperature $\Delta T$ from the surface to the depth of interest — and is given by $\alpha \Delta T$. By assuming that the rocks are linear elastic materials, we can again use Hooke's law to determine the stress induced in the rock needed to prevent this strain. This is the same in all directions in the horizontal plane and is given by $E\alpha\Delta T$. Thus, the state of stress in the crust generated by the overburden, basin geometry and geothermal gradient is characterised by Equations (2.44)–(2.46);

$$\sigma_z = z\rho g, \tag{2.44}$$

$$\sigma_y = [\sigma_z/(m-1)] + Ee_y + E\alpha\Delta T, \tag{2.45}$$

$$\sigma_x = [\sigma_z/(m-1)] - Ee_x + E\alpha\Delta T. \tag{2.46}$$

It can be seen from these equations that, in a non-tectonic environment, the state of stress in the crust is determined by the depth and the intrinsic properties of the rocks. The vertical (i.e., overburden) stress is controlled by the depth and the mean density of the overlying rocks, Equation (2.44), and the horizontal stresses are determined by the vertical stress, Poisson's ratio, Young's modulus, the coefficient of thermal expansion and the temperature change.

### 2.11.5 Possible fracture patterns in a layered crust

If we consider the same tectonically relaxed geological environment assumed in the construction of Figure 2.126(d), but one in which the overburden is made up of horizontal layers of different rock types rather than a uniform rock unit, then the simple fracture patterns shown in this Figure, i.e., vertical extensional fractures generated above the depth where $(\sigma_1 - \sigma_3) < 4T$ and shear failure below it, no longer apply. The tensile strength of each rock type will be different and therefore the depths at which fractures change from extensional to shear will also be different.

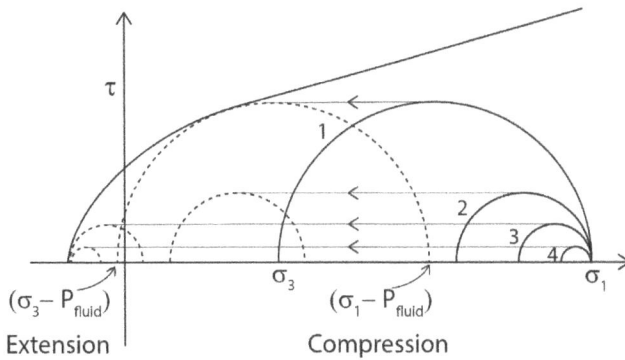

Figure 2.128. The influence of fluid pressure on the stresses in four different layers at some depth $z$ in the crust. Each layer has its own stress field represented by the Mohr circles 1–4. For simplicity, a single failure envelope has been drawn. In reality, each of the rocks will have its own failure criterion.

In addition, because the Poisson's ratio, Young's modulus and coefficient of thermal expansion for the different rock types will be different, it follows from Equations (2.45) and (2.46) that the induced horizontal stress will also be different and that each layer will have a different value of differential stress. Both these factors indicate that, at any particular depth, some layers may develop extensional fractures and others shear fractures.

The stress states in a series of layers made up of four different rock types (1, 2, 3 and 4) at a depth $z$ in the crust are shown as four Mohr stress circles (full lines) in Figure 2.128. Each layer is subjected to the same overburden stress but will have a different horizontal stress according to Equations (2.45) and (2.46). It is apparent from these equations, that adjacent layers at the some depth can have very different stress states depending on their physical properties. However, none of these stress states will cause brittle failure as none touch either of the failure envelopes, i.e., neither the straight, shear failure envelope of the compressive stress regime nor the parabolic, extensional failure envelope of the extensional stress regime.

It is clear from the discussion relating to the construction of Figure 2.126 that the stress state in the Earth's crust is generally compressive. However, as we have noted earlier, rocks in the upper part of the Earth's crust are frequently pervaded by extensional fractures. In order to account for this, geologists have invoked the effect of fluid pressure and the related concept of effective stress (see Section 2.5). It is argued that the fluid pressure ($p$) opposes the lithostatic stress ($\sigma$) and that the resulting stress, i.e., the effective stress, is given by ($\sigma-p$). The effect of the fluid pressure on the lithostatic stresses within the different layers is shown in Figure 2.128. Each Mohr circle is moved to the left, i.e., towards the failure envelopes, by an amount $p$. In this way, fluid pressures within the crust can cause stable lithostatic stress states (represented by the solid Mohr circles) to become unstable and lead to fracturing (the dashed Mohr circles). See also Figure 2.129.

However, whether or not this fluid induced failure occurs depends upon the magnitude of the fluid pressure generated in any particular layer during burial and this will depend to a large extent on its permeability $K$. Thus, beds with a low permeability are likely to generate high fluid pressures and are therefore also more likely to develop fractures than more

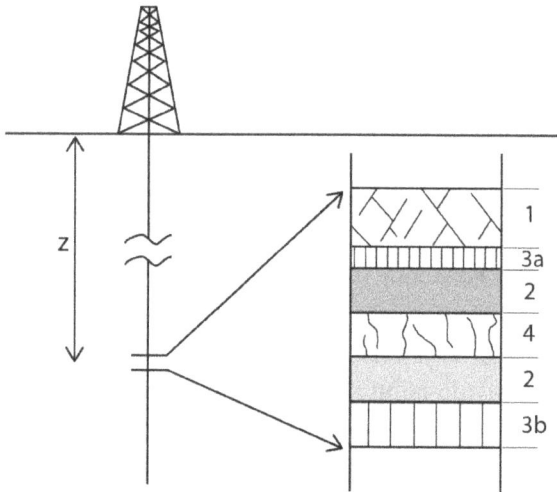

Figure 2.129. The potential variability of the type and organisation of fractures in a sequence of beds at some depth '$z$' in the crust. The rocks 1–4 have stress states given by the four Mohr's circles shown in Figure 2.128.

permeable units. It follows that, at a particular depth in the crust, some beds may remain unfractured whilst other, less permeable beds will develop extensional or shear fractures, depending upon the Poisson's ratio which controls the differential stress.

In addition, and as is discussed in Section 2.9, the organisation of any extensional fractures, i.e., whether they form regular arrays of straight, vertical fractures or less organised arrays, will also be determined by the differential stress. Figure 2.95 in Section 2.9 shows the range of possible organisations, from straight, vertical fractures to randomly oriented fractures which lead to brecciation of the rock, as the differential stress varies from around four times the tensile strength of the rock to zero. Other parameters that influence the organisation of extensional fractures have been discussed in Section 2.8 and include the layer thickness and the fracture toughness. The layer thickness determines the fracture spacing, and the fracture toughness determines whether the fractures are regularly spaced or tend to cluster into fracture corridors.

## 2.11.6 Variable stress states and therefore variable fracturing at the same depth

It follows from the above discussion that the expression of brittle failure in the rock can be extremely variable even when generated by the simplest of stress fields acting on flat-lying beds. Figure 2.129 is a schematic representation of the fractures that might develop in layers made from the four rock types whose stress states are shown by the four Mohr stress circles in Figure 2.128. Layers of rock type 1 have a high differential stress and therefore will develop conjugate shear fractures. Layers of rock type 2 have a high differential stress but their high permeability prevents the fluid pressures building up sufficiently for the formation of any fractures. Layers of rock type 3 have a differential stress less than $4T$ (where $T$ is the tensile strength of the rock) which results in the formation of extensional fractures whose spacing is determined by the layer thickness (cf. layers 3a and 3b, Figure 2.129). Layers of rock type 4 have a very low differential stress and the extensional fractures that form are therefore poorly aligned.

It should be noted that, in the above discussion regarding the overburden stress in the crust and its impact on fracturing, the influence of the intermediate principal stress ($\sigma_2$) has not been considered. The vertical section through the crust shown in Figure 2.126(d) is the $\sigma_1 - \sigma_3$ plane. The section containing the intermediate and minimum principal stresses is the horizontal plane, i.e., bedding planes. In Section 2.4, the influence of $\sigma_2$ on fracturing was considered and it was shown that its effect on fracture orientation, i.e., the strike of the fracture, was determined by its relative magnitude with respect to $\sigma_3$. When there is a significant difference between the two (i.e., when the differential stress $(\sigma_2 - \sigma_3)$ is large), the fractures have a consistent and regular strike, see Figure 2.24. As this differential stress decreases, the strike regularity degrades until the point where the two stresses are the same. Under this condition of no differential stress, the fractures develop with random strike and the result is the formation of polygonal fracture arrays.

*****

Also, in the above discussion, the fractures predicted to form in the various rock units making up the crust are the result of a single deformation (in this example caused by the overburden load). As a result, only one fracture set or at most a conjugate set of shear fractures will form. However, the type of fractured rock mass that frequently confronts an engineering geologist has generally been subjected to several different stress fields over geological time, each of which may have developed its own fracture set. The superposition of several fracture sets produces a fracture network which will have a major impact on the properties of the rock mass in which it is situated and upon the distribution of stress. In order to address the problem of establishing the likely geometry of a fracture network within a rock and of determining the resultant distribution of stress, the interaction between stress and fractures needs to be examined. This is considered in the following Section.

## 2.12 THE EVOLUTION OF FRACTURE NETWORKS — FRACTURE ANALYSIS

### 2.12.1 Introduction

Many rock masses exposed at the Earth's surface are pervaded by fractures. These may be the result of one episode of deformation which has resulted in the formation of a single set of fractures, but more commonly the rocks have been subjected to several different stress regimes during their evolution and therefore contain several fracture sets. These sets combine to produce a fracture network whose properties are sensitive to its geometry which, as discussed below, is determined by the order in which the different fracture sets are superimposed. The bulk properties of a fractured rock mass, such as strength, permeability and porosity, are often dominated by the properties of the fracture network and the intrinsic rock properties are often of second order importance. It follows that an understanding of the geometry of a fracture network and the way it evolved over geological time is highly relevant to both the structural geologist and rock engineer.

In Section 2.3, as Figure 2.9 we included the photograph of the limestone platform which is now included below as Figure 2.130, although the white numbers indicating the sequence of fracture set development were not included in Figure 2.9 version. It can be seen in Figure 2.130 that, by studying which fractures terminate against other fractures, the temporal sequence of the fracture pattern development can be deduced. Moreover, the key aspects of the sequence of imposed stress fields can also be deduced. This type of 'back analysis' study is profitable in the geological context and the understanding of the fracture array in terms of which fracture sets are likely to be the most through-going is most helpful to support a rock engineer's site investigation and any associated numerical modelling of the rock mass.

Because of the intimate relation between the fractures that form and the stress fields that generated them, structural geologists can use the number and orientation of the different fracture sets making up a fracture network to determine the number and orientation of the various stress fields that have affected the rock over the geological period during which the prevailing conditions favoured the formation of brittle (rather than ductile) deformation. If the relative age of the various fracture sets can be

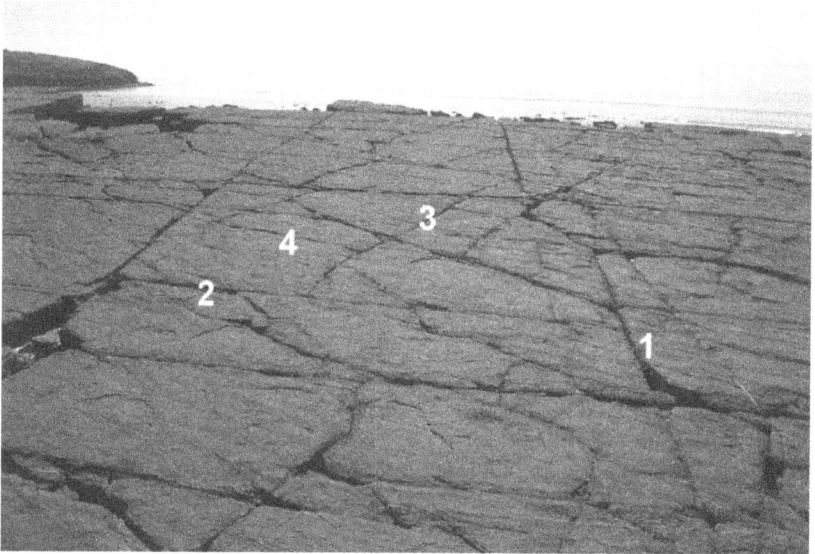

Figure 2.130. Fractured limestone rock mass, South Wales, UK. The white numbers indicate the sequence in which the fracture sets occurred.

determined, then the evolution of the stress regime that affected the rocks can also be determined. In the following Section, the mechanical principles that enable the relative age of fractures to be established are outlined. These provide the basis for 'Fracture analysis', i.e., a technique that can be used to determine the order of formation of the fracture sets making up a fracture network and therefore of the associated evolution of the stress regimes.

The fracture network shown in Figure 2.131 contains several sets of fractures. Three of the most apparent have been labelled A, B and C. Before reading through the following Section which outlines the rationale and technique used in 'fracture analysis' and, given the fracture set labelling in Figure 2.130, the reader is invited to speculate on the relative age of the three sets in Figure 2.131.

## 2.12.2 The effect of pre-existing fractures on stress orientation

The local reorientation of the regional stress field in the vicinity of an open fracture tip is shown in Figure 2.132(a). Because the orientation of

Figure 2.131. A fracture network made up of several fracture sets. The right-hand diagram shows the orientation of three of the sets, labelled A, B and C.

any fracture is determined by the orientation of the causative stress field, the orientation of any fracture that propagates into this local perturbed stress regime will be modified. The influence of the stress variation around the fracture tip on the orientation of a later fracture formed in the same stress regime that generated the original vertical fracture is shown in Figure 2.132(a).

Extensional fractures form normal to the maximum extensional stress, $\sigma_3$, i.e., parallel to the maximum principal compressive stress, $\sigma_1$, whose orientation is indicated by the dashes in the Figure. It can be seen that in the vicinity of the fracture tip these are normal to the open fracture. Consequently, a fracture propagating into this zone is rotated and intersects the fracture at 90°. This accounts for the linking of closely spaced fractures of the same fracture set as they propagate towards each other when the fracture tips begin to overlap, Figures 2.132(b)–2.132(d).

The mechanical interaction between two fractures is not restricted to fractures of the same fracture set. When a new fracture set is superimposed on an earlier set as a result of a later geological stress regime, the early fractures are likely to locally modify the orientation and impede the propagation of the later set. The interference of the early set with the late

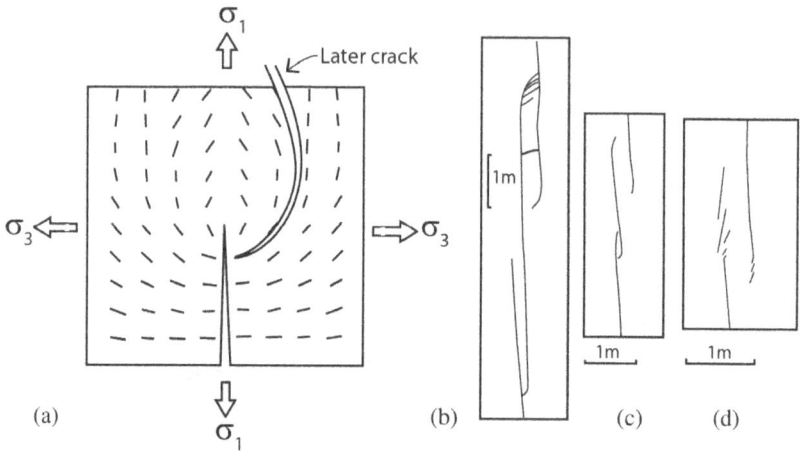

Figure 2.132. (a) The stress distribution around the tip of an open fracture. The maximum extension ($\sigma_3$) acts horizontally. The influence of the tip stresses on a later fracture generated during the same stress regime is shown. (b)–(d) Natural examples of closely spaced fractures of the same set linking as their tips overlap. (b)–(d) after Cruikshank *et al.* (1991).

set is particularly marked if the early fractures are open during the formation of the second set, either as a result of the orientation of the new stress field or because of the existence of a fluid pressure within the rock. Because the fracture sides of such open fractures are free surfaces, they cannot support a shear stress. Consequently, the stress field in the rock must reorient itself in the vicinity of the fracture so that the principal stresses are parallel to or normal to the fracture wall.

As noted, Figure 2.132(a), this deflection can be seen by its effect on subsequent fractures which will swing into an orientation either parallel to or normal to the fracture wall. The impact of an early set of fractures on the orientation and continuity of a later set is illustrated in Figure 2.133, where it is shown on two scales. In (a), a small-scale fracture to the SE of the coin is rotated through an angle of ~40° as it approaches an early open fracture which is parallel to the upper and lower edges of the photograph, so that it intersects this fracture at 90°. Similarly, as the fracture approaches the fracture on the left-hand side of the photograph, which is parallel to the edge of the image, a similar rotation occurs so that the fractures meet

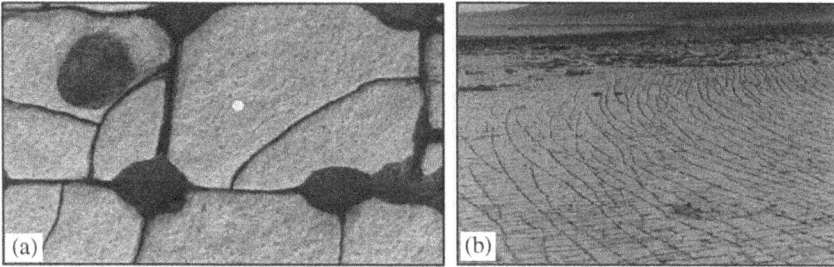

Figure 2.133. (a) The deflection of later fractures by early (horizontal) fractures. The change in stress orientation caused by the early open fracture is recorded by the curving of the later fractures as they approach and eventually meet the early fractures. (b) Stress and fracture deflection on a larger-scale than in (a). Both images from Lilstock, Bristol Channel coast, SW England.

at 90°. It is apparent from this Figure that, in addition to causing the late fractures to rotate, the early fractures also act as barriers which prevent the propagation of later fractures. Thus an early set of fractures will tend to deflect and impede the propagation of later fractures.

In Figure 2.133(b), a set of extensional, fractures in a Liassic carbonate bed at Lilstock on the Bristol Channel coast is shown, curving towards a large-scale normal fault situated in the mid distance and delineated by the seaweed. The curving and abutting of late fractures against early fractures enables the relative age of the different fracture sets making up a fracture network to be determined. This enables a 'fracture analysis' to be carried out which provides a detailed description of the evolving stress regimes during the formation of the fracture network.

### 2.12.3 Fracture analysis

Using the criteria developed in the preceding Section, it is possible to put the different fracture sets making up a fracture network into their chronological order. Figure 2.134 shows a cartoon of a fracture network made up of three fracture sets. The long, continuous fractures which do not abut against other fractures (fracture set A) are the earliest formed fractures. Later formed fractures are impeded by this set and end against it. It can be seen that fractures of set B abut against A and in addition, in the vicinity of fractures of set A, are deflected into an orientation normal to A. As noted above, this indicates that fractures of set A were open during the formation of fracture set B because then the principal stresses close to the fracture would have been either parallel or normal to the fracture surface. Fractures of set C abut against fractures belonging to sets A and B — indicating that they are the youngest set. Assuming that the three sets of fractures are extensional fractures, the fractures represent the $\sigma_1$–$\sigma_2$ plane of the stress field associated with their formation, and $\sigma_3$ was oriented normal to the fractures. The $\sigma_3$ orientations linked to the formation of the three fracture sets are indicated in Figure 2.134(b). It can be seen that in this example the fracture network was built up during a systematic anti-clockwise rotation of the regional stress field.

In this illustrative example of the 'fracture analysis' of a fracture network, the late fractures rotate into an orientation normal to the early fractures as they approach them. As noted above, this indicates that the early fractures were open during the formation of the later fractures and therefore unable to sustain a shear stress parallel to their walls. High fluid pressures can ensure that the early fractures remain open — regardless of the orientation of the later stress fields. However, inspection of natural

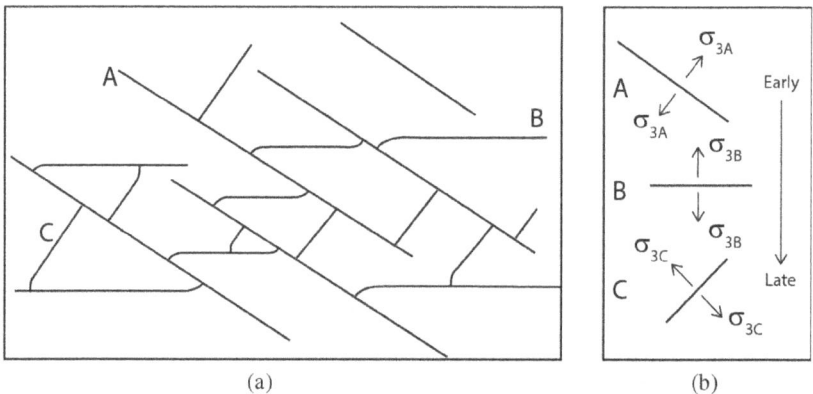

Figure 2.134.   (a) A cartoon of a fracture network illustrating the features that can be used to determine the relative age of the individual fracture sets making up the network. (b) Shows the individual fracture sets making up the network, together with their associated stress orientations and their relative age.

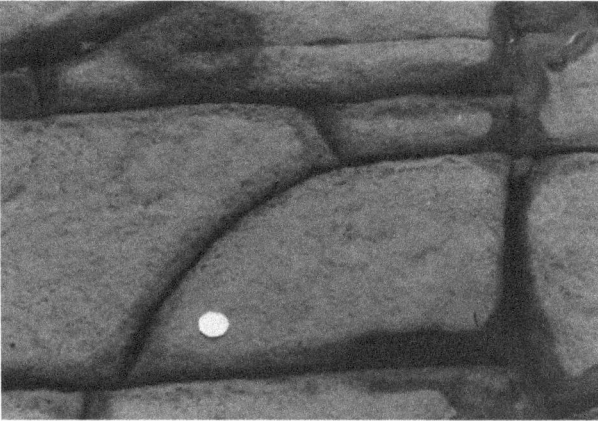

Figure 2.135. The fracture to the left of the coin curves towards and abuts against an earlier fracture below the coin. Note that it does not meet the fracture at 90°.

fracture networks indicates that early fractures are not always open during the formation of later sets. For example, the curved fracture to the left of the coin in Figure 2.135 abuts against the fracture which is sub-parallel to the bottom edge of the photograph. It is evident that the orientation of the curved fracture has been influenced by the fracture it ends against but it is also clear that, in contrast to the fractures shown in Figure 2.133(a), it does not meet this fracture at 90°. This indicates that, at the time of formation of the curved fracture, the earlier fracture was not completely open and was therefore able to sustain a shear stress parallel to its surface.

As the cohesion along an early fracture increases, its ability to impede the propagation of a later fracture decreases and it is instructive to consider what parameters determine whether or not a fracture will propagate across the early fracture. This problem is particularly relevant to the evolution of fracture systems in layered, sedimentary rocks. During burial and diagenesis of a sedimentary succession, the original ductile sediments become lithified and begin to respond to the overburden load in a brittle manner. The stress state in the crust in a non-tectonic environment and the fractures that form in response to this stress have been discussed in Section 2.4, where it was argued that in the upper part of the crust where the differential stress is small, vertical extensional fractures will form. The bedding planes act as an important pre-existing horizontal fracture

(a)                                                    (b)

Figure 2.136.   (a) Strata-bound fractures cutting Devonian sandstones, Bude, Cornwall, England. (b) Through-going fractures cutting Carboniferous limestones, Tytherington, Gloucestershire, England.

set that can impede the propagation of the vertical fractures. If the bedding planes inhibit joint propagation, then instead of the development of through-going joints, Figure 2.136(b), the fractures become strata-bound, Figure 2.136(a). Which of these two options develops has a major impact on the resulting fracture system and therefore on the bulk properties of the fractured rock mass. It is therefore important to understand what controls whether a fracture approaching an earlier fracture (or bedding plane) will be arrested by that fracture or whether it will propagate across it.

The high tensile stress that characterises the region just ahead of the fracture tip is shown in Figures 2.132(a) and 2.137(b) and it is in response to this stress that the fracture propagates. Figure 2.137(a) shows a vertical extensional fracture in a relatively strong layer (high Young's modulus, $E_2$). If the bedding plane separating this layer from the overlying layer with the same Young's modulus has a high cohesive strength (Figure 2.137(b)), then it will be able to sustain the build-up of tensile stress as the fracture approaches it — thus enabling the fracture to propagate and cross the bedding. If, in contrast, the bedding plane has a low cohesive strength then the tensile stress in front of the fracture tip driving fracture growth will be dissipated by slip along the bedding plane as shown in Figure 2.137(a). The fracture will therefore be unable to grow and will stop at the bedding. The cohesive strength may be an intrinsic property of the bedding plane or it may be related to the frictional resistance to sliding determined by

Amonton's law; this states that the frictional resistance to slip is determined by the product of the normal stress across it and the coefficient of sliding friction, $\mu$, i.e., $\tau = \sigma\mu$.

Because the normal stress is sensitive to fluid pressure, $\tau = (\sigma - p)\mu$, it is possible for a particular bedding plane to effectively inhibit fracture propagation on some occasions (i.e., when the fluid pressure is high) and allow fracture propagation across it on others (i.e., when the fluid pressure is low).

Another factor that controls whether or not a fracture will propagate across a bedding plane is the contrast in material properties of the beds, Figure 2.137(c). The magnitude of the tensile stress generated in front of a propagating fracture is determined by the strength of the material through which it is passing; the stronger the material the larger the stress needed to propagate the fracture. Thus if a fracture attempts to pass from a relatively weak rock into a stronger rock, propagation is hampered as the fracture hits the interface to a higher modulus rock. The tensile stress available for fracture growth is likely to be too small and the fracture will

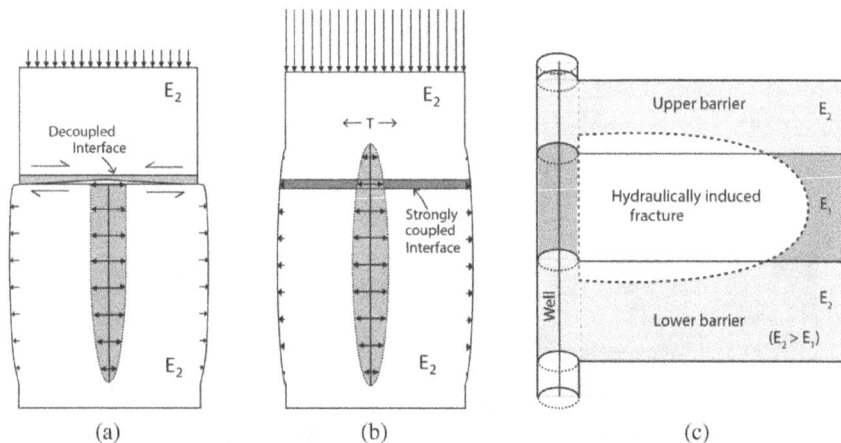

Figure 2.137. A schematic illustration of two mechanisms able to impede opening mode fracture propagation: (a) bed slippage which prevents fracture propagation and (b) no slippage which facilitates propagation. The elastic properties of the two adjacent layers are the same. (c) Vertical hydraulic fractures generated in a weak layer by the local application of a high fluid pressure. They are prevented from propagating vertically by the occurrence of the adjacent stronger layers. From Casabianca and Cosgrove (2012).

therefore stop. This phenomenon has important consequences when rocks at depths are subjected to high fluid pressures in order to stimulate fracture formation. A well is drilled to the horizon of interest (horizon $E_1$ in Figure 2.137(c)) and the borehole lined with a casing, except in $E_1$. This allows the application of high fluid pressure to be applied specifically to this horizon. The pressure is increased until hydraulic fractures form. These can propagate vertically (both upwards and downwards) but may be constrained when they hit stronger layers (i.e., layers with a higher Young's modulus) as illustrated in the Figure.

It follows that fractures would be expected to propagate from a strong layer into a weak layer, but not the other way round. However, it is commonly observed in sedimentary successions, that fractures dominate the strong rocks and do not generally propagate into the adjacent weaker rocks and an example where joints have been constrained within a strong limestone bed by the adjacent weaker shales is shown in Figure 2.138. This apparent anomaly can be explained by arguing either that weak bedding planes separate the layers which allow the fracture propagating stress to be dissipated, Figure 2.137(a), or that the weaker rocks behave as ductile rather than brittle materials at the time of fracturing. As discussed in Section 2.8, it has been argued that one of the mechanisms for the formation of extensional fractures in competent beds interlayered with weaker horizons is that the ductile flow of the weaker material during layer-normal compression generates a tensile stress within the competent layer which can lead to brittle failure.

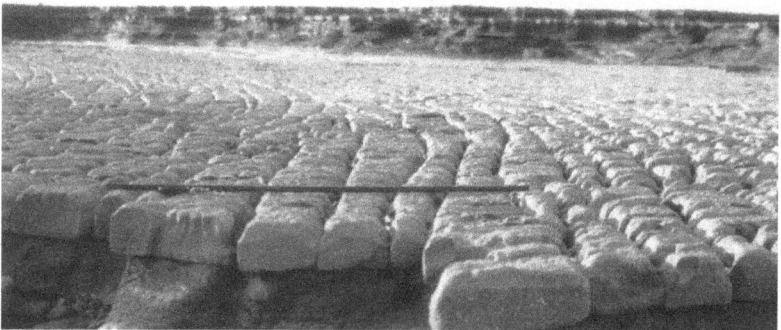

Figure 2.138.   Strata-bound joints constrained to the relatively strong limestone bed by the weak adjacent shales. Liassic beds from Lilstock, Somerset, UK.

An understanding of the mechanical interaction between early and later fractures enables the relative age of the different fracture sets making up a fracture network to be determined and thereby the chronology of stress fields linked to the brittle deformation of the rocks, see Figure 2.134. It was noted that the earliest fracture set will be characterised by long, continuous fractures — reflecting the uniform stress regime imposed by the regional stress field responsible for the fracturing. Subsequent regional stress fields will be locally modified by the pre-existing fracture sets and this will be reflected in the form of the associated fractures which will be less continuous and which will curve into and abut against the earlier fractures. As more fracture sets are added to the fracture network any subsequent regional stress field will be progressively more and more degraded within the fracture network and the resulting fractures less continuous and more irregular.

This is apparent from an inspection of the fracture network shown in Figure 2.139. Three or four clear fracture sets can be recognised. The earliest are the long continuous fractures, labelled 1 in Figure 2.139(c). Using the criteria discussed above, the next set of fractures to form was set 2, followed by set 3. Note that there are only a few members of set 3, indicating that it is 'under-saturated'. As expected, the final set to form, set 4, is made up of short fractures which abut against all the earlier sets. The fracture sets making this fracture network appear to evolve in response to an anticlockwise rotating regional stress field. Any regional stress field currently acting on such a network is likely to be significantly degraded. This has been convincingly demonstrated using numerical modelling. The development of local areas of low differential stress within the fracture network shown in Figure 2.139 is indicated by the local formation of polygonal fractures, region A.

## 2.12.4 The importance of the order of superposition of fracture sets on the geometry of the resulting fracture network

It was noted in Section 2.8.3 that, when a fracture set forms in a rock, there is an initial period when the density of fractures is low (i.e., when the beds are under-saturated with fractures) and where the fracture spacing is random. As the fracture density increases, the fractures begin to self-organise

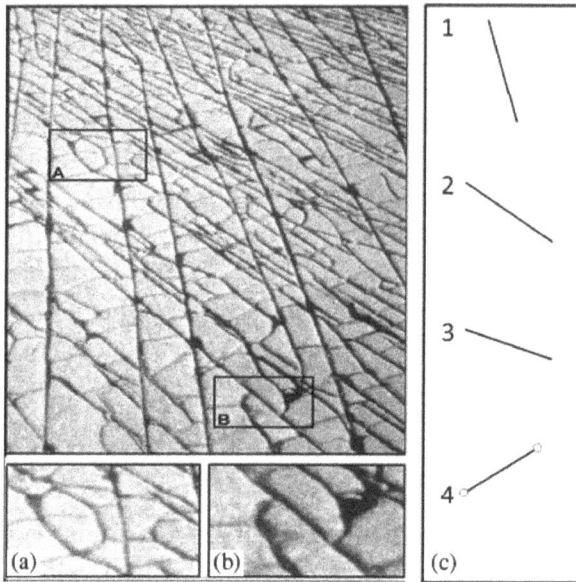

Figure 2.139.   A fracture network cutting a limestone bed. Using the criteria discussed in the text, (the abutting and curving of later fractures against older fractures, see insert B), the relative age of the fracture sets (1 → 4) can be determined. They show that the network developed in an anticlockwise rotating regional stress regime. The progressive degradation of the stress field within the network as it develops is illustrated by the late fractures forming local polygonal fracture arrays, inset A.

and a regular spacing develops which is often related to the thickness of the bed in which the fractures are forming. This represents the condition of fracture saturation. In the above discussion of the formation of a fracture network by the superposition of several fracture sets, the impact of early fractures on the later fractures was emphasised. The earliest fractures are long and continuous and the length and regularity of successive fracture sets progressively decreases. Thus, when populating a numerical model with fracture data collected in the field (which may record fracture density, length, orientation etc.), it is important to bear in mind the tendency for fractures of a particular set to be uniformly spaced and the mechanical interaction that occurs between fractures of different fracture sets. In this way, not only will the model honour the data collected, but the fracture network generated in the model is more likely to be geologically realistic.

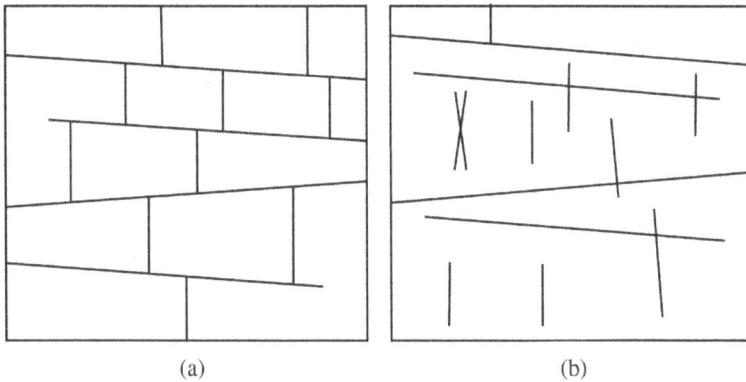

(a)                                                    (b)

Figure 2.140.  (a) A model populated with two fracture sets. The model is sensitive to the order of superposition of the two sets and respects the natural spacing that develops between fractures. (b) A model populated stochastically with two fracture sets. No consideration is given to the relative age of the two fracture sets, the spacing of the sets or the fracture interaction. Consequently, the resulting fracture network in (b) is geologically unrealistic.

Extreme examples of introducing identical fracture data to a model are shown in Figure 2.140. In 2.140(a) a regular spacing of the fractures in each set is used and the mechanical interaction of the two fracture sets considered. The result is a geologically realistic fracture network. Neglecting to constrain the data with this understanding and populating the model with fractures in a random, stochastic manner, leads to a geologically unrealistic network, Figure 2.140(b). Both networks shown in this Figure contain identical fractures with identical orientations; however their geometries and therefore the properties they impart to the rock are very different. Although the number of fracture sets, the number of fractures, the density of the fractures and the mean lengths of the fractures in both models in Figure 2.140 are identical, model (a) which has the geometry of a natural fracture network has a greater connectivity and therefore greater capacity for fluid flow than model (b). Its impact on bulk rock strength will also be different to that of the fracture network shown in 2.140(b).

It can be seen from this simple illustration that, unless the fracture network used in a model has a geologically realistic geometry which requires a knowledge of (i) the order in which the fracture sets making up the network were superimposed (which controls fracture length) and (ii) the fracture spacing of each set which is determined by the degree of fracture saturation, and layer thickness, the predicted rock mass properties are likely to be unrealistic — even if the field data used relating to orientation and length are correct and complete. It follows that the more accurately the geological evolution of a fracture network is known the more accurately its geometry can be established and the better the rock mass can be modelled and its bulk properties determined.

## 2.13 THE BULK PROPERTIES OF FRACTURE NETWORKS AND THE INFLUENCING FACTORS

### 2.13.1 Introduction

As already noted, the properties of a fractured rock mass (e.g., its strength and permeability) are often dominated by the fracture network, rather than by the intrinsic properties of the rock. It can be shown that these 'bulk' properties are determined by a variety of factors including the following.

— The order in which the various fracture sets making up the network have been superimposed.
— The detailed geometry of the individual fracture sets making up the fracture network, i.e., their fracture spacing, their tendency to cluster, their length and orientation with respect to the other fracture sets.
— The scale of interest with respect to the scale of the fracture network.
— The history of fluid flow through the fractures over geological time. This may result in dissolution and weathering of the rock which may weaken the rock mass and increase permeability, or in the precipitation of minerals that infill the fractures resulting in an increase in bulk strength and decrease in the fracture-induced permeability.
— The present day stress regime.

These will be discussed in turn.

## 2.13.2 The influence of the order of superposition of fracture sets on network geometry

It was shown in Section 2.12 that early fractures inhibit the propagation of later fractures which typically stop (abut) against them, Figure 2.141(a). This observation provides the basis for 'fracture analysis' whereby the relative ages of the various fracture sets making up a fracture network can be established and used to determine the evolution of the stress regimes through geological time.

In Figure 2.141(a), the early fractures are the sub-horizontal fractures and it can be seen that they impede the propagation of the later, sub-vertical fractures which abut against them. If the order of formation of the two fracture sets is reversed, then the sub-vertical fractures will be long and continuous and the later, sub-horizontal fractures will be shorter and abut against these longer fractures, Figure 2.141(b). The impact of the order of superposition of the fracture sets on the geometry of the fracture network and therefore upon its properties is apparent. It is clear that permeability is greatest in the horizontal direction in the network shown in Figure 2.141(a) and in the vertical direction in the network shown in Figure 2.141(b).

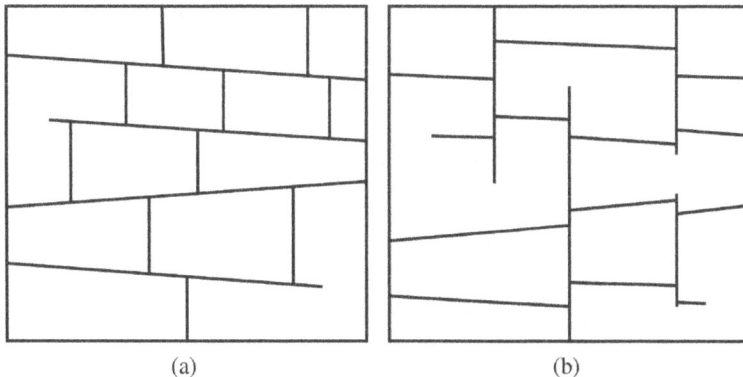

(a)  (b)

Figure 2.141. Two fracture networks each made up of two fracture sets. The spacing and the fracture length for each set is the same in both diagrams. In (a), the horizontal fractures formed first, and, in (b), the vertical fractures formed first. It can be seen that the resulting networks have very different geometries and therefore different properties.

### 2.13.3 The influence of fracture spacing, length and orientation on network geometry and mechanical properties

In addition to the importance of the order of superimposition of the fracture sets on the geometry of the resulting fracture network, the detailed geometry of the individual fracture sets making up the network, i.e., fracture spacing, length and orientation with respect to the other fracture sets, also impacts on the geometry and therefore properties of the network. This is illustrated in Figure 2.142 which shows three fracture networks $a_1$, $b_1$ and $c_1$ constructed by the superposition of two fracture sets inclined at 90° to each other.

The number of fractures in the fracture sets and their mean spacing is the same for all three models, but the lengths of the fractures increase from $a_1$ to $c_1$. The fracture networks were tested for connectivity between the upper and lower boundary and any connectivity recorded (see $a_2$, $b_2$ and $c_2$). Three patterns of connectivity emerged: namely 'no connectivity' for $a_2$, 'channelised connectivity' for $b_2$, and 'pervasive connectivity' for $c_2$. The critical length of the fractures that causes the non-connected network to change to one in which the connectivity is channelised is determined by the relative orientation of the two sets and their fracture spacing. This sudden switch from an unconnected to connected system can be represented by the step function shown in Figure 2.143. In contrast, once the channelised connectivity has been established, there is a gradual increase in connectivity as the lengths of the fractures increase and the connectivity becomes progressively more and more pervasive. In the example shown in Figure 2.142, the sudden switch from non-connected to connected occurs when the length of the fractures reaches around 0.065 of the distance $L$ between the upper and lower boundary of the model, i.e., the width of the block of material across which the property (connectivity) is being determined. It is important to note that the critical length associated with the switch in connectivity is scale-dependent.

### 2.13.4 The scale-dependence of bulk fracture network properties

The scale-dependence of bulk fracture network properties can be demonstrated using Figure 2.142 in which it can be seen that smaller sub-areas

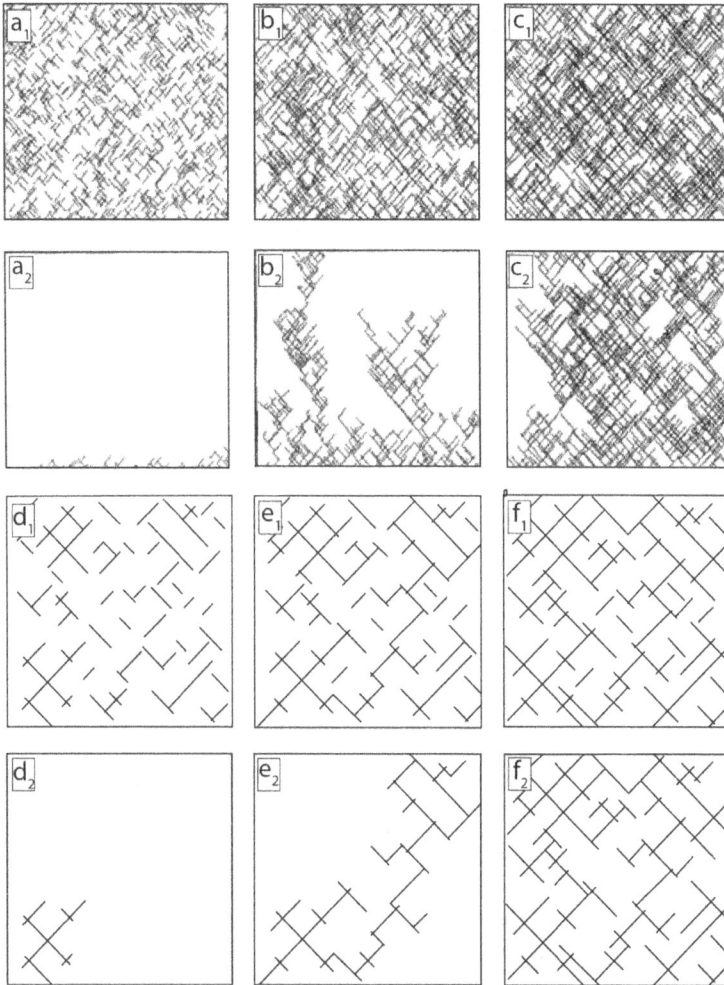

Figure 2.142. $(a_1)$–$(c_1)$ show three fracture networks containing the same number of fractures and formed by two fracture sets. The length of the fractures increases from $(a_1)$ to $(c_1)$. The corresponding connectivity across the squares from top to bottom is indicated in $(a_2)$, $(b_2)$ and $(c_2)$. From Jolly and Cosgrove, (2003). A simplified model containing relatively few fractures is illustrated in $(d_1)$, $(e_1)$ and $(f_1)$ to show the change in connectivity with increase in fracture length more clearly. The connectivity of each stage in the evolution of the fracture network is shown in $(d_2)$, $(e_2)$ and $(f_2)$. Courtesy of D. Sanderson.

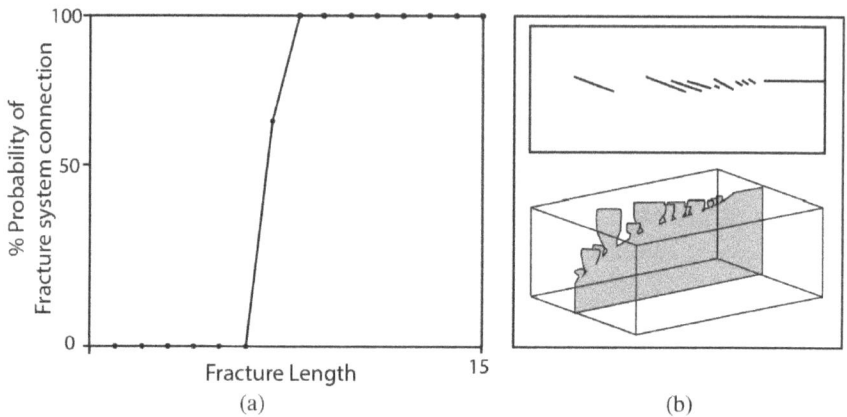

Figure 2.143.   (a) The critical effect of changing the fracture length on connectivity. From Jolly and Cosgrove (2003). (b) Plan and block diagram showing the out-of-plane linking of *en echelon* fractures to a parent fracture.

where the connectivity is channelised or in which there is no connectivity can be identified nested within the fracture network defined as pervasive ($c_2$). Similarly, if larger areas are considered, it is equally likely that local pervasive patterns of connectivity may be found to form part of a larger, channelised pattern.

In the context of determining the properties of a fractured rock mass when the scales of the volume of interest is 20 km × 20 km and the fracture network is as shown in $c_1$, it is likely that the properties can be represented by 'bulk' properties. Thus, the bulk hydraulic conductivity will adequately describe the fluid flow through the system and could be used to model water movement in an aquifer of this size. Conversely, when a much smaller region of rock is considered (say, for example, a block of a few 100 m by a few 100 m as may be the case for a potential nuclear waste repository) the scale of the network with respect to the block of interest is such that it is likely to contain only a few important fractures in terms of the project objective. Here the concept of 'bulk' properties becomes invalid and the rock mass properties will be dominated by local fractures and fracture patterns which may be non-connected, channelised or pervasive, depending on the location. Clearly, the relative scale of the fracture network and the volume of interest are of first order importance in determining whether the bulk rock properties (i.e., a general property of the rock mass) can be applied or whether the local properties are likely to dominate and the local system has to be treated specifically. When the volume of rock involved in a study falls between these two extremes, as for example many oil reservoirs and rock engineering projects, it is unclear whether the system should be modelled using bulk properties or whether local properties dominate.

It should be noted that the fracture networks in Figure 2.142 are considered in two dimensions and it is assumed that the individual fractures continue in the third dimension without any change in length or orientation. A study of natural fractures demonstrates that this is generally not the case and that fractures do vary along their strike, frequently connecting with other fractures of the same set, Figure 2.132, or with a parent fracture as illustrated in Figure 2.143(b). It follows that connectivity and conductivity values determined from two dimensional outcrops are likely to be underestimates of these properties. Conversely, because of the interaction of the fractures in the strike direction, the estimate of the bulk rock strength based on two dimensional outcrops is likely to be an over estimate. Having assessed whether the appropriate properties to use for a particular project are bulk or local, the problem of how to determine (i.e., quantify) them remains. One possible method is that of numerical modelling.

## 2.13.5 The influence of the current stress regime on fracture network conductivity

Another factor that influences the conductivity of fractures, in addition to their length, is their orientation with respect to the current stress field. This phenomenon has long been recognised in the hydrocarbon industry where it is often observed that similar fracture sets in a reservoir have very different conductivities and that these differences can be accounted for by considering the orientation of the sets with respect to the present day stress field. Fractures orientated parallel to the maximum compression direction, $\sigma_1$, and therefore at right angles to $\sigma_3$, are the most conductive and those parallel to $\sigma_3$ i.e., normal to $\sigma_1$, the least. This is well illustrated by the Scandinavian Shield where the conductive fractures are sub-vertical and striking NW–SE, i.e., parallel to the current maximum principal compression (the result of the Atlantic ridge push and of the Alpine collision) of ~25 MPa and normal to minimum principal compression.

Other factors that influence the conductivity of fractures relate to the properties of the fluids that move along them. As noted in the following Section, these fluids can have a positive or negative impact on permeability and can cause the permeability to change over geological time. A fracture may be currently conductive or may have been originally conductive but subsequently in-filled by mineralisation linked to the flow of mineral charged fluids along it. In addition to filling the fractures with minerals, fluids may modify the fracture and its ability to conduct fluids in a variety of other ways. These include widening of the aperture by solution of the wall rocks (particularly common in limestones) and alteration or weathering of the wall rock. This may either increase the permeability or decrease it as, for example, when the alteration of feldspars in the wall rock generates kaolin and other clays.

### 2.13.6 The superposition of planar deformation zones and their influence on rock properties

The factors listed in the Introduction to this Section that can influence the bulk rock properties resulting from the fracture set or fracture network it contains are not restricted to fracture networks, but also apply to the complete spectrum of structures that can form in a rock — as the conditions under which the rock is deformed change from brittle to ductile. These include:

1. Discrete fractures, either extensional or shear,
2. Individual shear zones, either brittle or ductile,
3. Shear zone networks and
4. Pervasive fabrics.

The majority of these structures (1–3 in the list above) represent localised zones of deformation and all of them result in a change in the mechanical properties of the rock in the region where the structures form. It is therefore important to know of their occurrence, location, orientation and scale at any engineering site where they have a potential influence on design and construction. In addition, because rocks are frequently exposed to more than one episode of deformation, it is likely that, just as in the simple example of the superposition of two sets of extensional fractures shown in Figure 2.141, the various deformation features making up the brittle/ductile spectrum are also likely to be found superimposed.

It is therefore appropriate to consider the effect of the superposition of two or more of these structures on the internal geometry of a rock mass in order to understand how they will affect its rock mechanics properties. Figure 2.144 shows an original, undeformed, homogeneous, isotropic rock (a), the protolith, subjected to a stress which, depending on the prevailing conditions, will produce one or other of the deformation features shown in (b)–(f). In regions of multiple deformation, any one of these features can replace the protolith and experience a second deformation which will result in one of the structures (b)–(f) being superimposed upon it. An interaction matrix showing the range of possible superpositions is shown in Figure 2.145.

Three specific examples of the superposition of such structures are considered in the following Sections. They are the result of the superposition of

— A brittle deformation on a localised ductile deformation,
— a brittle deformation on a pervasive ductile deformation and
— a ductile deformation on a ductile deformation.

In the construction of the matrix shown in Figure 2.145, two important simplifying assumptions have been made. The first is that the two deformations are coaxial, i.e., the orientation of the principal stresses remains the same and, secondly, that the pre-existing structure behaves passively during the second deformation. In the following discussion of the three superpositions (1)–(3) listed above, the effect of relaxing this second assumption is considered.

### 2.13.6.1 *Brittle deformation superimposed on a localised ductile deformation*

Figure 2.146 shows the superposition of brittle structures (tensile or shear fractures, Figures 2.146(b) and 2.146(c) respectively) on localised ductile

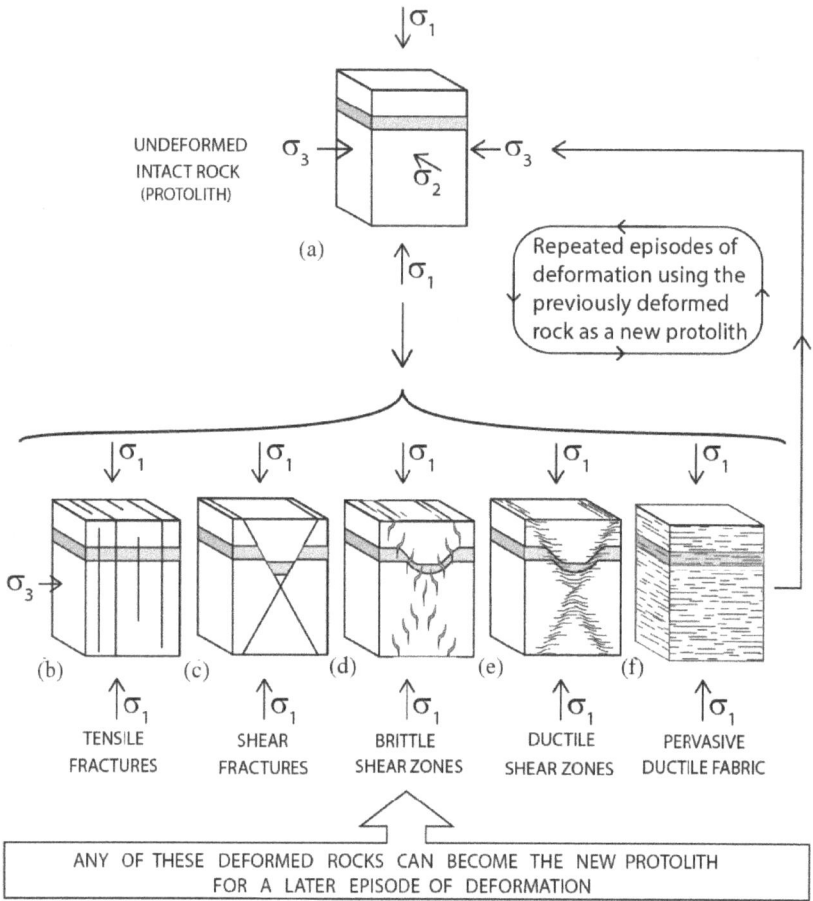

Figure 2.144. The deformation of the original, homogeneous, isotropic rock, (a) the protolith, can produce any of the deformational structures (b)–(f), depending on the conditions under which it deformed. In areas where the rocks have undergone multiple episodes of deformation, any of the deformation states shown in (b)–(f) can act as the new 'protolith' for subsequent deformational events.

deformational structures, Figure 2.146(a). As noted earlier, it is assumed that the stress field that generated the early ductile shear zones and the later brittle fractures had the same orientation. A further assumption made in the construction of Figure 2.145 is that the earlier structures remain mechanically passive during the second deformation. However, the initial ductile shear zones are localised planar zones of deformation in

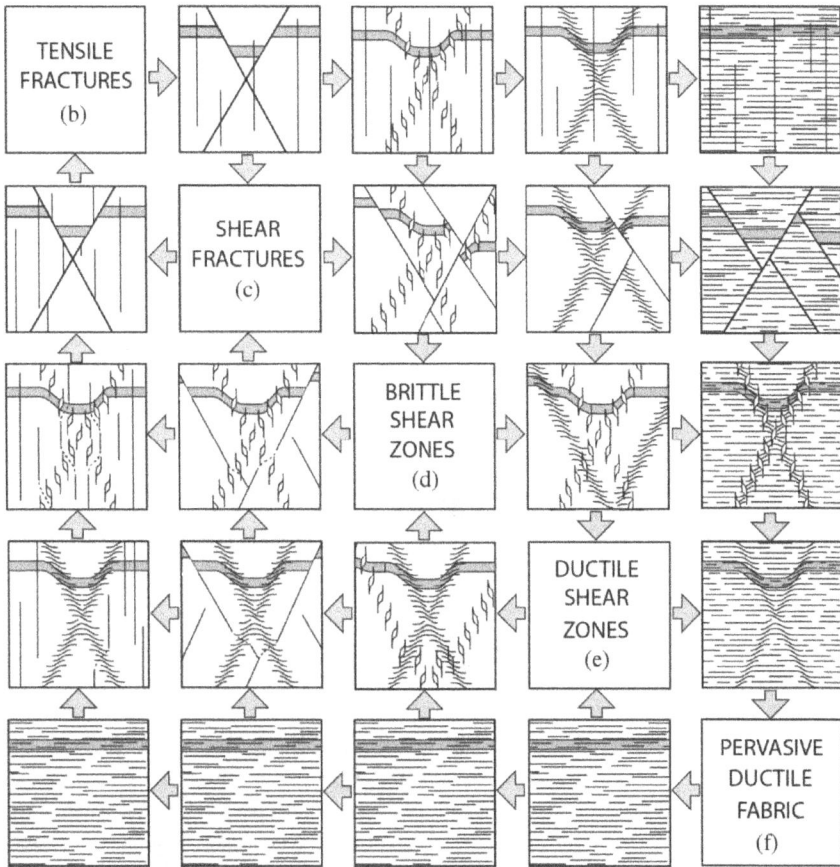

Figure 2.145. An interaction matrix illustrating possible deformation overlay sequences consisting of two deformational events. The leading diagonal terms (top left to bottom right) are the individual deformation modes illustrated in Figure 2.144(b)– 2.144(f). The off-diagonal terms indicate the superposition of one mode in the leading diagonal on another mode in the leading diagonal, following the clockwise direction of the arrows. Note that each off-diagonal box represents a single event: e.g., the fourth box in the second row of the matrix represents shear fracturing imposed on a ductile shear zone.

which a ductile fabric has developed. These zones will have different properties to the un-sheared country rock: the shear zones are likely to be weaker and therefore can act as sites where later brittle deformation will preferentially occur, Figure 2.146(d). They may also have a higher

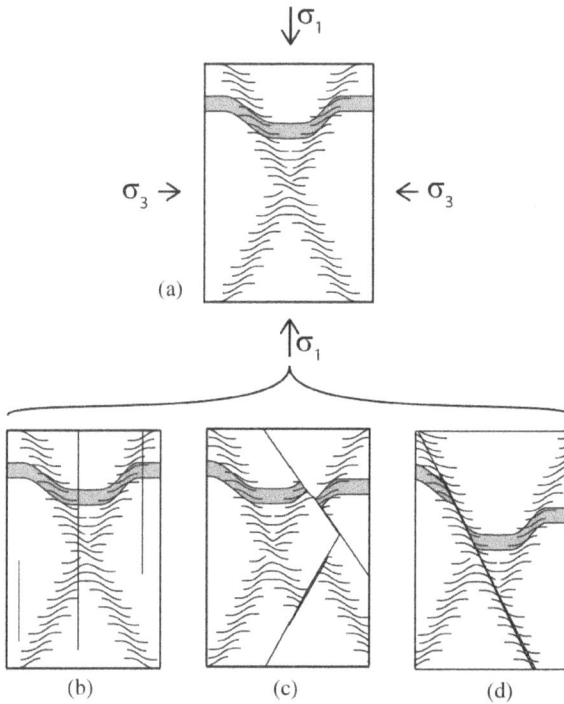

Figure 2.146. The superposition of brittle structures on localised ductile shear zones (a). If the shear zones are not appropriately orientated with respect to the later fractures (b) or, if they have approximately the same mechanical properties as the country rock (c), they are mechanically passive and do not influence the orientation or location of the later fractures. When they are weaker and appropriately oriented, (d), they can be reactivated at stress levels lower than those required to generate new fractures.

permeability than the surrounding rock with the result that fluids have selectively migrated along them, leading to an alteration of the rock and a further reduction in strength.

Clearly, if the shear zones are not weaker than the surrounding rock and/or if they are not advantageously orientated with respect to the orientation of the newly forming brittle structures, the location and orientation of the brittle structures will not be influenced by them, Figure 2.146(b) and c. Alternatively, if the shear zones are weakened zones and are appropriately orientated with respect to the new fractures, then it is likely that the new fractures will form along them, Figure 2.146(d).

## 2.13.6.2 *Brittle deformation superimposed on a pervasive ductile deformation*

The superposition of brittle deformation on a pervasive ductile fabric is shown in Figure 2.147 which shows two possible orientations of the maximum principal compression with respect to the pre-existing ductile fabric, perpendicular in (a)–(c) and at an angle in (d) and (e).

When the principal compression is at right angles to the pre-existing rock fabric, no slip can occur along the fabric and it remains mechanically passive during the imposed brittle deformation. Depending on the magnitude of the differential stress ($\sigma_1 - \sigma_3$) either tensile or shear fractures can form, Figures 2.147(b) and c, respectively. Conversely, in Figures 2.147(d) and e, the orientation of the rock fabric is such that the applied stress can cause the rock to deform by slip along the fabric. In this case the fabric is mechanically active and has a major impact on the mode of deformation, inhibiting the formation of the brittle structures.

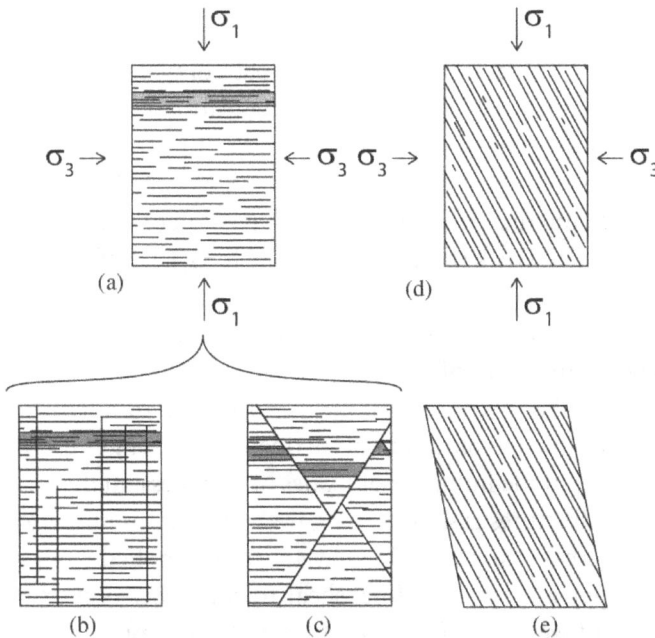

Figure 2.147. The possible consequences of the superposition of brittle failure on a pervasive ductile fabric, See text.

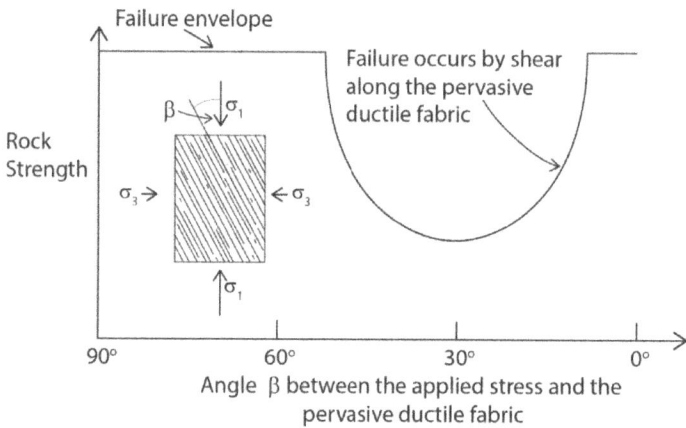

Figure 2.148. The influence of the orientation of a pervasive planar fabric on the rock strength. Along the straight portions of the failure envelope deformation occurs by brittle failure. Shear along the fabric can occur for the range of stress orientations corresponding to the dip in the failure envelope.

Which of these two modes of deformation occurs is determined by the relative orientation of the applied principal compression and the rock fabric. The two potential failure modes, i.e., the formation of new fractures or re-shear on the pre-existing fabric, are shown in Figure 2.148, see also Section 2.6.7. As noted in Figure 2.147, when the applied compression is at a high angle to the rock fabric (the straight portion of the failure envelope in Figure 2.148), the rock strength is relatively high, i.e., deformation occurs by the formation of new fractures which traverse the fabric. As the principal compression becomes less inclined to the rock fabric (in the example shown between the range of 5 and 50°) shear becomes possible. This occurs at a lower stress than that required for the formation of fractures resulting in a distinct, U-shaped minimum occurring in the strength profile.

An example of the influence of planes of weakness on the location and orientation of fractures is shown in Figure 2.149. In this numerical model, the planes of weakness are inclined at 45° to the applied compression. It can be seen that, although the final shear fracture does occur sub-parallel to the planes of weakness, Figure 2.149, Step 87, it is preceded by the formation of small-scale extensional fractures that eventually link to produce a through-going zone of shear failure, see Sections 2.10.4 and 2.10.5.

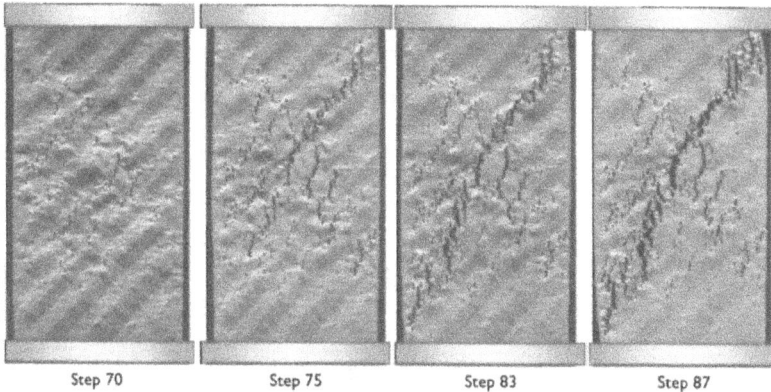

| Step 70 | Step 75 | Step 83 | Step 87 |

Figure 2.149. Numerical model showing the influence of a planar rock fabric on rock matrix microstructural failure and the localisation of macroscopic deformation (from Tang and Hudson, 2010).

### 2.13.6.3 *Ductile deformation superimposed on a ductile deformation*

If a ductile deformation is superimposed on a rock and if the result is the formation of a pervasive ductile fabric, then any previous structure (whether brittle or ductile) will be obliterated and the rock will exhibit a uniform ductile fabric. If, as in the example shown in Figure 2.150, the previous structure was a pervasive ductile fabric, then there are two types of response.

These are:

— The formation of a new, pervasive fabric. This can occur either by the enhancement of the pre-existing fabric if the orientations of the stresses are the same for the two deformations (Figure 2.150(a)) or by the obliteration of the early fabric and the generation of a new fabric at a different orientation if the stress orientations for the two deformations are different.
— Alternatively, the early fabric can be mechanically active during the later deformation and this can result in the development of a localised, rather than a pervasive, ductile deformation, Figure 2.150(b).

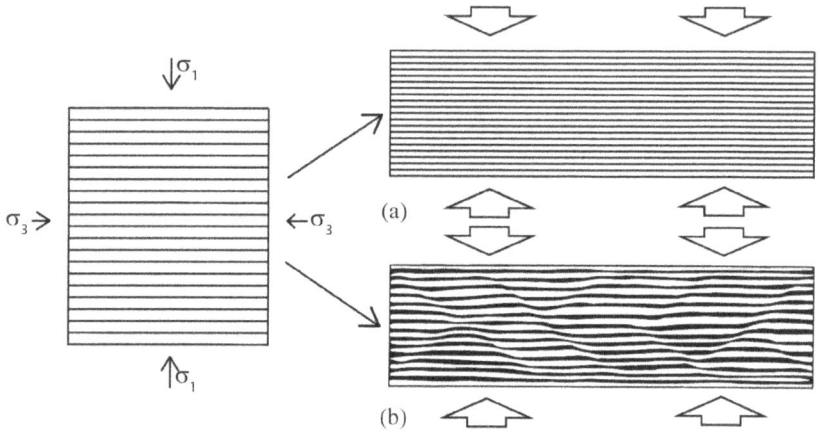

Figure 2.150. Superposition of ductile deformation on ductile deformation. In (a), the early fabric behaves passively and is enhanced by the later deformation. In (b), the early fabric is mechanically active and responds to the later deformation by developing conjugate normal kink bands.

The geometry of the structures that form is determined by the orientation of the applied stress with respect to the pre-existing rock fabric and the magnitude of its mechanical anisotropy. The suite of structures that can form is shown in Figure 2.151. It can be seen that, if the pre-existing ductile fabric has a high mechanical anisotropy, the later ductile deformation is localised into planar bands known as 'kink bands'. If the anisotropy is high and if the stress orientation for the first and second deformations are the same, then conjugate normal kink bands will form as illustrated in Figures 2.150(b) and 2.151(h) and 2.151(i).

If the principal compression of the second deformation is normal to the first, i.e., if it acts along the induced ductile fabric, then, depending on the mechanical anisotropy of the fabric, the resulting ductile structures will be either folds with axial planes normal to the principal compression (Figure 2.151(a).) or conjugate kink bands (box folds) with axial planes inclined to this compression (Figure 2.151(c)). It can be seen from this Figure that a complete spectrum of structures exists between these two end-members as the mechanical anisotropy of the fabric gradually increases.

*****

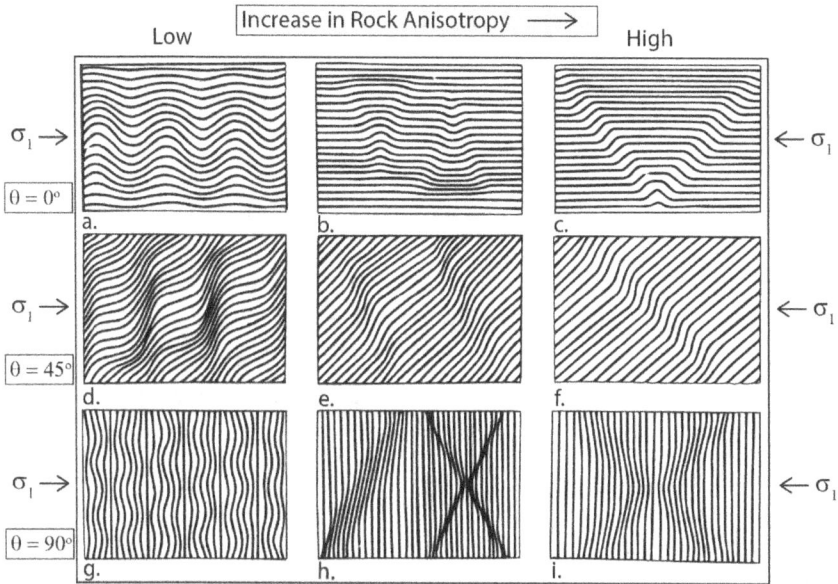

Figure 2.151. The suite of structures that can form in a planar, anisotropic fabric when compressed at different angles to the fabric. $\theta$ is the angle between $\sigma_1$ and the rock fabric. (a→c) compression parallel to the fabric, (d→f) at 45° to the fabric and (g→i) normal to the fabric. Note that the structures become progressively more localised as the mechanical anisotropy increases. (From Cosgrove, 1976).

The examples of superimposed deformation modes illustrated and discussed are just three examples of a complete range of possible superpositions of two of the five individual structures shown in Figure 2.144(b)–2.144(c). The complete range of such possible superpositions is illustrated in the interaction matrix in Figure 2.145. To recapitulate, the five deformation modes are represented in the boxes along the leading diagonal (from top left to bottom right) in this Figure. All possibilities of one deformation mode superimposed on another are then illustrated in the 20 off-diagonal boxes. For example, the consequence of tensile fracturing being superimposed on shear fracturing is shown in Box (1, 2), i.e., Row 1, Column 2. Conversely, the consequence of shear fracturing being superimposed on tensile fracturing is shown in the complementary off-diagonal Box (2, 1), i.e., Row 2, Column 1 — and this same matrix component locational principle is applied to all the other deformation mode superimpositions.

The specific deformation overlays that have been discussed here can be located in this diagram. For example, brittle deformation, tensile and shear, superimposed on a pervasive ductile deformation is illustrated in Row 1, Column 5 and Row 2, Column 5, i.e., in Boxes (1, 5) and (2, 5), respectively. Similarly, the superposition of brittle deformation, tensile and shear, on localised ductile deformation is shown in Boxes (1, 4) and (2, 4) respectively.

Note that the superposition of ductile deformation on a pervasive ductile deformation represented by Box (4, 5) requires special consideration because, as discussed in Section 2.13.6.3, if the imposed deformation results in a pervasive ductile fabric, then any previous structures will be obliterated because the mineral constituents of the rock recrystallise in response to the later deformation. (It will be recalled that in this diagram it is assumed that, unlike the situation represented in Figure 2.151, the early structures do not interact mechanically with the later structures.) The orientation of the new fabric is then controlled totally by the new orientation of the associated stress field with the fabric forming normal to the maximum compressive stress, $\sigma_1$.

The Figure 2.145 matrix diagram considers just one deformational event superimposed on another and assumes that the stress orientations in both cases are the same, i.e., with $\sigma_1$ acting from top to bottom of the page and $\sigma_3$ acting across the page. In many rocks, many episodes of ductile and brittle deformation have affected the rock, and, in general, the stress orientations linked to the different events are unrelated and therefore are likely to have been different and had different magnitudes. Such a change in the orientation of $\sigma_1$ would result in more complex deformation patterns than those shown in Figure 2.145.

In addition, and in terms of the compilation of the matrix in Figure 2.145, as noted above, it has been assumed for the purposes of illustration that the later structures have been superimposed passively on the earlier structures. However, this assumption is also over-simplistic because the formation of the earlier structure destroys the homogeneous and isotropic properties of the protolith and generates a mechanically heterogeneous and often anisotropic rock mass. As discussed in Sections 2.13.6 (1–3) the possibility therefore arises that the early structures will not behave passively during subsequent deformation and may influence the orientation and location of any superimposed structure and can even result in the formation of new types of structures not illustrated in Figure 2.144, such as the conjugate normal kink bands, shown in Figure 2.150(b) and Figure 2.151(h) and i. If a pervasive ductile fabric is compressed parallel to the fabric and, if the pressure and temperature conditions are appropriate, folds can form, Figure 2.151(a)–2.151(c).

However, the graphical overview provided by the interaction matrix is helpful in considering the wide range of structures generated by superposition of the deformational modes illustrated in Figure 2.144. Similar interaction matrices could be generated for considering more complex multiple superpositions according to other user requirements. An understanding of this potential complexity leads to a deeper understanding of the likely distribution of deformational features within the rock mass and hence provides a clearer appreciation of how these deformational structures are likely to impact on its rock mechanics properties. In addition, the interaction matrix approach can be used for a variety of engineering application, as is discussed in detail in Hudson (1992 on rock engineering systems, and 2015 in the context of rock engineering risk).

### 2.13.7 Scale-dependent properties of fracture networks, other deformation features, and the link with engineering

The scale dependence of bulk fracture network properties has been discussed in Section 2.13.4. However, the importance of the relative scale of an excavation at an engineering site with respect to the geological structure formed within the local rock mass is not restricted to fracture networks but also applies to the complete spectrum of structures that can form in a rock as the conditions under which the rock is deformed change from brittle to ductile, Figure 2.144(b)–2.144(f). These include:

— Discrete fractures, either extensional or shear. These may be a single fracture set made up of extensional fractures, or a fracture network made up of a conjugate set of shear fractures linked to a single deformation event, or two or more superimposed extensional fractures linked to different deformation events.
— Individual shear zones, either brittle or ductile.

— Shear zone networks.
— Pervasive fabric.

A series of schematic diagrams showing the various planar deformational structures and excavations of different sizes in a rock mass is shown in Figure 2.152. It can be seen that, depending on the location, orientation and size of the excavations with respect to the structures, a variety of potential excavation effects and instabilities can arise, some of which are described in the following sub-sections. The large circles represent tunnels and the smaller circles represent boreholes. Note that, depending on the relative scales, the excavation can occur in intact rock or be intersected by one or more fractures/structures. Also, note that the small box within the shear zone network box, Figure 2.152(g), represents the individual shear zone box drawn directly above it, i.e., the shear zone network will be on a much larger scale than the individual shear zone.

### 2.13.7.1 *Intact rock*

The intact rock case, Figures 2.144(a) and 2.152(a), is assumed to contain no discontinuities. Accordingly, there is no major issue of scale when encountering the intact rock in a borehole or tunnel, although a minor scale effect can occur because the elastic modulus and strength of intact rock are known to decrease with the increasing size of the sample being considered. The intact rock may, however, be anisotropic and inhomogeneous. The former can be established by determining if the material properties are directionally dependent and the latter can be accommodated by dividing the rock into domains within which the material properties are essentially constant. Rock failure can occur if there are high *in situ* stresses concentrated around the excavation causing spalling.

### 2.13.7.2 *Discrete fractures*

When brittle deformation results in the formation of discrete fractures, Figures 2.144(b) and 2.144(c), the individual fractures often form part of approximately uniformly spaced fracture sets, either one set of extension fractures, or conjugate sets of shear fractures, see Section 2.8. All are likely

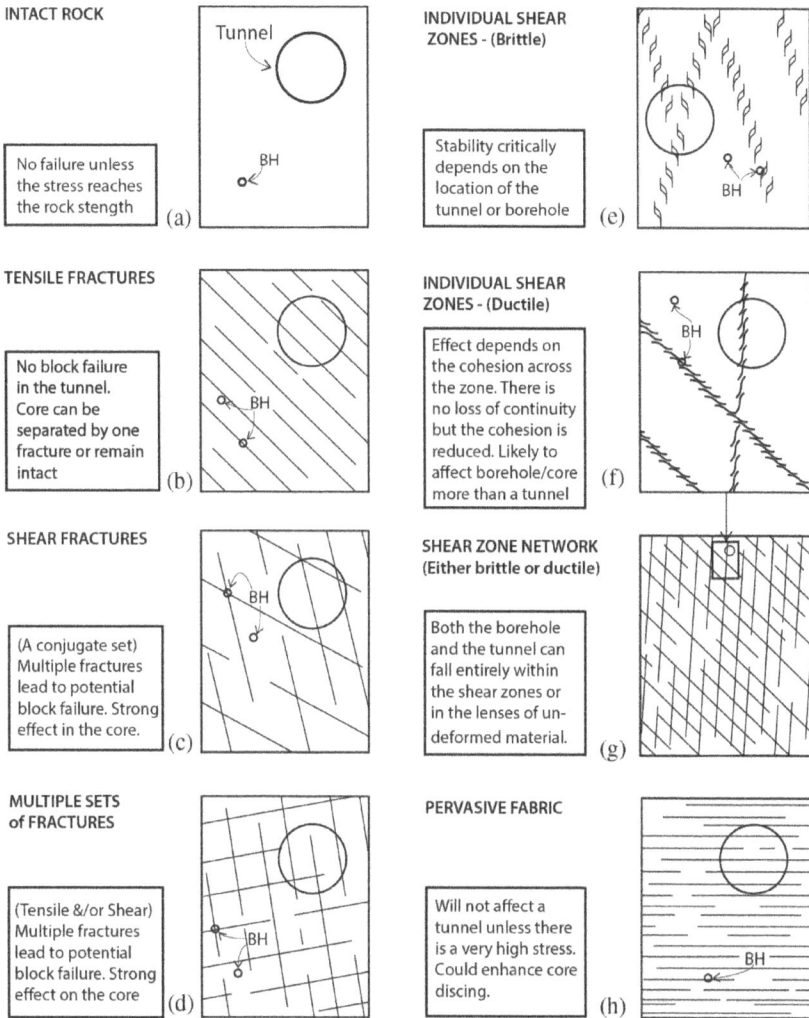

**INTACT ROCK**

No failure unless the stress reaches the rock stength (a)

Tunnel

BH

**TENSILE FRACTURES**

No block failure in the tunnel. Core can be separated by one fracture or remain intact (b)

BH

**SHEAR FRACTURES**

(A conjugate set) Multiple fractures lead to potential block failure. Strong effect in the core. (c)

BH

**MULTIPLE SETS of FRACTURES**

(Tensile &/or Shear) Multiple fractures lead to potential block failure. Strong effect on the core (d)

BH

**INDIVIDUAL SHEAR ZONES - (Brittle)**

Stability critically depends on the location of the tunnel or borehole (e)

BH

**INDIVIDUAL SHEAR ZONES - (Ductile)**

Effect depends on the cohesion across the zone. There is no loss of continuity but the cohesion is reduced. Likely to affect borehole/core more than a tunnel (f)

BH

**SHEAR ZONE NETWORK (Either brittle or ductile)**

Both the borehole and the tunnel can fall entirely within the shear zones or in the lenses of un-deformed material. (g)

**PERVASIVE FABRIC**

Will not affect a tunnel unless there is a very high stress. Could enhance core discing. (h)

BH

Figure 2.152. An illustration of the impact on stability of the relative scale of the excavation and the geological structures shown in Figure 2.144. BH in the sketches refers to a borehole and the larger circles represent a tunnel.

to have an adverse effect on excavation stability. The case of discrete fractures is the one most commonly addressed in rock mechanics analyses with reference to the formation of blocks that can fall or slide into the excavation, as illustrated in Figure 2.153(a). Considerable research has been

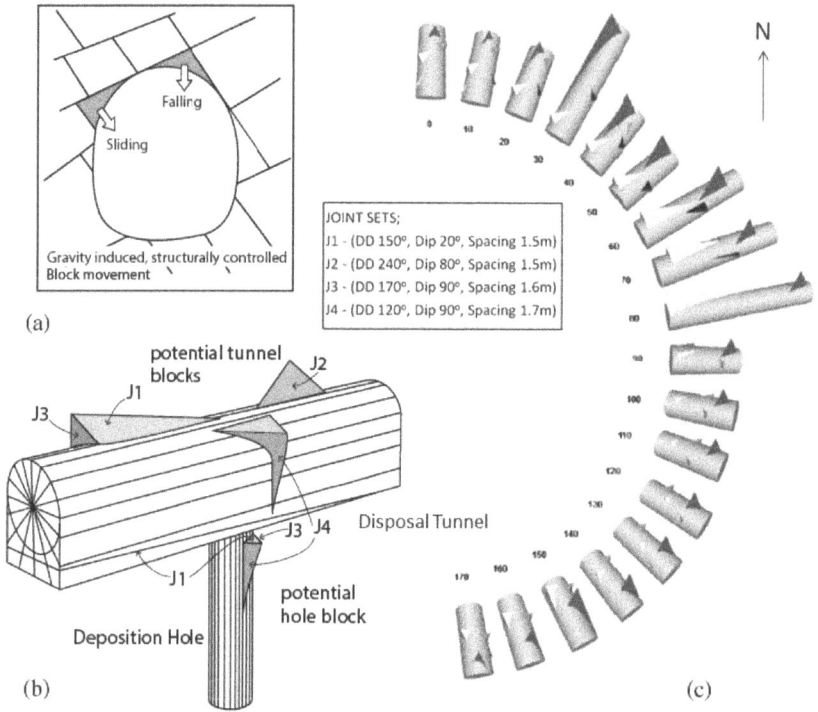

Figure 2.153. The influence of discrete brittle fractures on excavation stability. (a) A general sketch showing the formation of rock blocks that could potentially slide or fall as a result of the construction of a tunnel within a fractured rock mass. (b) A similar illustration of rock blocks that can form around a nuclear waste disposal tunnel and deposition hole formed in a rock mass with several fracture sets (from Rautakorpi *et al.*, 2003). (c) Shows the kinematic feasibility of rock blocks falling or sliding into a tunnel for a specific rock mass with four joint sets, J1–J4, and for a horizontal tunnel excavated in different azimuth direction (after Johansson *et al.*, 2002).

undertaken to predict the formation of such blocks as a function of the geometry and orientation of the fracture sets and the orientation of the excavations, Figure 2.153(b), e.g., Goodman and Shi (1985). Figure 2.153(c) illustrates the blocks that can fall into a tunnel for given sets of fractures for all tunnel azimuth directions.

In a fractured rock mass, two of the main factors in tunnel design are designing to avoid damage due to concentration of the *in situ* stresses and to avoid rock block fallout. As a general principle, a tunnel will be less susceptible to stress damage when it is driven parallel to the major principal stress. However, the direction in which the tunnel will have the least block fallout may not be so clear, see Figures 2.153(b) and 2.153(c). In a rock engineering project where there is flexibility in the orientation of tunnels, e.g., an underground radioactive waste repository, the designer needs to find the direction that provides the optimal compromise between the two factors of stress concentration and block fallout.

Another factor is the impact of the relative scales of the excavation and the fracture system. This is shown in Figures 2.152, e.g., (b)–(d). Depending on their orientation with respect to the fractures, boreholes may penetrate intact rock and be unaffected by the fractures. In contrast, tunnels, whose scale may be comparable to the spacing of the fractures, will intersect the fractures with the resulting possibility of instability, i.e., the large circles in Figures 2.152(b)–2.152(d).

### 2.13.7.3 *Individual shear zones*

Under this heading, both brittle and ductile shear zones are included (Figures 2.144(d) and 2.144(e), as illustrated in Figures 2.152(e) and 2.152(f)). In this case, because the degree of strain localisation is high, as with the discrete fractures discussed above, then depending on the spacing of the shear zones, both the small and large excavations could be located completely in the intact rock. If the excavation does intersect a shear zone then, as illustrated in Figures 2.152(e) and 2.152(f), the smaller excavation could be totally within the shear zone and the situation would be analogous to that of a core being drilled in a pervasive fabric, (Figures 2.144(f) and 2.152(h)), which characterises the internal part of a ductile shear zone with the consequent problems of potential instability discussed below.

In the case of the larger excavation, the stability will depend on the orientation and location of the deformation structure intersection, i.e., whether it is in the roof, sidewall or floor of the excavation, and on the direction in which the shear zone is approached. For example, it is better for construction if the tunnel first intersects a gentle dipping zone in the floor (i.e., the zone is dipping towards the tunnel approach), rather than in the roof first (i.e., the zone is dipping away from the tunnel approach), because in the latter case the presence of the structure is not evident until it is first intersected in the roof — a situation that could result in unexpected roof collapse.

### 2.13.7.4 Shear zone networks

The influence of the scale of the excavation relative to the deformational feature is illustrated in Figure 2.152(g) for the case of the shear zone network. Like the conjugate shear fractures (Figure 2.144(c)), the conjugate shear zones that define the shear zone network form in response to a single stress field (Figures 2.144(d) and 2.144(e)). In terms of stress failure, the relative scale of the excavation and the shear zone network will not significantly affect the likelihood of rock spalling due to excessive stresses. The scale affects whether the excavation is located entirely within the intact rock or the shear zone and the likelihood of block failure (i.e., a block formed by the shear zone boundaries and the excavation surface) depends critically on the scale. In the case of the larger excavation, the probability of a kinematically admissible block is naturally much greater than for a borehole. Note that the small box in Figure 2.152(g) represents the Figure 2.152(f).

### 2.13.7.5 Pervasive fabrics

Figure 2.152(h) illustrates excavations made in a rock with a sub-planar pervasive fabric. In this case, the scale of the excavation relative to the scale of the rock fabric is not likely to be significant. However, the orientation of the excavation surface relative to the foliation orientation is important. When an excavation is made in a rock mass, the *in situ* stress is deflected around the excavation, with the maximum principal stress being parallel to the excavation boundary — see the white arrows in Figure 2.154. If this stress component is large enough, the rock can fail. However, the strength of the rock is determined by the relative orientation of the excavation boundary stress and the rock fabric, as has been already illustrated in Figure 2.148, so that the conditions for rock failure vary around the opening. In the case of a pervasive fabric, there is little scale effect, so long as the size of the fabric elements in the foliation is much smaller than the excavation dimensions.

Figure 2.154. The effect of an excavation intersecting a pervasive planar rock fabric, i.e., foliation, on the potential location of rock failure at the excavation periphery. This potential is a function of both the magnitude of the rock stress acting parallel to the excavation boundary and the orientation of the foliation relative to the excavation boundary, as indicated by the white arrows and the shading in the individual rock samples, respectively.

The potential impact of the various structures illustrated in Figure 2.144 on the stability of excavations discussed above makes it clear that their recognition during any site investigation is of first-order importance. Site investigations are frequently carried out by surface mapping and borehole analysis. The following discussion highlights the difficulty of detecting, recognising and characterising the geometry and properties of these complex planar zones of deformation on the basis of borehole intersections and the resulting borehole logs and cores.

## 2.13.8 The identification of a deformational structure at outcrops and in tunnels and boreholes

In the preceding part of this Section, a variety of tectonically-induced, sub-planar structures that can form in a rock mass have been described. These have been summarised diagrammatically in Figure 2.144 which shows the structures to be part of a complete spectrum ranging from individual fractures representing extreme examples of strain localisation through to a pervasive ductile flattening fabric which is the result of a perfectly uniform distribution of strain. All structures in this spectrum are likely to affect the rock mechanics properties of the rock, sometimes in a way that may impact adversely on the construction of excavations within the rock mass. It is therefore important for rock engineering to establish how these structures can be recognised so that the effects on any proposed excavation can be predicted.

The rock structures can be identified at an engineering site either by

— direct observation from outcrops at the surface, within tunnels and in boreholes and via cores,
— remotely using aerial and satellite images to detect structures outcropping at the surface, or
— using geophysical techniques to investigate their presence within the rock mass at depth.

The ease with which the various structures can be recognised in outcrop relates directly to the relative size of the structure and the outcrop, tunnel or borehole surface available for inspection. The more localised the structure is, the smaller the outcrop needed to convincingly identify it. Thus, a major diffuse shear zone would be more difficult to detect in a tunnel than localised fractures. Having identified a structure in an outcrop, an additional factor needs to be considered: namely the extent of the structure, i.e., how far into the rock mass it continues.

When there is a considerable outcrop area in and around a potential engineering site there is ample opportunity to study the fractures and other deformational features in detail, at least over the essentially two dimensional outcrop surface, and to identify their nature. However, the 'outcrop' surfaces made available by the excavation of tunnels and the drilling of boreholes to provide borehole logs and core surfaces are much more restrictive. It was noted in the discussion related to Figure 2.152 that, depending on the relative scales of excavation and deformational structure, some sub-planar structures will appear as discrete features in a borehole or tunnel and others will not. However, this problem is not one that necessarily concerns the rock mechanics characterisation. For example, a tunnel cutting a major ductile shear zone will penetrate a region where a pervasive ductile planar fabric exists in the rock. This will be readily identified in the tunnel and, although it may not be apparent that this is not a pervasive ductile fabric of the type shown in the last sketch of Figure 2.152, but part of a localised ductile shear zone, from the point of assessing its rock mechanics significance, this is only of second order importance.

It is clear from the discussion above that the more diffuse the structure is the more difficult it is likely to be to identify within the relatively small exposed surfaces provided by boreholes and tunnels. Indeed, the problem of comprehensively identifying sub-planar deformational structures in boreholes and tunnels, establishing correlation between two or more observations, and determining the extent of the structures has been recognised as one of the major problems facing both geologists and rock mechanics engineers.

### 2.13.8.1 *The expression of the different deformational structures in tunnels and boreholes*

The intact protolith, Figure 2.144(a), and the pervasive ductile fabric, Figure 2.144(f), are ideally homogeneous and will show no variations in their properties apart from local variations. The discrete extensional and shear fractures, Figures 2.144(b) and 2.144(c), will be easily recognised on borehole and tunnel walls and it may be possible to differentiate between the two on the basis of displacement and surface features. It will be recalled from Section 2.3, that the displacements characterising extensional fractures are dominantly perpendicular to the fracture surfaces and those for shear fractures are dominantly parallel to these surfaces. The displacement direction for both types of fractures may be obtained from the displacement of markers within the rock, or, in the case of shear fractures from kinematic indicators such as slickensides, crystal fibres and other features that form sub-parallel to the fracture (Sections 2.7.4 and 2.7.5).

Both extensional fractures (joints) and shear fractures (faults) may be either uniformly distributed (spaced) or may occur in clusters (see Section 2.8.4). Thus, depending on the size of the excavation with respect to the width of any fracture cluster, the fracture spacing may appear approximately uniform or clustered. The expression of shear zones, both ductile and brittle (Figures 2.144(d) and 2.144(e)) in a borehole or tunnel

Table 2.1. The ability to characterise the geometrical and mechanical properties of the deformational features in a rock mass.

| Deformational feature | Size of representative feature for identification and characterisation relative to sampling size, especially a borehole | Ease/Difficulty of characterisation |
|---|---|---|
| Pervasive fabric — foliation | Core size | Easy because pervasive feature and evident in borehole cores |
| Networks of shear zones | Requires large exposure for complete identification | Extremely difficult |
| Andersonian shear zones | | |
| Discrete extensional fractures | Some information from core intersections; more information from exposures | Medium |
| Intact rock, unaltered, and altered | Core size | Easy because pervasive feature and evident in borehole cores |

will also depend upon their relative scales. Thus, the expression in an excavation may be that of a relatively uniform shear fabric characterized by the various second order structures associated with shear zones (*en echelon* tension gashes, shear zone fabrics, etc.) or intact rock punctuated with zones of intense shear fabric.

In Table 2.1, we summarise the comments already made in terms of the ease or difficulty of identifying the specific deformational structures in outcrops, boreholes and tunnels. It should be noted that, following the sequence of structures as listed in the Table, which corresponds to the suite of features in Figure 2.144, the identification is easy → difficult → easy, a trend that has been foreshadowed by the bottom arrow of the earlier Figure 2.1 in Section 2.1 for the rock mechanics characterisation capability. Other aspects of identifying the different deformational structures in boreholes and tunnels are the influence of the borehole orientation relative to the particular structure and the scale of observation. These two aspects are discussed in the following Section.

### 2.13.9 The ability to characterise the geometrical and mechanical properties of the deformational features from a borehole or tunnel

Figure 2.155 shows a fracture zone in the Pre-Cambrian crystalline basement rock of Finland. The sub-horizontal, whitish grey lines on the rock face in the background are the 'half-barrels' remaining from the blastholes used to excavate the tunnel and the bold white lines indicate the portions of the brittle deformation zone that would be intersected if boreholes were to be located in these positions in order to characterise the zone in a site investigation.

In theory, the influence of the direction of the borehole on the deformation zone width intersected can be evaluated through the bias correction discussed by Terzaghi (1965) and Priest and Hudson (1976). However, this correction assumes that the width and geometry of the deformation

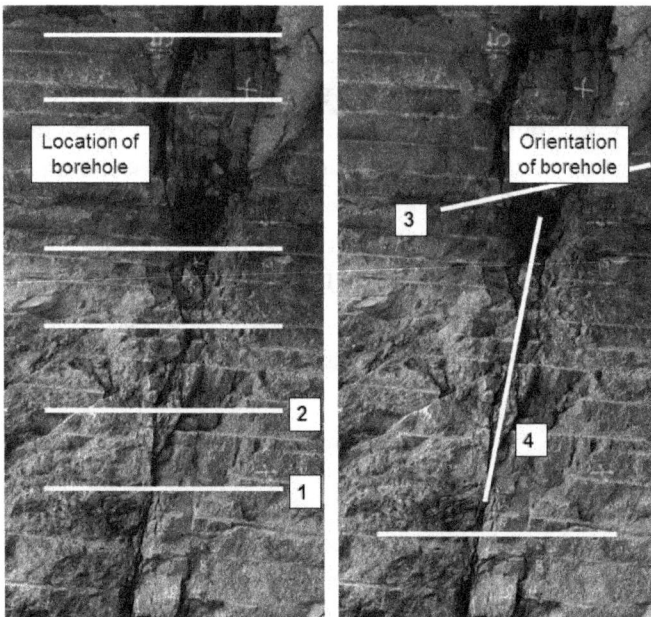

Figure 2.155.   A fracture zone cutting the Pre-Cambrian shield of Finland. The white lines show the orientation and position of hypothetical boreholes drilled to investigate the geometry of the zone.

zone remains constant along its length, an assumption that is clearly inappropriate for deformation zones of the type shown in Figure 2.155 which are typical of many natural fractures. This Figure illustrates the important effect of the apparent deformation zone geometry (width, fracture intensity, etc.) as observed in the borehole core caused by the location and orientation of the borehole through such a fracture zone. The left and right diagrams in the Figure illustrate the effects of the 'location' and 'orientation', respectively, of a borehole with respect to a deformation zone on the apparent width of the zone when observed in the core. For example, boreholes 1 and 2 on the left-hand side of the Figure would indicate a relatively thin fractured zone; whereas, boreholes 3 and 4 on the right-hand side of the Figure would indicate a greater thickness. It is evident from Figures 2.144, 2.145, and 2.155 that the ability to characterise the geometrical and mechanical properties of the deformational features from data obtained from a borehole or tunnel will depend strongly on their type, as summarised in Table 2.1.

A summary of the ability to characterise a rock mass containing the different deformational features described is that it is relatively easy at both ends of the strain localisation spectrum, i.e., for intact or strongly foliated rock without discrete fractures. It is difficult to characterise the remaining range of structures, whether they occur in isolation or association, because of the large size of the required rock sample — necessary both for deformational feature identification and characterisation of the bulk rock properties. In these cases, for characterisation it is necessary to use either an empirical or numerical modelling approach in which the bulk properties are determined from modelling the constituent components.

*****

The motivation for the above discussion has been to determine whether or not it is possible to assess the rock mass properties at an engineering site from surface exposures, boreholes and tunnel monitoring data. As noted above, this is easier for certain structures than for others. Such an on-site study would lead to assessing the effects of the geological features on the excavations, thus providing coherent support for design, Feng and Hudson (2011) and Hudson and Feng (2015). A key aspect of these effects is the influence of the relative sizes of the deformational features and the excavations. The scale of observation of the deformational features and the impact that this has on the associated interpretation has been noted, Figure 2.152, and it has also been pointed out that there is an analogous circumstance in considering the relative size of an excavation intersecting the deformational features. This concept is illustrated in the series of diagrams presented in Figure 2.152 which considers a circular excavation intersecting the main deformational types shown in Figure 2.144.

Although a circular excavation has been used for the illustrations, the same principles apply for any shaped excavation. Also, the diagrams illustrate cases where the interaction between the deformational features and the excavation is where the excavation axis is perpendicular to the profile cross-section of the feature, i.e., parallel to the strike of the shear zone network and parallel to the foliation. It should be remembered that this is a special case and that the excavations will generally be at various orientations to the deformational features, resulting in more complex interactions.

The structures shown in Figure 2.144 are the result of a single deformation. However, many rocks have experienced more than one deformation, each of which is likely to generate one or other of the structures shown in this Figure and which will be superimposed on earlier structures. This is shown diagrammatically in Figure 2.145. It should be noted that this is a simplified diagram in that there are only two deformation events, the principal stresses generating the two sets of structures are in the same orientation ($\sigma_1$ vertical) and the mechanical interaction between the early structures and late structures has been ignored. Clearly, the more general situation in which there are more than two deformation events, where the deformations are non-coaxial and where the pre-existing structures are mechanically active, will lead to rock masses with a complex structure and with heterogeneous mechanical properties. However, it may be possible to characterise the rock mass with general, bulk properties if the size of the block under consideration is large enough. Alternatively, if the block size is relatively small with respect to the scale of the rock structures, its properties are likely to be determined by the local properties of a few major shear zones and require a more specific analysis.

It is important to recognise the type, size and distribution of the major structures affecting a potential engineering site and to assess whether it is likely to be the bulk or local properties that best represent the behaviour of the rock mass. Although the task of determining these properties is formidable: a clear understanding of the geometry, orientation and location of the major structures combined with an appreciation of the relative scale of these structures with respect to the proposed excavation is a crucial first step. Moreover, this structural geology input will enable the site investigation content, any numerical analyses, the design process and monitoring during excavation to be based on a sound appreciation of the host rock mass. There are also ramifications in terms of *in situ* stress measurements which are discussed in the next Section.

## 2.14 THE TIMING, SCALE AND DISTRIBUTION OF STRESS IN THE EARTH'S CRUST AND ITS INFLUENCE ON FRACTURING

### 2.14.1 Introduction

The stress fields that affect rocks can be conveniently divided into three groups namely:

— **Past stress fields** — those that have been dissipated in the crust by generating deformation (either ductile or brittle) in the geological past.
— **Present day stress field** — linked to current overburden and plate motion, i.e., a currently active stress in the crust which can have an important impact on the stability of a fractured rock mass and upon its properties, e.g., conductivity.
— **Residual stress** — the stress field locked into the rock and only released during exhumation, i.e., on removal of the confinement provided by the overburden and surrounding rock.

These will be briefly considered in turn.

## 2.14.2 Past stress fields

Many rocks in the Earth's crust have been subjected to several major stress regimes over their long geological history. These stresses are generally dissipated as the rock responds by deforming (straining) in either a ductile or brittle manner. However, structural geologists can determine the stress history of a rock by the analysis of the structures that form in response to the individual stress regimes making up this history. Although structures induced during ductile deformation can have an important impact on the mechanical properties of a rock (e.g., the formation of a pervasive cleavage or schistosity), the structures that generally have the most important effect on the geomechanical properties of a rock mass are fractures. The technique of 'fracture analysis', by which the chronology of the fracture sets making up a fracture network (together with the network geometry and the orientation of the various stress regimes responsible for its formation) can be established, has been described in Section 2.12.

The evidence for the ancient stress fields that have affected the crust is often preserved as fractures and these indicate clearly that, although fractures do form in response to regional stress fields of the type associated with overburden stresses and plate motion, the heterogeneity of the crust can result in the degradation of these stress fields into smaller regions. It is observed that the stress fields responsible for the generation of fractures range from those that occur on a regional scale to stress fields that are extremely localised. An example of a regional stress field generating a fracture system is illustrated in Figure 2.156 which shows jointing in the Appalachian plateau caused by the collision between ancestral Africa and the North American plate approximately 300 million years ago. This collision formed the central and southern Appalachian Mountains and the pattern of jointing shows that the vector of the collision was oriented ~NW–SE.

However, the crust is heterogeneous, i.e., it contains a variety of rock types with different properties and geometries (e.g., sedimentary layering, circular or irregular igneous intrusions etc.) and these heterogeneities cause local perturbations of the regional stress fields. This is illustrated in Figure 2.157 which shows a map of the Spanish Peaks in Colorado, i.e., two

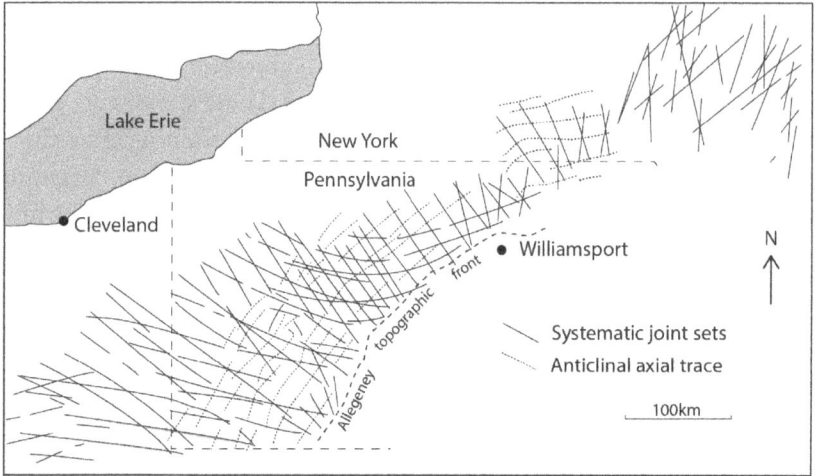

Figure 2.156. A fracture system covering several thousand square kilometres caused by the regional stress field generated by the collision between ancestral Africa and the North American plate in the Carboniferous period (~300 mya). From Engelder (1993).

volcanic necks situated just to the east of the Rocky Mountain front. During their intrusion, the stress field into which they were intruded was made up of the regional stress field linked to the formation of the Rocky Mountains (an ~E–W horizontal compression) and a local stress field generated around the two necks.

Fracturing occurred in response to this combined stress field and the fractures became in-filled with magma to form dykes. The distribution of the dykes is shown in Figure 2.157(a) and the pattern shows a remarkable correspondence to the $\sigma_1$ stress trajectories predicted from the stress analysis in which the two stress fields are combined, Figure 2.157(b).

The distribution of deformation around a buckle fold developed in a homogeneous, isotropic layer such as an unbedded sandstone or limestone is shown in Figure 2.158. The regional, horizontal, compressive stress field that generated the fold is modified by the local stresses which develop in the hinge regions. Above the neutral surface (i.e., the surface within the folded layer that experiences no deformation), the layer is stretched and below the neutral surface it is compressed. The fractures that form in these local regimes are shown in Figures 2.158(c) and 2.158(d).

(a)

(b)

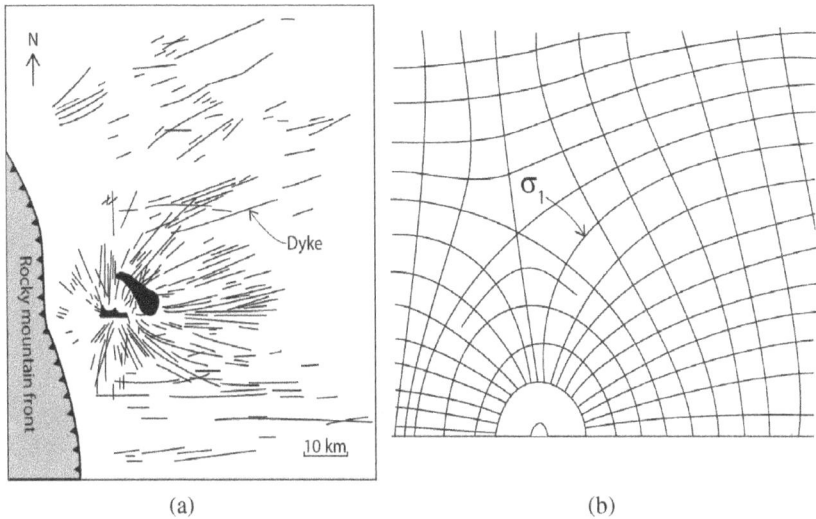

Figure 2.157. (a) Map of the Spanish Peaks, two volcanic necks and the associated dyke swarm, Colorado, USA. (b) Stress trajectories resulting from the superposition of the stress system around a volcanic centre on a regional E–W compressive stress field. After Ode (1957).

The formation of these fractures during folding can cause a dramatic increase in the permeability of the hinge regions and a considerable weakening of the rock mass and thus increases its permeability. The fracturing of fold hinge regions in this manner is typical of many folded areas and provides an example of the synchronous formation of ductile and brittle structures.

Another example of a pre-existing structure locally modifying a later stress field is that of a fracture. This modification of the stress causes fractures that formed, (i) later during the same stress regime or (ii) during a later stress field, to change their orientation as they approach the earlier fractures, Figure 2.159. This forms the basis of 'fracture analysis' which is discussed in detail in Section 2.12.3. In that Section, the gradual degradation of later stress fields caused by the evolving fracture network generated by the successive superposition of individual fracture sets was discussed.

This is illustrated in Figure 2.160 which shows the early fracture set made up of long, continuous, straight fractures and the gradual reduction in length and regularity of the later sets as progressively more and more

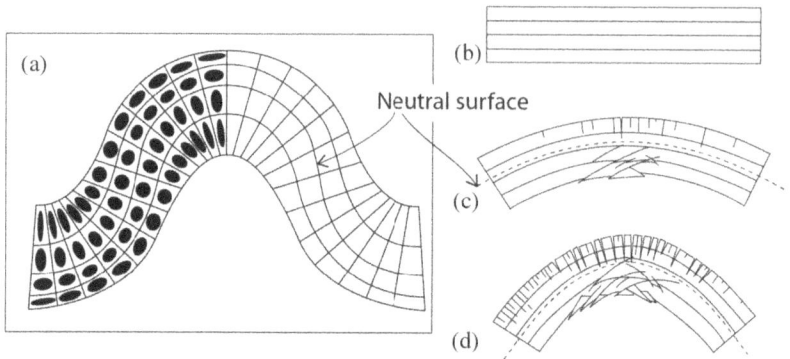

Figure 2.158. (a) The local extension and compression in the outer and inner arcs of a folded layer generates local stress fields which can form local fracture patterns, (c) and (d). From Ramsay (1967).

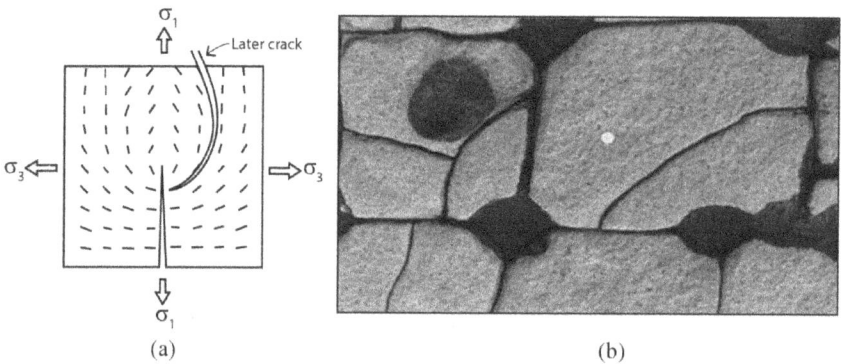

Figure 2.159. (a) Stress modification around a fracture tip and its impact on the trajectory of a later fracture. (b) Natural fractures in a limestone pavement from Lilstock, Somerset, showing the deflection of later fractures as they approach and abut against earlier fractures.

fracture sets were added to the point where eventually the regional stress field is so affected by the fracture network that the resulting fractures bear little relation to it.

The breakdown of the regional stress field by the fracture network that exists in the rock has been illustrated by numerical modelling and the results of some of these studies are shown in Figure 2.161.

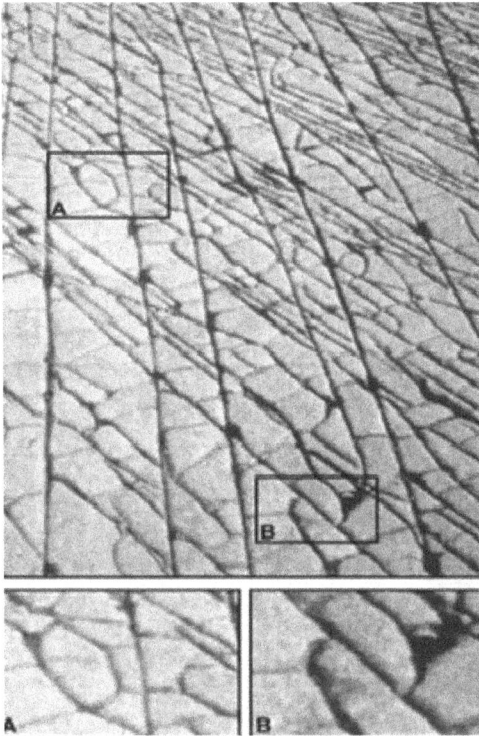

Figure 2.160. Fracture networks formed in the Liassic limestone of Lilstock, Somerset. The fracture sets become progressively less regular and less continuous as the fracture network develops — as a result of the increase in disturbance of the regional stress field.

Figure 2.161 from Valli *et al.* (2011) shows a vertical and horizontal section through a fractured rock mass in which a nuclear repository is to be constructed. The impact of the fractures on the regional stress field can be seen. The effect is to locally reorient and change the magnitude of the applied stress field. It is interesting to note that the impact of the fractures on the regional stress field generally reduces with depth. This is because, as the overburden and horizontal stress increase with depth, Figure 2.23, the normal stress across the fractures also increases. This enables the fractures to sustain progressively larger shear stresses, thus reducing the need for the regional stress to reorient in order to keep the shear stresses along the fractures low. It should be noted that the break-up of the stress field by the fracture network shown in these computer outputs is scale

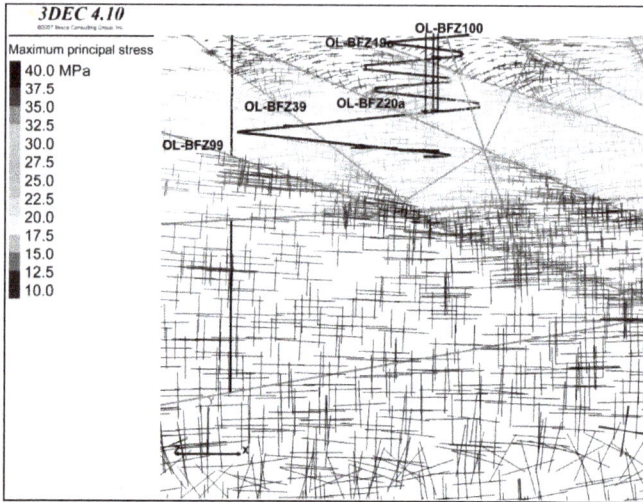

(a)

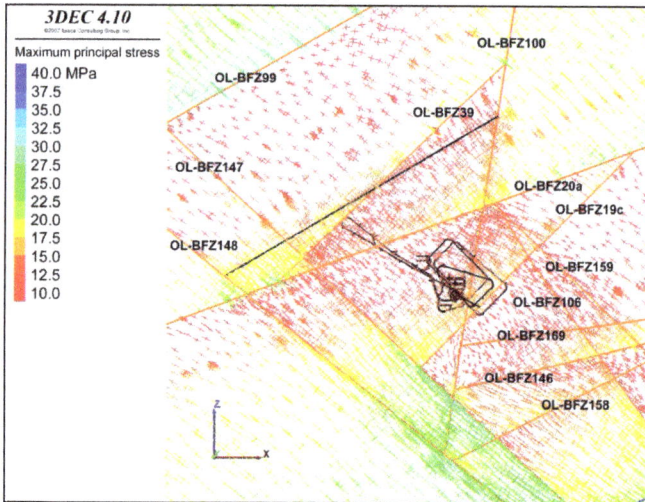

(b)

Figure 2.161.   Numerical modelling of the stress distribution in a fractured rock mass within which a nuclear waste repository is to be constructed in Finland. Diagram (a) is a vertical section and diagram (b) a horizontal section. The regional stress field applied is that caused by the present day plate motion in which the maximum principal compression acts approximately NW–SE with a magnitude of 25 MPa. The dramatic effect of the brittle deformation zones (annotated lines) on the magnitudes and orientations of the local maximum principal stress values is evident (from Valli *et al.*, 2011).

independent. The image in the bottom diagram of Figure 2.161 could be a few centimetres across or tens of kilometres.

Because of this interaction between the regional geological stress field and the fractures within the rock, the magnitude and orientation of the local stress fields can be very varied, particularly in the vicinity of a fracture. An appreciation of this raises questions about the significance of local stress measurements taken within a fractured rock mass with regard to both the far-field geological stress field that is acting on the rock and the state of stress elsewhere within the excavation site. In the same way, the presence of a fracture network also results in the rock properties becoming heterogeneous and, as with stress measurements, the significance of a locally determined rock property (e.g., strength, stiffness and hydraulic conductivity) with regard to the property of the rock mass as a whole is difficult to assess. However, the advent of advanced computer modelling and simulation, such as the 3DEC computer outputs in Figure 2.161, provides a method of approaching the links between the local and regional stress fields which has not been possible until relatively recently.

### 2.14.3 Present day stress field

All rocks in the crust are currently subjected to a stress field. This may just be the result of an overburden load when the rocks are situated in what is termed a 'tectonically relaxed region', i.e., one in which plate motion has little or no impact on the state of stress, or the rock may be affected by plate motion. This current stress regime may or may not be of sufficient magnitude to generate failure and the formation of a fracture set, but it will have an impact on the properties of any fracture network the rock contains. In particular, and as noted in Section 2.13, it will have an influence on the conductivity of the various fracture sets making up the network. All other things being equal, fractures normal to the current maximum principal compression, $\sigma_1$, will be less conductive than those parallel to it. In addition, as noted in Chapter 3, Principle 13, the orientation of the present day stress in the crust has a major influence on the planning of tunnel and excavation orientations.

## 2.14.4 Residual stress (the potential for future rock fracture formation)

There is evidence that rocks can contain within them 'locked-in stresses' relating to earlier stress regimes and that these stresses have the capacity to generate fractures during exhumation, i.e., during the removal of the confinement originally provided by the overburden and surrounding rock. They have been termed residual stresses and were first recognised when creep tests were being carried out on a series of Coal Measures rocks from South Wales. The results of one of the tests are shown in Figure 2.162. In describing this experiment, Price (1966) notes that at the instant the specimen was loaded it underwent instantaneous elastic shortening but, as can be seen from the graph in this Figure, almost immediately after the load was applied, the specimen began to expand. The rate of expansion was relatively rapid at first but, after the first seven days, settled down to approximately 0.7 microstrains ($\mu s$) per day.

When, after a period of 70 days, the specimen was unloaded, it expanded showing instantaneous and time-elastic recovery to a length 350 $\mu s$ longer than its original length. It remained unloaded for a further 70 days during which time it continued to lengthen (see Figure 2.162). The specimen was then subjected to the same vertical load that had been applied during the first loading cycle and again showed instantaneous elastic shortening. However, instead of expanding as it had in the previous cycle of the test, the specimen showed 'normal' primary and secondary creep, Figure 2.162, and when the specimen was subsequently off-loaded and elastic recovery was complete, it was some 80 $\mu s$ shorter than it had been immediately prior to the second application of load.

Price emphasises that the experimental procedure in both loading cycles was identical and that, consequently, the difference in the time–strain data must be attributed to some fundamental change in the test specimen. He argued that there was a release of strain energy stored in the rock specimen which occurred during and after the first period of loading, and that it was this release of energy which brought about the increase in dimensions of the specimen. The mechanism of storing this energy can be illustrated with reference to the experiment illustrated in Figure 2.163. The

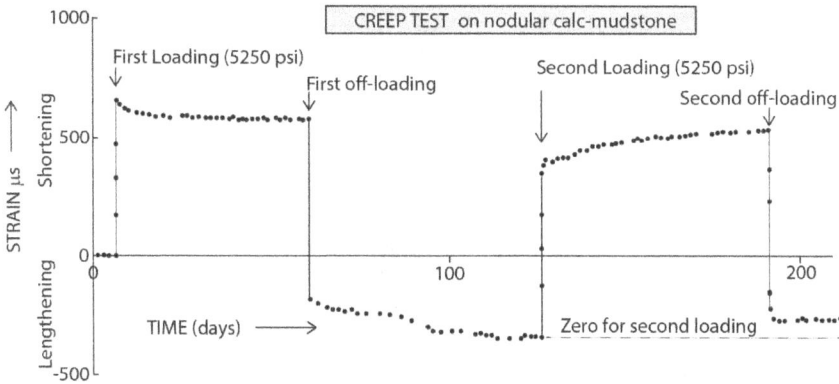

Figure 2.162.   Time–strain data obtained from a specimen of nodular muddy limestone showing expansion during and after the first loading and the normal time — strain behaviour during the second loading cycle. From Price (1966).

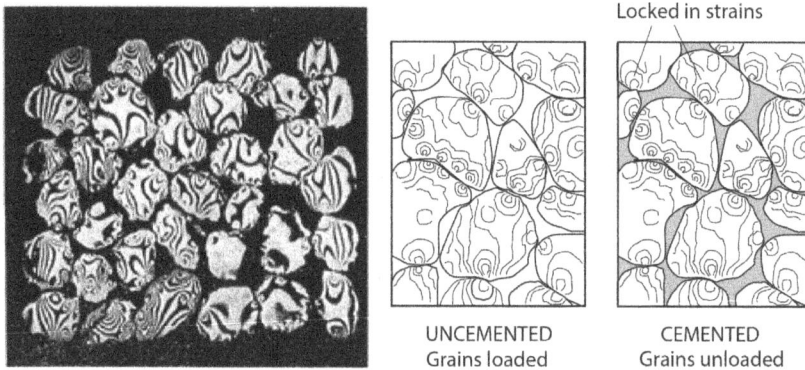

Figure 2.163.   The heterogeneous distribution of stress in a perspex model of a granular material caused by point loading at grain contacts, from Price (1966) and Friedman (1968).

grains of the sediment are represented by perspex analogues which are photoelastic and therefore show the stresses within them when viewed through polarised light. It is argued that during deposition and burial the vertical load on the grains increases and generates point contact stresses as illustrated. The elastic strains generated by these stresses are maintained by the overburden stress whilst the sediment is cemented. If the rock is subsequently brought to the surface by exhumation and the overburden stress

Figure 2.164. Exfoliation fractures in the granite tors of Slieve Binnian, Mourne Mountains Ireland. Photograph courtesy of Paddy Dillon.

removed, the cement tends to prevent the grains from expanding again locking the elastic strains into the rock.

However, as the rocks approach the surface, these residual stresses are often of sufficient magnitude to fracture the rock or the processes of weathering remove some of the confining cement resulting in the release of some of this stored strain energy and the development of sub-horizontal fractures which parallel the topography. These fractures are termed exfoliation fractures and are particularly well developed in massive igneous bodies such as granites, Figure 2.164 and in exhumed areas characterised by unstratified rocks. Exfoliation fractures are extensional fractures and occur in the upper 150 m of the crust. They tend to decrease in intensity with depth, Figure 2.165.

It should be pointed out that, in addition to the formation of new, sub-horizontal fractures, the process of stress release can also lead to the reactivation of geologically ancient fractures, both gently and steeply dipping. Small block rotation linked to this process could lead to the reactivation of ancient, steeply dipping structures. [Further pictures of the quarry shown in Figure 2.165 can be found in Chapter 5.]

Figure 2.165. Horizontal exfoliation fractures developed in the Grabinex granite quarry in Strzegom, Poland. Note the decrease in intensity (i.e., the increase in spacing) with depth. The quarry has exploited several of these fractures to act as the quarry floor at various stages in the evolution of the quarry.

The concordance between topography and exfoliation joints noted above has been used by Jahns (1943) to determine the amount of erosion caused by glaciation. He presents representative cross sections through the Chelmsford granite in Massachusetts, U.S.A., showing the relation between the exfoliation fractures and the pre-glacial topography, indicated by the dotted lines in Figure 2.166. The impact of glacial erosion can be clearly seen. More recently, Ziegler *et al.* (2014) noted that the granitic rock mass of the upper Aar valley (Grimsel area, Switzerland) contains distinct generations of exfoliation fractures, which formed during different stages of the Pleistocene, sub-parallel to distinct glacial valley palaeo-topography.

The bulk of the exfoliation joints show prominent, fractographic features: (1) radial plumose structures with distinct plume axes; (2) arrest marks superimposed by plumose striations; and (3) gradually developing *en echelon* fringe cracks, Figure 2.59. Multiple arrest marks reveal that exfoliation joints formed incrementally and, together with the absence of hackle fringes, suggest stable, i.e., subcritical fracture conditions.

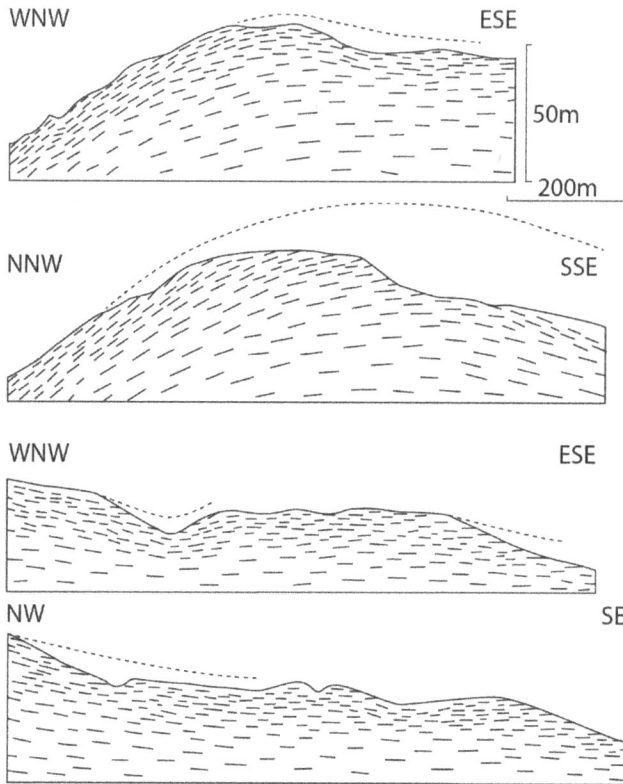

Figure 2.166. Representative cross sections of Oak Hill and adjacent hills in the Chelmsford granite N.E. Mass., U.S.A. showing the relationship between the exfoliation fractures (dashed lines) and the present-day topography. The dotted lines show the pre-glacial topography. (After Jahns, 1943.).

*****

This concludes our summary of the structural geology principles within the context of the 'structural geology–rock engineering' combined approach. The rock mechanics principles are presented in the next Chapter and these are followed in Chapter 4 by an illustrative case example (the Clifton Suspension Bridge in the UK) of the structural geology–rock engineering synthesis as applied to a real engineering stability problem.

# CHAPTER 3
# ROCK MECHANICS PRINCIPLES

## 3.1 INTRODUCTION

Following the previous Chapter on the structural geology principles, we now present a summary of rock mechanics principles. As explained in Chapter 1, the purpose of this book is to describe the synthesis of structural geology and rock mechanics, both for structural geology analyses and rock engineering design. So, in addition to the structural geology principles, it is necessary for this synthesis to also provide an understanding of rock mechanics principles. Accordingly, in this Chapter the basis of rock mechanics is presented in the form of 50 compact principles with associated illustrations — as a complement to the structural geology principles presented in Chapter 2. Readers interested in the development of rock mechanics over the 50-year history of the International Society for Rock Mechanics (ISRM) are referred to Brown (2012a).

## 3.2 FIFTY ROCK MECHANICS PRINCIPLES

### Principle 1:  Rock is a natural material

Rock is a natural material: its properties cannot be specified as with a fabricated material; the properties have to be measured on site or in the laboratory or back-calculated from rock behaviour during excavation.

Figure 3.1.   Jointed limestone, UK. The width of the rock mass illustrated is ~1.5 m across. Note that the joints are approximately normal to the bedding which dips steeply to the left.

The geological history of a rock mass will determine several key characteristics, as has already been illustrated in Chapter 2. These include:

— the presence and mechanical characteristics of the discontinuities (bedding and pre-existing fractures, Figure 3.1),
— the inhomogeneity (different properties at different locations),
— the anisotropy (different properties in different directions), and
— the hydraulic properties.

As a result of these various planar fabrics and other more complex textures, rock masses are not Continuous, Homogeneous, Isotropic and Linearly Elastic (CHILE) materials: they are Discontinuous, Inhomogeneous, Anisotropic and Not Elastic (DIANE) materials. Although it is tempting to study a rock mass as a CHILE material (because the analysis is simpler), the DIANE characteristics must be taken into account in rock engineering design. Furthermore and for underground engineering, the rock mass is pre-loaded — because of the vertical stress component due to the weight of the rock above and the horizontal stress components induced by this vertical load and by tectonic plate movement (see Section 2.4.1, Figure 2.23 and the related text). However, the advent of advanced numerical analysis applied to computer modelling and simulation has now made the inclusion of many of these features possible in analyses, noting the example in Section 2.14.

A common type of anisotropy in rocks is that represented by a planar layering or fabric. Such materials are termed transversely isotropic materials, see Principle 16 in this Chapter, and fall into two categories, namely primary and secondary. Bedding planes (Figure 3.1) are an example of the former being linked to the process of sedimentation which generates the rock. The second group of transversely isotropic rocks form later as a result of tectonism. This group includes fracture sets which form as a result of tectonism if the conditions are appropriate for brittle deformation, and rock cleavage or schistosity which develops when the rock responds in a ductile manner to the applied tectonic stress, see Section 2.10 for a more detailed explanation.

These ductile fabrics form as a result of the recrystallisation and alignment of the minerals making up the rock. Unlike the fracture planes mentioned above, which form in the upper crust where the temperature is relatively low and the rock behaves in a brittle manner, ductile fabrics require the application of a stress and an increase in temperature in order to form. However, whether or not the fabrics develop is determined by the mineralogy of the rock. For example, a sandstone is made up primarily of quartz grains ($SiO_2$). These have a stubby, tabular shape and therefore will not produce a good cleavage in the rock even when the crystals become aligned in response to a differential stress and high temperature. In contrast a mudstone containing clay minerals which have a platey geometry will respond to deformation by aligning within the rock so that they are all parallel. This imparts a macroscopic, planar fabric to the rock known as cleavage. The intensity of the fabric depends on the degree of deformation and metamorphism. Low grade metamorphism (300°–450°C and 200–1000 MPa) causes the clay minerals to recrystallise as micas which align within the rock to form a slate. As the metamorphic conditions increase in intensity, the mica grains increase in size and the rock becomes progressively coarser grained, passing through what is described as a schist, and finally a gneiss.

## Principle 2: Intact rock

The nature of the intact rock will depend on its geological type, as discussed in Chapter 2 and the Principle 1 text above, i.e., whether sedimentary, igneous or metamorphic and the degree of weathering/ alteration to which it has been subjected. Rock mechanics started with detailed studies of the deformation and failure of intact rock and the engineering properties of intact rock are functions of the rock microstructure (Figure 3.2) which in turn is a function of the geological formation and history (see Principle 1).

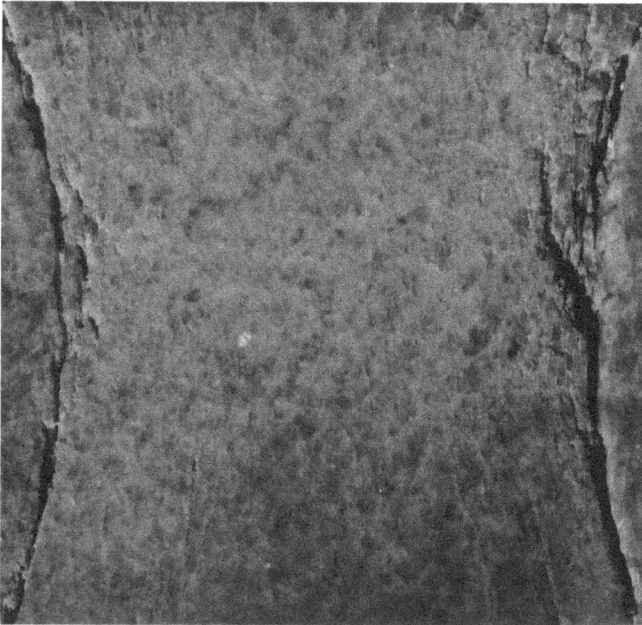

Figure 3.2.  Cut and polished cross-section of a 1:1 H:W ratio (100 mm) Georgia Cherokee marble specimen loaded vertically in compression to an advanced state of failure.

## Principle 3: The complete stress–strain curve for intact rock

When testing intact rock specimens in the laboratory to simulate the effects of stresses and to determine the rock strength properties, it is necessary to consider whether the *in situ* loading conditions are 'soft' or 'stiff'. Figure 3.3(a) illustrates these two conditions. Any structure built at the ground surface and hence exerting a vertical load on the ground will continue to exert that load if the ground fails. This is termed a 'soft' loading system. On the other hand, although the excavation of a tunnel will alter the local stress conditions and create stress concentrations (Figure 3.3(a)), the surrounding rock will generally not continue to move into the tunnel unless it is weak. This is termed a 'stiff' loading system.

Therefore, when testing rock specimens in the laboratory, it is better to use a testing machine that simulates the 'stiff' condition, rather than one

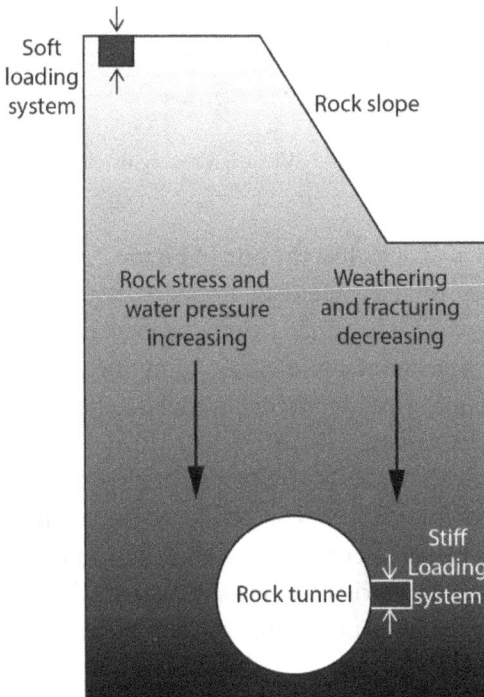

Figure 3.3(a).   Soft and stiff *in situ* rock loading conditions.

Figure 3.3(b).   Victorian testing machine.

that is inherently 'soft', e.g., like the one in Figure 3.3(b) — which is inherently soft because of the system of levers used to apply the load. The required stiffness can be obtained simply by increasing the stiffness of the testing machine and reducing any soft components, or by the use of a servo-controlled machine which can simulate the stiffness through an electronic feedback system.

Thus, using a stiff or servo-controlled testing machine, the complete stress–strain curve for intact rock in uniaxial compression can be obtained (Figure 3.3(c)). The test is conducted with axial strain as the independent variable (the controlled variable) and axial stress as the dependent variable (the measured variable). The complete stress–strain curve represents the structural collapse of the rock microstructure from initial loading to complete disintegration. The most widely-used characteristics of the complete stress–strain curve are the elastic modulus ($E$, measured at 50% maximum

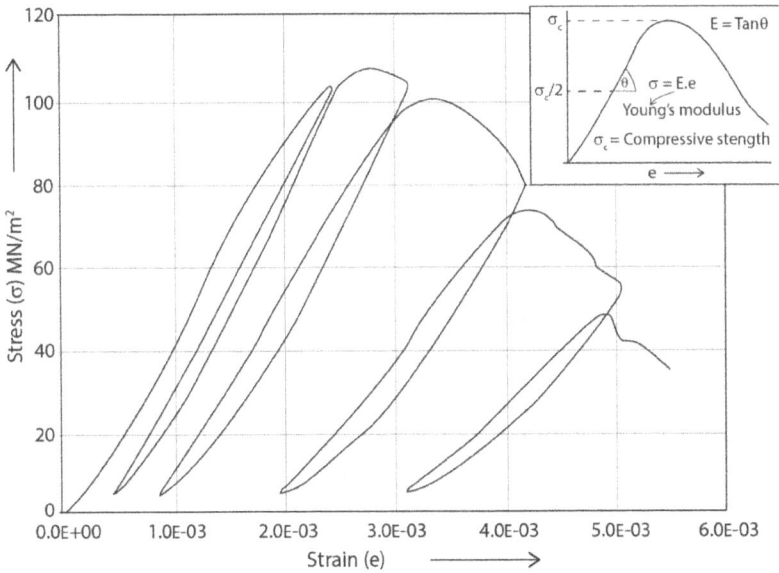

Figure 3.3(c). Complete stress–strain curve for marble with unloading and reloading cycles (from École Polytechnique Fédérale de Lausanne (EPFL) rock mechanics laboratory, Switzerland).

stress) and the compressive strength ($\sigma_c$, the maximum stress sustained). After the peak stress, the rock specimen suffers increased microstructural degradation as the applied strain increases and the capacity of the specimen to sustain the stress decreases, Figure 3.3(d).

Numerical models of the type shown in Figure 3.3(d) show the formation and organisation of the individual extensional microfractures and provide an insight into the mechanism by which macroscopic shear and extensional fractures develop. This is discussed in detail in Sections 2.10.3 and 2.10.4.

Figure 3.3(d) contains images from a computer simulation (using the RFPA — Rock Failure Process Analysis code) of a uniaxial compression test on a rock sample. There is minor axial splitting damage occurring up to Image (e) at 100% peak stress, i.e., at the compressive strength, and then further increasing damage in Images (f) → (j) as the complete stress–strain curve descends in the failure region. Note how the multitude of axial splitting microcracks gradually develops into a macroscopic shear zone.

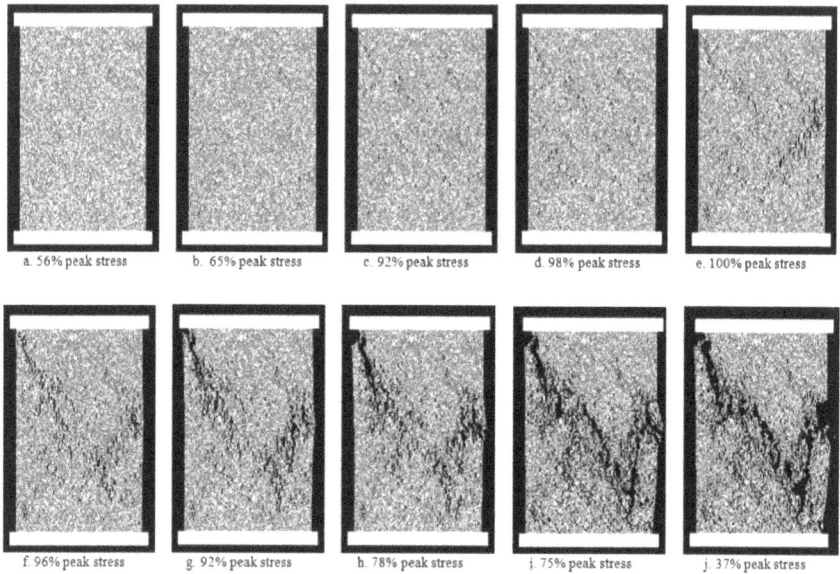

Figure 3.3(d).  Computer modelling of rock specimen microstructural degradation as the strain is increased during uniaxial compression applied vertically, from Tang and Hudson (2010).

## Principle 4:  Measured rock properties are not material properties *per se*

The complete stress–strain curve for intact rock, e.g., Figure 3.3(c), depends on the specimen geometry, loading conditions of the test and the environmental conditions, and it will be recalled from Section 2.3.2 and Figure 2.12 that the compressive strength of a rock is also influenced by the confining stress under which the strength test is carried out. However, a material property does not depend on these factors. So, neither the compressive strength nor the tensile strength are actual material properties — because they both depend on the specimen geometry and the loading conditions of the test.

In Figure 3.4, the results of a computer simulation of testing rock specimens in uniaxial compression are illustrated. It can be seen that the compressive strengths obtained for each height:width (H:W) specimen ratio are different, the associated complete stress–strain curves are different, and

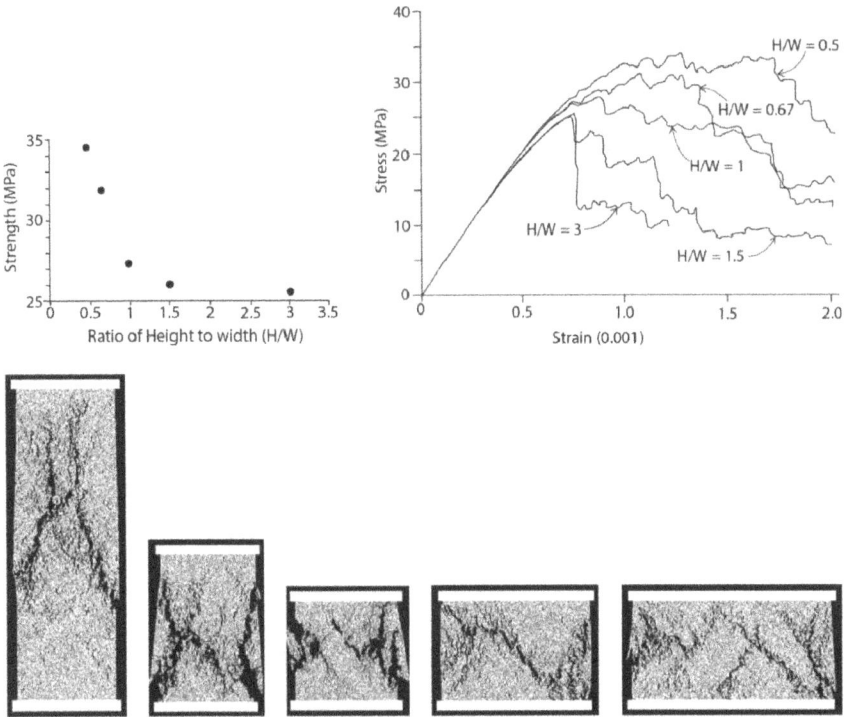

Figure 3.4. Computer simulation (using the RFPA code) of axially loaded rock specimens having the same microstructure but different H:W ratios. Each specimen exhibits a different compressive strength (top left graph) and a different type of complete stress–strain curve (top right graph). From Tang and Hudson (2010).

the detailed specimen failure modes are different. Thus, the compressive strength values obtained are not material properties but properties of the rock and the testing system, and hence are experimental properties. A material property is just that, a property of the material: its value should not depend on how it is obtained.

## Principle 5: Effect of confining pressure

When a confining pressure is also applied during a compression test on intact rock, the rock will exhibit either brittle or ductile behaviour. In brittle behaviour, the stress decreases after the compressive strength has been

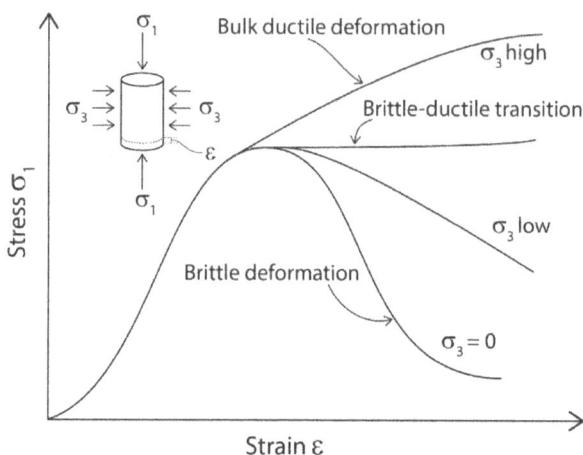

Figure 3.5.    The effect of confining pressure ($\sigma_3$) on the stress–strain curve for an intact rock specimen. Depending on the confining stress, some samples yield brittle deformation plots, and others yield plots characteristic of bulk ductile deformation.

reached (e.g., Figure 3.3(c)); in ductile behaviour, the stress continues to increase (Figure 3.5). The confining pressure associated with the brittle-ductile transition is, for example, 0 MPa for rock salt, 20–100 MPa for limestone, and more than 100 MPa for sandstone and granite.

It should be noted that in this discussion the classification of rock into either brittle or ductile is based on the macroscopic behaviour of the rock sample which is captured by the stress–strain curves shown in Figure 3.5. In Section 2.10 which outlines the geologist's approach to the brittle-ductile transition, the classification is linked to the microscopic behaviour of the rock. These two appropriate but different approaches to the transition are thus discussed in this Section and also in Sections 2.10.3 and 2.10.5.

## Principle 6: Rock failure criteria

The most widely used failure criteria for intact rock are the Griffith, Navier–Coulomb (as used in Chapter 2, Section 2.3) and the Hoek–Brown failure criterion. The Griffith criterion considers the energy required by a propagating crack in terms of an initial crack length. The Navier–Coulomb criterion considers the cohesion and angle of friction associated with shear

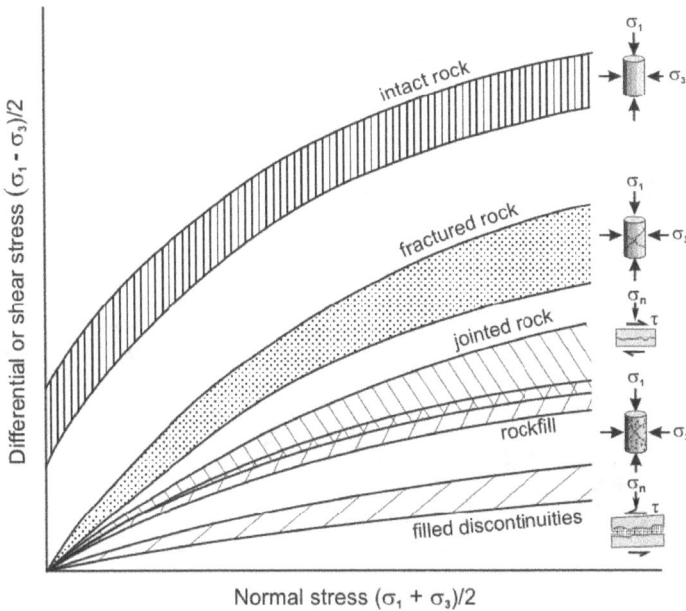

Figure 3.6. Differential shear stress versus normal stress at failure (prepared by N.R. Barton). Cover diagram for the 'Orange Book' published by Springer Science in 2014 and which contains all the ISRM Suggested Methods generated in the period 2007–2013, including several on rock failure criteria.

failure. The Hoek–Brown criterion is an empirical criterion using two parameters which can be estimated from the rock description.

Many other failure criteria for intact rock have been developed, see the ISRM Suggested Methods papers in Issue 45 of the Rock Mechanics and Rock Engineering journal published in 2012 and in the ISRM 'Orange Book' (Ulusay, 2014). The failure of fractured rock can be complex and the failure envelopes are different for different rock circumstances, as shown in Figure 3.6.

## Principle 7: The presence of rock discontinuities

In most practical cases, the properties and engineering behaviour of rock masses are governed by the discontinuities — which are any breaks in the mechanical rock continuum. These can occur on a variety of scales, from

Figure 3.7(a). Bedding planes dipping gently to the right and fractures sub-normal to the bedding in limestone strata. Such discontinuities often act as channels for water flow which has resulted in a reduction of their mechanical integrity.

micro-fissures to fissures to bedding planes to joints and faults (see Chapter 2 and Figures 3.7(a)–3.7(c) below). The bulk properties of a fractured rock are scale-dependent, see Figure 3.7(b), Section 2.13.4 and Figure 2.142 and the related discussion, plus Section 2.13.7.

The most important discontinuities for engineering are faults or other shear features. This is because, both during the formation of a shear fracture and during its subsequent evolution, there is a progressive degradation of the surrounding rocks and the generation of a zone of deformation, a fault zone, see Figure 3.7(c), plus Sections 2.10.6.1 and 2.10.6.2, particularly Figures 2.118 and 2.119. Joints which have been created by normal tensile stresses can be significant as well. Note that discontinuities have little or zero tensile strength.

Shear failure is not restricted to purely brittle deformation and, as discussed in Section 2.10 on the brittle-ductile transition, localised shearing can occur over a wide range of conditions, see Figure 2.10.1. As the deformation moves from purely brittle towards ductile, the shear

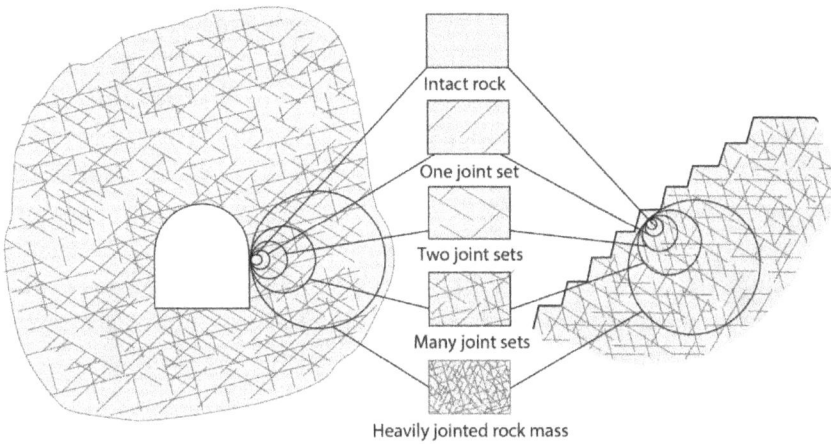

Figure 3.7(b). The transition from intact rock to heavily jointed rock with increasing sample size, from Hoek and Brown (1997).

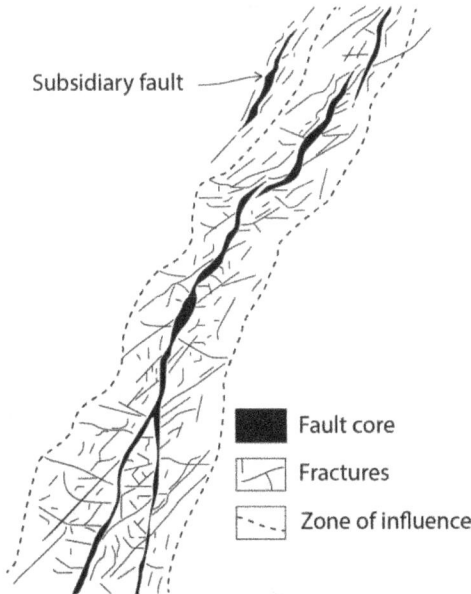

Figure 3.7(c). The geometry of larger discontinuities such as a fault zone can be complicated by their internal structure. (From the Swedish Nuclear Fuel and Waste Management Co., SKB, Sweden).

instability widens into a shear zone. Although these zones may not contain discrete planes of failure, they often develop intense ductile fabrics (see Figure 2.70(b)) which are planar and which allow hydrothermal fluids and/or weathering agents to invade the zones. This results in the alteration and weathering of the rock mass along the shear zone, processes that can cause considerable weakening. These ductile fabrics within shear zones are discussed in Section 2.7.5 where their formation and their use as kinematic indicators are considered, and in Section 2.10.2 where it is shown that shear zones containing such fabrics are part of a spectrum of geological structures which lies between the brittle and ductile end members.

## Principle 8: The characteristics of rock discontinuities

Ten main characteristics are used to describe discontinuities: spacing, orientation, persistence, roughness, aperture, number of sets, block size, filling, wall strength, and seepage. These are illustrated in Figure 3.8(a) below.

Unfortunately, it is difficult to obtain the ten characteristics shown in Figure 3.8(a) because most information about a potential rock engineering site comes from boreholes, Figure 3.8(b). Some of the characteristics

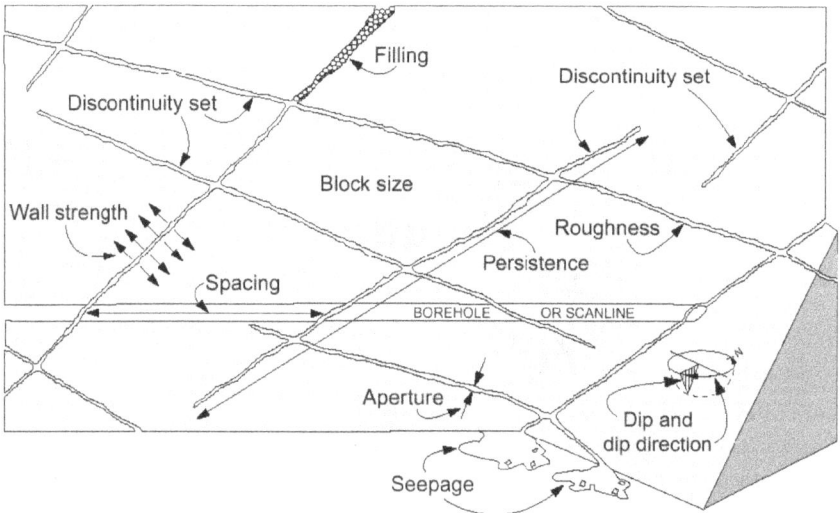

Figure 3.8(a).  The 10 key rock fracture characteristics for engineering purposes.

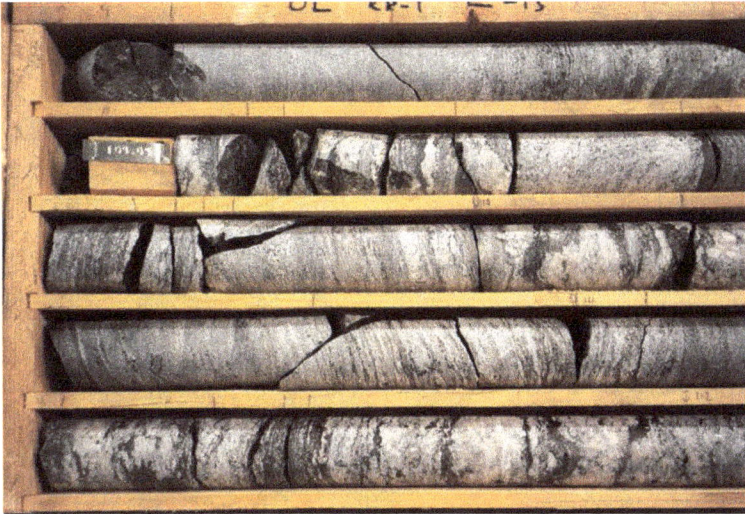

Figure 3.8(b). Portion of a corebox containing rock core from an exploratory borehole during a site investigation in a rock mass — illustrating the difficulty of obtaining the characteristics shown in Figure 3.8(a).

(such as the fracture spacing and the dip and dip direction of the fractures) can readily be obtained from the borehole core, but others have been affected by the drilling process. For example, although fracture aperture can sometimes be measured from a core, the confining stress under which the fracture existed in the *in situ* fractured rock mass is no longer present when the fracture is in the corebox. Consequently, the measured aperture is likely to be significantly greater than the actual aperture when the fracture is *in situ*.

## Principle 9: The Rock Quality Designation (RQD)

The most widely used rock mechanics parameter to describe discontinuity occurrence for engineering purposes is the RQD. This is the percentage of pieces that are greater than 100 mm or 4 inches in a borehole core or lengths along a scanline on the rock surface, Figure 3.9. The RQD can be related to the discontinuity frequency if the nature of the discontinuity spacing histogram is known. For a negative exponential distribution of spacings along a scanline (which is often the case, see Section 2.8.3) the

RQD can be estimated as RQD $= 100e^{-0.1\lambda}$ $(0.1\lambda + 1)$, where $\lambda$ is the fracture frequency and the reciprocal of the fracture spacing in metres.

Note that a negative exponential distribution of fracture spacings can be a result of an apparent random distribution of spacings, which in turn can be the result of the superimposition of different fracture sets, each with its own specific distributions of spacings — which is why the fracture spacings often appear to be randomly distributed along the borehole core or along a scanline on a rock exposure, see Section 2.8.3.

## Principle 10: Discontinuities tend to occur in parallel sets

Because discontinuities tend to occur in sets (of parallel or sub-parallel discontinuities, Figures 3.10(a)–3.10(c) and Figures 2.24, 2.28, 2.33 and the related discussions) the discontinuity frequency value is different along lines in different directions through a rock mass. It follows that the RQD (Principle 9 above) will also be different in different directions through the rock mass.

## Principle 11: Rock discontinuity persistence

The persistence, or extent of a discontinuity, is an important characteristic for many engineering considerations, such as the modulus of deformation of the rock, the degree to which rock blocks are formed, and the hydraulic connectivity of the discontinuity network (Figure 3.8(a)). The roughness, aperture and filling of the discontinuities are also important for the mechanical and hydrological features.

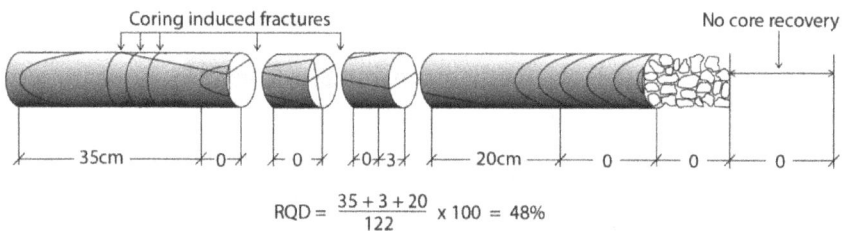

$$RQD = \frac{35 + 3 + 20}{122} \times 100 = 48\%$$

Figure 3.9. Example of an RQD estimation from a core sample of total length 122 cm, from González de Vallejo and Ferrer (2011).

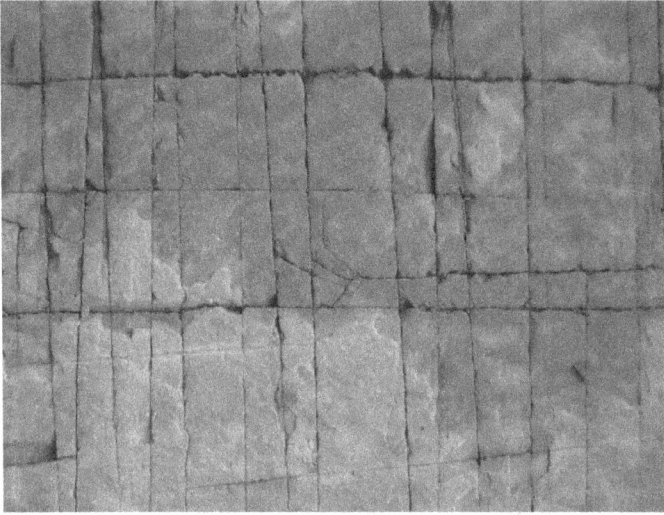

Figure 3.10(a). Two orthogonal sets of sub-parallel fractures in Carboniferous limestone, South Wales, UK. Image represents a width of ~1 m.

Figure 3.10(b). Several sets of sub-parallel fractures in a Carboniferous stratum, South Wales, UK.

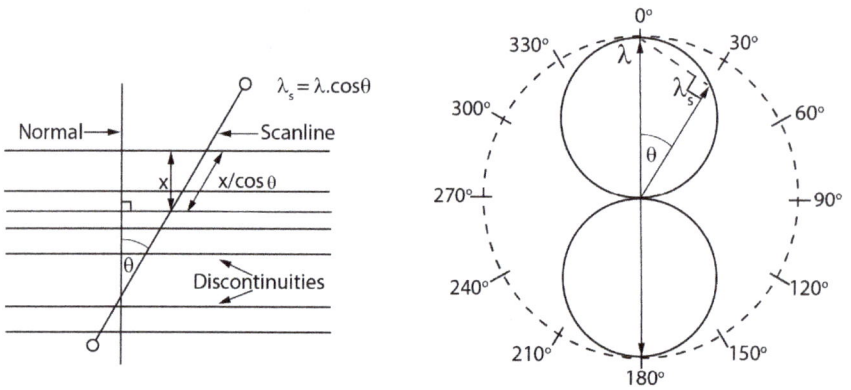

Figure 3.10(c).  The discontinuity frequency/m ($\lambda_s$) will be different along scanlines at different angles to a fracture set with frequency $\lambda$ where this refers to the frequency along a line perpendicular to the fracture set. Knowing the orientation and individual frequency of a fracture set, its frequency along a scanline or borehole at any orientation can be calculated. The left diagram shows how the spacing value depends on the orientation of the scanline with respect to the discontinuities; the right diagram show the locus of the fracture frequency as the scanline is rotated through 360° for the case of a single fracture set. The same procedure can be used in 3D for a rock mass containing several fracture sets by adding the components from each fracture set and plotting the overall fracture frequency on a stereographic projection, see the example in Figure 3.10(d).

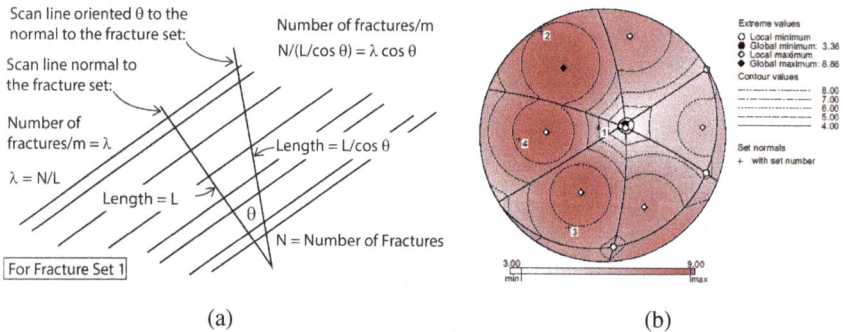

(a)

(b)

Figure 3.10(d).  (a) Resolution of the fracture frequency for one set. (b) Example of a lower hemisphere stereographic plot of the fracture frequency along all directions through a rock mass containing four discontinuity sets (prepared by J.P. Harrison).

## Principle 12: The stiffness and strength of rock fractures

The main mechanical properties of a fracture for engineering analyses are the stiffness and strength. The stiffness is considered via the normal (i.e., perpendicular) elastic stiffness plus the two elastic shear stiffnesses (upper part of Figure 3.12(a)). The strength is specified by the shear strength, i.e., the cohesion, C, and angle of friction, $\phi$, the tangent of which is the coefficient of sliding friction, see Equation (2.11) (lower part of Figure 3.12(a)). The discontinuity may have essentially no tensile strength, and is also often assumed to have little or no cohesion. The angle of friction is a complex combination of the basic friction angle, the strength of the asperities and the discontinuity roughness. Note that 'adhesion' refers to resistance against a tensile stress; but 'cohesion' refers to resistance against a shear stress.

The influence of fracture infill can be significant and the vein material will clearly have an effect on the elastic and failure properties of the

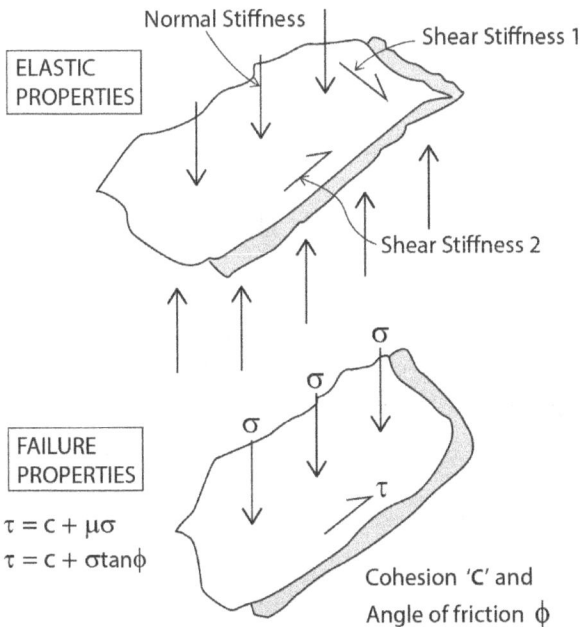

Figure 3.12(a).   The normal and shear 'elastic' stiffnesses of a rock fracture (upper diagram); the cohesion and friction failure properties of a rock fracture (lower diagram).

Figure 3.12(b).    Infilled fracture in a granodiorite rock core. The lineations which can be seen faintly at the top of the core are the result of *in situ* shear sliding along the fracture.

fracture. It might be expected that the infilling of a fracture with minerals to produce a vein would increase its strength. Indeed, this is generally so for the tensile strength which, as noted above, for a barren fracture tends to be very low. However, this is not always the case for shear strength. The effect of the vein material is to infill the asperities along the fracture walls which, in a barren fracture, tend to inhibit shear and therefore increase its shear strength. Once the vein has formed, the strength of the fracture will be determined by the shear strength of the vein material. Thus, if a fracture is infilled with a weak mineral such as epidote, chlorite or calcite, it is likely to have a lower shear strength than the original barren fracture.

An example where this has occurred is shown in Figure 3.12(b) which shows a granodiorite core cut by a vein of epidote and chlorite within which there is clear evidence of shear. The vein is sufficiently thick to ensure that the asperities along the fracture play no part in determining its shear strength — which is determined solely by the shear strength of the infilling material.

## Principle 13: *In situ* rock stress

A rock mass contains a pre-existing natural stress state, the *in situ* stress, which is caused by the overburden weight and tectonic processes, see Sections 2.4.1 and 2.4.2. As explained in Chapter 2, Section 2.2, the quantity 'stress' is not a scalar nor vector quantity, but a second order tensor quantity which has to be characterised by six independent values, usually the magnitudes and directions of the three principal stresses. These rock stresses are mainly caused by the overburden and current tectonic activity but old, residual stresses can also be present, see Section 2.14.4. In an underground rock engineering project, the natural *in situ* stresses are perturbed by the creation of openings — which increases the magnitudes of the stress components in some regions and decreases it in others, see Section 2.3.3, particularly Figures 2.15 and 2.16.

In Figure 3.13(a), the peripheral stress component around a mine opening (which has been caused by both vertical and horizontal stress components, see Figure 2.15) has been magnified by the creation of the opening to such an extent that the rock around the opening has been significantly fractured, a process known as spalling. In Figures 3.13(b) and

Figure 3.13(a).   Rock splitting due to high stresses concentrated in the roof around a mine opening (Pyhäsalmi zinc and copper mine, Oulu Province, central Finland).

Figure 3.13(b).   Rock spalling due to high stresses in the wall of a tunnel at the Jinping II hydroelectric project in China (from Hudson and Feng (2015)).

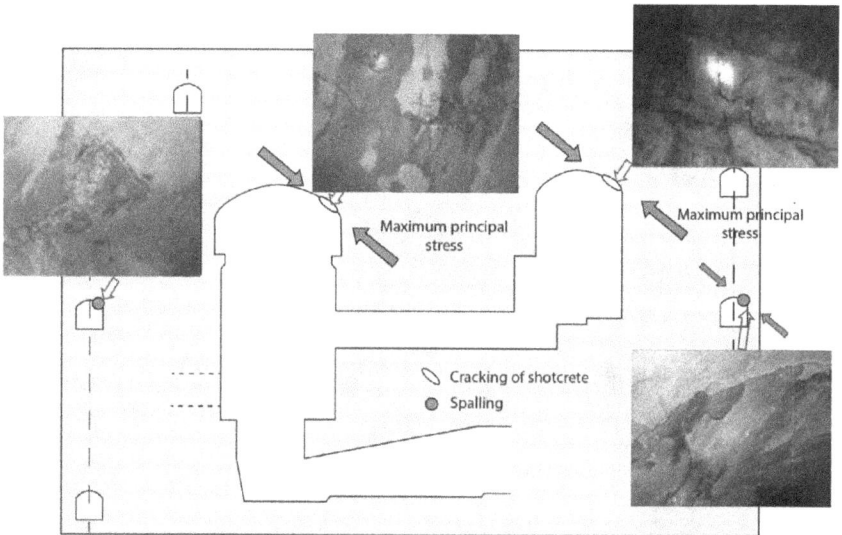

Figure 3.13(c).   Stress concentrations and associated damage at different locations in the excavation-peripheral walls of the electricity generating chambers at the Jinping II hydroelectric project in China. The dark arrows indicate regions of concentrated stress (from Hudson and Feng (2015)).

Figure 3.13(d).   Rock discing in a borehole core caused by a high *in situ* rock stress and by the process of drilling.

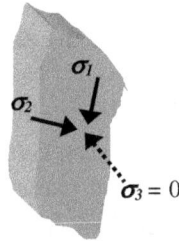

$$\begin{bmatrix} \sigma_1 & 0 & 0 \\ 0 & \sigma_2 & 0 \\ 0 & 0 & 0 \end{bmatrix}$$

Figure 3.13(e).   At all unsupported rock surfaces, whether natural or artificial, the principal stresses are orientated perpendicular and parallel to the surface. This is because the shear stresses along the surface are zero — and this is the definition of a principal stress plane.

3.13(c), the same process has caused fracturing around one of the tunnels and in the electricity generating chambers of the Jinping II hydroelectric project in China. In Figure 3.13(d), the concentrated *in situ* stress caused by the drilling of a cored site investigation borehole in a highly stressed rock mass has created a series of regular, induced fractures, a process known as core discing.

The stress state at unsupported rock surfaces, whether natural or artificial, is as shown in Figure 3.13(e). The pre-existing, natural, principal

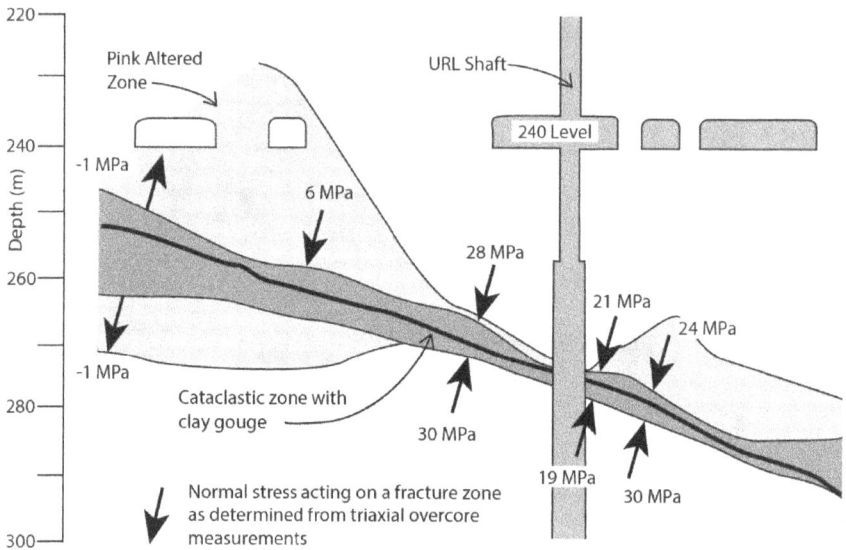

Figure 3.13(f). Stresses normal to a fault measured at the Underground Research Laboratory in Canada.

stresses are altered because the stress component normal to the surface becomes a principal stress — because there can be no shear stress on the surface, and this is the definition of a principal stress plane (see Sections 2.12.2 and 2.2.3 respectively). In general, the major principal stress acting parallel to the plane will then be greater than the natural, pre-existing major principal stress. The presence of a fault, can have a similar effect (see Figure 3.13(f)). Although the stress state along the fault will vary because of the idiosyncrasies of the fault geometry, aperture and stiffnesses, the ability to sustain a shear stress parallel to the fault will generally be considerably less than that of the intact rock. This will be particularly so at times during the geological history or during the development of an engineering site when the fluid pressure, $p$, within the fault is increased. The normal stress, $\sigma_n$, acting across the fault plane is reduced to an effective stress, $\sigma_n - p$, reducing its ability to support a shear stress (see Principle 15 and Section 2.5 and Figure 2.30). This causes a deflection of the stress trajectories in the adjacent rock mass until, in the limit, when the fluid pressure equals the normal stress across the fault and its tensile strength, the fault acts as a free surface and the principal stresses in the adjacent rock become normal and parallel to the fault plane, Figure 3.13(e) and Chapter 2, Section 2.12.2, Figures 2.132 and 2.133 and the related discussion.

## Principle 14:  Measuring and plotting *in situ* rock stress

There are four main methods of measuring the *in situ* stress: the flat jack, hydraulic fracturing, the USBM type of overcoring torpedo, and the CSIRO type of overcoring gauge. The first two methods use hydraulic fluid; the second two use strain gauges. The CSIRO type of gauge is the most reliable and hydraulic fracturing can be used significant distances from 'person–access'. It is the rule, rather than the exception, that the maximum horizontal stress component is greater than the vertical stress component — because of tectonic plate movement. Fractures have a significant effect on the local principal stress magnitudes and directions (see Figure 2.161), so the *in situ* stresses are expected to vary at the project location.

In addition to the complications such as those illustrated in Figure 3.14(a), the fact that rock stress is a tensor quantity with six independent components makes it difficult to illustrate the changes in rock stress at a point in the rock mass during the development of an underground engineering project in which rock is removed. In Figure 3.14(b), the changes in the directions and magnitudes of the three principal stresses, $\sigma_1$, $\sigma_2$ and $\sigma_3$, for a hypothetical case are shown: the paths of $\sigma_1$, $\sigma_2$, and $\sigma_3$ across the hemispherical projection are shown on the left, remembering that at each step they must remain mutually orthogonal; and, on the right, the changing magnitudes of the principal stresses are indicated.

Remember that a rock stress field can only be described through the specification of six independent quantities because the stress tensor has six independent components. Thus, a statement such as, "The rock stress is 6 MPa" has no meaning without further qualification.

## Principle 15:  The effective stress in a rock mass

Water can be present in the pores of the intact rock and in the discontinuities. As described in Chapter 2, Section 2.5.3, the water pressure in a rock mass is subtracted from the normal stress components of the stress tensor

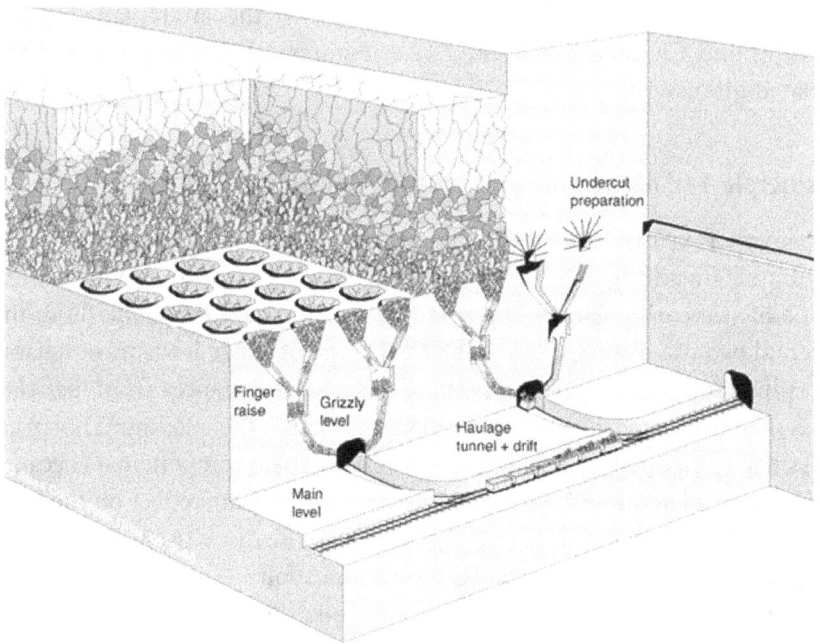

Figure 3.14(a). The transmission of the rock stresses through a block caving mining project will be complex and difficult to model.

A hemispherical projection allows evolution of orientation to be shown but not magnitude. However, at any given stage of the stress path the principal stresses can be seen to be orthogonal.

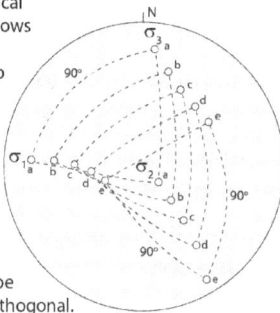

Complete visualisation requires a composite diagram. Here, evolution of the stress path is again shown by the letters **a** through **e** but now the orientation and the magnitude are shown in terms of position on the hemispherical projection and height above it.

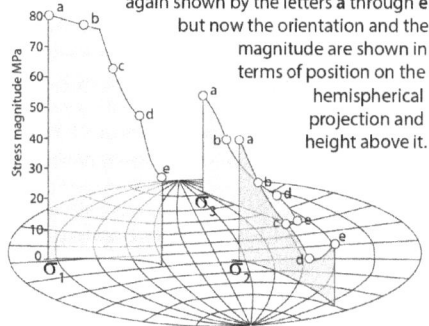

Figure 3.14(b). An attempt using plots on a stereographic projection to illustrate the changing orientations and magnitudes of the three principal stresses $\sigma_1$, $\sigma_2$, and $\sigma_3$ at a point in the rock mass during the development of an underground engineering project in which rock is removed (compiled with the assistance of J.P. Harrison).

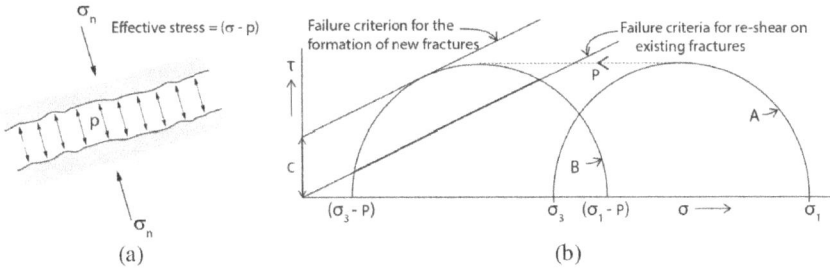

Figure 3.15. (a) Water pressure, P, acting inside a fracture and (b) the consequential movement of Mohr's circle to the left in $\sigma$–$\tau$ space. The original stress state defined by $\sigma_1$ and $\sigma_3$ is stable and unable to cause the formation of new fractures or slip on a pre-existing fracture. In contrast, the effective stress state can cause fracturing of the intact rock or slip along a pre-existing fracture. After González de Vallejo and Ferrer (2011).

to give the effective stress, Figure 3.15. The effective stress concept is used widely in soil mechanics but less so in rock mechanics for engineering — because rocks are fractured and do not generally have the regular particulate nature of soils. However, when the potential slip on a single fracture is being considered, the effective stress is a key concept, Figure 3.14 and Section 2.6.7.

The effect of a fluid pressure in a rock is to reduce the lithostatic stress ($\sigma_1$ and $\sigma_3$) to an effective stress (($\sigma_1 - P$) and ($\sigma_3 - P$)). This change in stress state is shown in Figure 3.15 and results in the Mohr circle representing the lithostatic stress being moved to the left by an amount P, the fluid pressure. In this way a stable stress field represented by Mohr circle A can be changed to an unstable stress state, Mohr circle B, capable of causing failure. If the rock is unfractured, new fractures will form when the effective stress satisfies the failure criterion for intact rock. If the rock contains suitably oriented fractures, then failure (slip) will occur on the fracture, see also Section 2.6.7, Figure 2.54.

## Principle 16: The elastic properties of rocks

Strain is a second order tensor quantity like stress. Assuming the rock is elastic, the six components of the strain tensor can be related to the six components of the stress tensor through the elastic compliance matrix. As shown in Figure 3.16, for complete isotropy (symmetry in all three

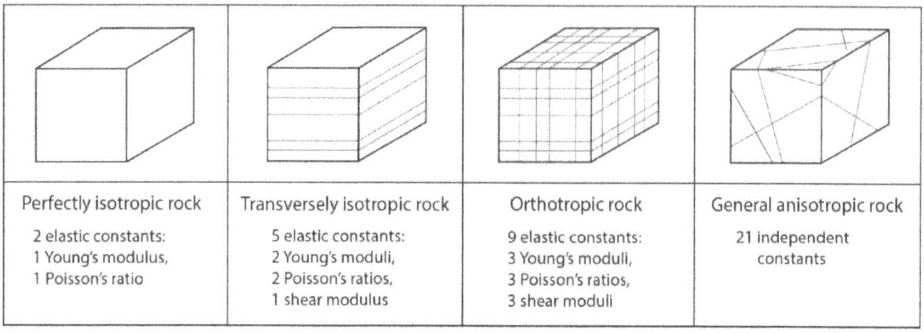

| Perfectly isotropic rock | Transversely isotropic rock | Orthotropic rock | General anisotropic rock |
|---|---|---|---|
| 2 elastic constants:<br>1 Young's modulus,<br>1 Poisson's ratio | 5 elastic constants:<br>2 Young's moduli,<br>2 Poisson's ratios,<br>1 shear modulus | 9 elastic constants:<br>3 Young's moduli,<br>3 Poisson's ratios,<br>3 shear moduli | 21 independent<br>constants |

Figure 3.16.   The four main types of isotropy/anisotropy considered in the theory of elasticity.

dimensions), this reduces to two elastic constants: the familiar Young's modulus and Poisson's ratio. However, for transverse isotropy (symmetry in two of the three dimensions), five elastic constants are needed: two Young's moduli, two Poisson's ratios and a shear modulus. For orthotropy, nine elastic constants are required: three Young's moduli, three Poisson's ratios, and three shear moduli. For complete anisotropy, the 21 independent constants of the elastic compliance matrix are required.

By far the greatest number of analyses related to rock engineering are conducted assuming the rock is isotropic. Sometimes the transverse isotropy parameters are used when the rock microstructure is clearly layered or when the rock mass has a single or dominant set of sub-parallel fractures. The orthotropy assumption is rarely used and, until recently, no one had ever estimated or measured the 21 elastic constants for a completely anisotropic rock. However, in a paper presented at the 2015 ISRM Congress held in Montréal, Canada and titled "Acoustic Emission Monitoring of the Fracturing Process of Migmatite Samples" M. Petružálek, T. Lokajíček and T. Svitek describe a microseismic technique using rock spheres by which this can be done. It is unlikely that the 21 independent elastic constants will ever be used in the practical design of any rock engineering project, but it is satisfying that the use of the full range of elastic symmetries can now be accessed.

## Principle 17:  Rock is not a perfect material; CHILE vs. DIANE

Principle 1 stated that rock is a natural material; one of the ramifications is that rock is not a perfect material. The 'ideal' rock mass is a CHILE

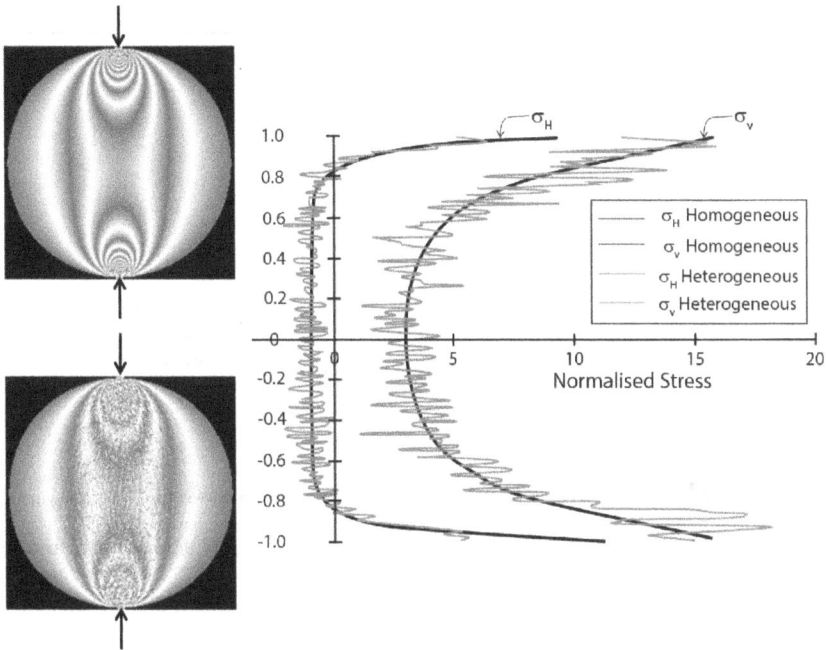

Figure 3.17(a). Left images: computer simulation comparison of the stress variation within a homogeneous and inhomogeneous rock specimen during the compressive loading of a rock disc (upper image homogeneous; lower image inhomogeneous). Right plots: the normalised stress distributions along the loaded diameter — the left-hand plot is the horizontal stress component; the right-hand plot is the vertical component. From Tang and Hudson (2010).

material: Continuous, Homogeneous, Isotropic and Linearly Elastic. The real rock mass is a DIANE material: Discontinuous, Inhomogeneous, Anisotropic and Not Elastic. Rock masses are discontinuous because they contain fractures. They are inhomogeneous and anisotropic because they are composed of different geological strata containing different discontinuity geometries at different locations and which have different properties in different directions. Rock masses are not elastic because the energy given to the rock mass during deformation cannot be recovered completely. A major shift occurred in the 1970s when modellers changed from simulating CHILE rocks to simulating DIANE rocks, Figure 3.17(a).

The reader is invited to speculate on how rock stresses might be transferred through the layered chalk in Figure 3.17(b) and the fused volcanic rock layers in Figure 3.17(c).

Figure 3.17(b).   Transverse isotropy in chalk strata composed of alternating hard and soft layers (Lower chalk, Isle of Wight, UK).

Figure 3.17(c).   Rock masses are discontinuous, inhomogeneous, anisotropic, and not elastic (volcanic rock, Jeju Island, South Korea).

### Principle 18: The deformation modulus of a fractured rock mass is directionally dependent

The deformability of a fractured rock mass results from deformation of both the intact rock and the discontinuities. Because the intact rock can be anisotropic and because the discontinuities occur in sets causing the discontinuity contributions to also be anisotropic, the deformation modulus of the rock mass will be different in different directions. An example illustration of the different elastic moduli in different directions is shown in Figure 3.18.

### Principle 19: Rock mass strength is a function of both the intact rock and the fractures

The strength of a rock mass will depend on whether failure occurs through the intact rock or along one or more fractures. This is illustrated

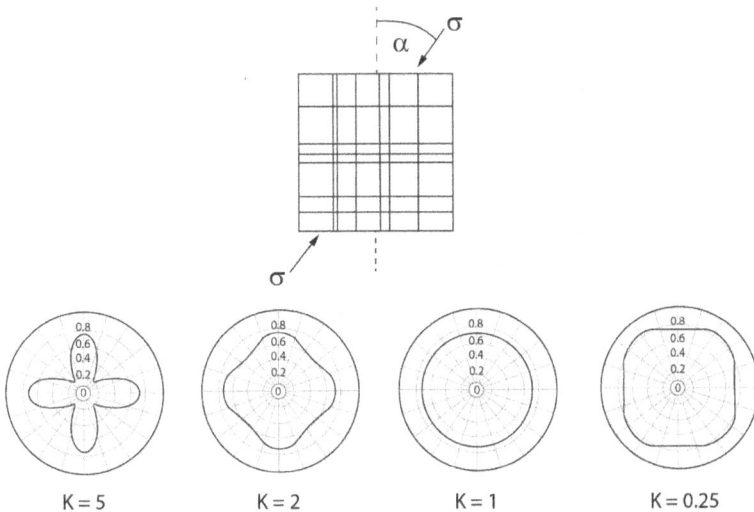

$$\frac{1}{E_m} = \frac{1}{E} + \lambda_1 s_{11}^1 \cos^4 \alpha + \lambda_1 s_{22}^1 \cos^2 \alpha \sin^2 \alpha + \lambda_2 s_{22}^2 \cos^2 \alpha \sin^2 \alpha + \lambda_2 s_{11}^2 \sin^4 \alpha$$

Figure 3.18.   Illustrative example of polar plots of the variation of elastic modulus in a rock mass containing two sets of fractures. $E_m$ is the modulus of the rock mass, $E$ is the modulus of the intact rock, the $\lambda$s are the fracture frequencies, the $s$ values are the moduli of the fractures (assumed to be the same for the two fracture sets), and the $K$ values below the images above are the ratios of the normal and shear fracture moduli. From PhD work by Wei (1988) at Imperial College London.

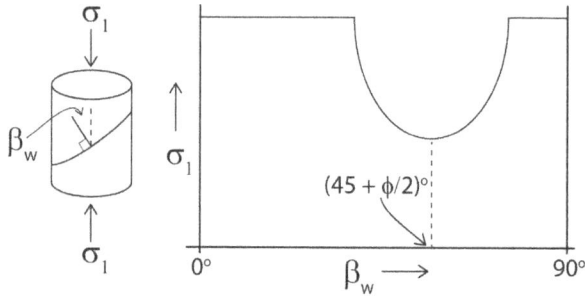

Figure 3.19. The overall strength of a rock specimen containing fractures at different angles to the applied stress $\sigma_1$.

in Figure 3.19 in which a cylindrical rock specimen containing a planar fracture (with its normal at an angle $\beta_w$ to the applied stress $\sigma_1$) is loaded to failure.

It can be shown that when the fracture is either sub-vertical or sub-horizontal, the inherent strength of the rock specimen is not altered by the presence of the fracture. However, between these orientations, at some $\beta_w$ angles, the fracture does affect the rock strength, with the greatest reduction being when the angle $\beta_w$ is $(45 + (\phi/2))°$ where $\phi$ is the angle of friction for the fracture, Figure 3.19. See also Figure 2.148. In a rock mass, the fracture will be subjected to a triaxial stress state and so the situation is then more complicated, and can be studied via the Mohr circle approach already presented in Chapter 2.

## Principle 20: Rock hydraulic conductivity/permeability

The ease with which a fluid flows through a rock mass is expressed by the hydraulic conductivity (when the fluid is water) and by the permeability (when the fluid is not water). Like stress and strain, hydraulic conductivity and permeability are second order tensors with six independent components and are usually characterised, like stress and strain, by the magnitudes and directions of the principal values. The hydraulic conductivity/permeability of fractured rock masses can vary greatly, Figure 3.20(a). The processes occurring within hydraulically active fractures over long time periods are important for some applications, Figure 3.20(b).

Figure 3.20(a). In general, water flows through the fracture network in a rock mass and so the hydraulic conductivity of a rock mass can be primarily a function of the fracture network and thus can be idiosyncratically variable. Note that in the fractured granodiorite pictured, water is only exiting through the single crack at the bottom right.

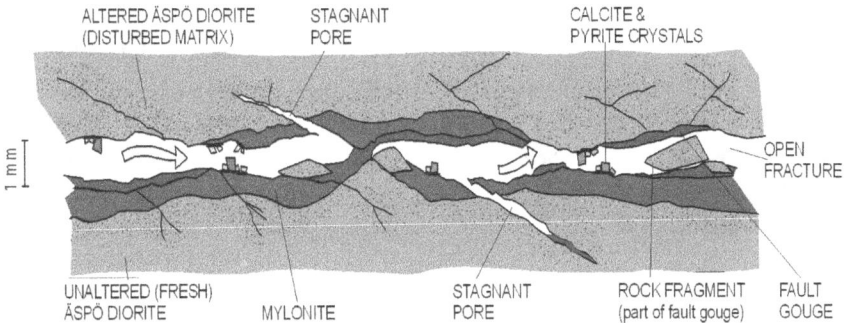

Figure 3.20(b). For the design of an underground repository for radioactive waste disposal, the potential flow of radionuclides through rock fractures and the long-term alteration to fracture properties by chemical processes are important subjects. From SKB, Sweden.

## Principle 21: The Representative Elemental Volume (REV)

The REV is an important concept for rock mass properties, e.g., the permeability of rock masses. In a rock mass sample, the number of fractures present is a function of the sample size, stabilising in average properties when the sample size is large enough, Figure 3.21.

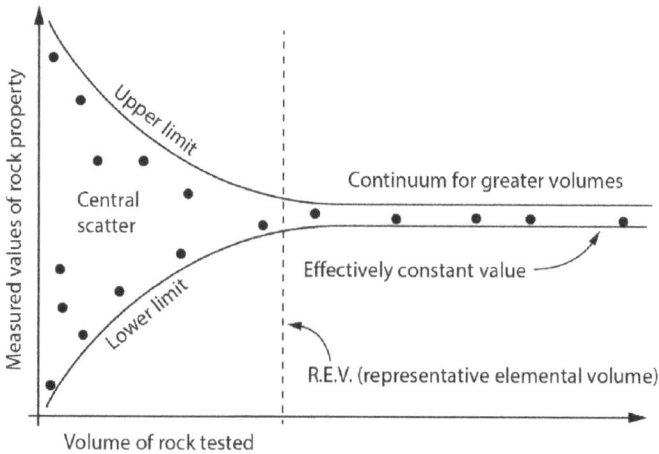

Figure 3.21.   The REV for a particular rock property is the sample size at which that property becomes essentially constant. The REV will be different for different properties, e.g., the 3D fracture density and the hydraulic conductivity tensor will have different REVs.

The REV is the rock mass sample size below which the property can vary significantly and above which the property is essentially constant, Figure 3.21. This REV concept applies to all properties governed wholly or partly by the fractures. Because the rock mass REV size is generally of the order of tens of metres and it is usually not possible to conduct meaningful tests directly on the rock mass, tests are conducted on the intact rock and the discontinuities separately and their significance for the rock mass properties evaluated (e.g., via numerical modelling).

## Principle 22: Rock testing

In the laboratory, the most frequently conducted tests are those for determining the compressive, tensile and shear elastic and failure properties. A biaxial compression test is shown in Figure 3.22(a) with $\sigma_1$ acting vertically and $\sigma_3$ acting horizontally. If this were to be extended to a pure triaxial test, there would be a $\sigma_2$ acting into the page.

There are many different configurations for applying force/stress or displacement/strain to a rock specimen and these can now be numerically modelled to provide enhanced understanding in order to improve the testing techniques and data reduction. The photograph in Figure 3.22(b)

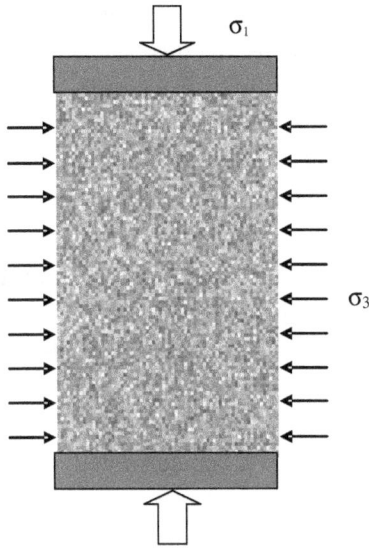

Figure 3.22(a).    Schematic of a biaxial compression test.

Figure 3.22(b).    The measurement of circumferential displacement in a uniaxial compression test (Figure 3.21(a)) as feedback in a closed-loop, servo-controlled testing machine. Photo from the SP Laboratory in Borås, Sweden and taken at the end of the test.

illustrates how a displacement transducer in the form of a chain can be used to measure the circumferential displacement during a uniaxial compression test — and to use this displacement as testing machine servo-control feedback in the test in order to be able to obtain the complete force–displacement curve (as in Figure 3.3(c)).

The rock properties obtained during such tests should, however, be regarded as experimental properties, rather than material properties *per se*, because the rock properties obtained via both laboratory and field testing vary according to the test arrangements. For example, the compressive strength of a rock determined from a laboratory test such as that shown in Figure 3.21(a) will vary according to the H:W ratio (Height to Width) ratio of the sample dimensions. A true material property does not vary as a function of the way in which it is determined, see also Principle 4.

## Principle 23: Testing methods

Given the caution expressed at the end of Principle 22 above and in order to inject coherency into establishing the properties of rocks, specific testing techniques are used and many of these have now been standardised (Ulusay and Hudson, 2007; Ulusay, 2014). Note, however, that different properties are required for different projects and, because there are many different rock engineering objectives, there can be no standardised overall site investigation procedure. The main organisations publishing rock testing methods are the ISRM and the American Society for Testing Materials (ASTM).

## Principle 24: Accuracy, bias, precision and resolution

The concepts of accuracy, bias, precision and resolution are useful when considering rock tests on intact rock and fractures.

— Accuracy is when the correct answer is obtained on the average;
— The bias is the difference between the sample mean and the actual mean;
— Precision is when the results are closely spaced (whether they are accurate or not); and

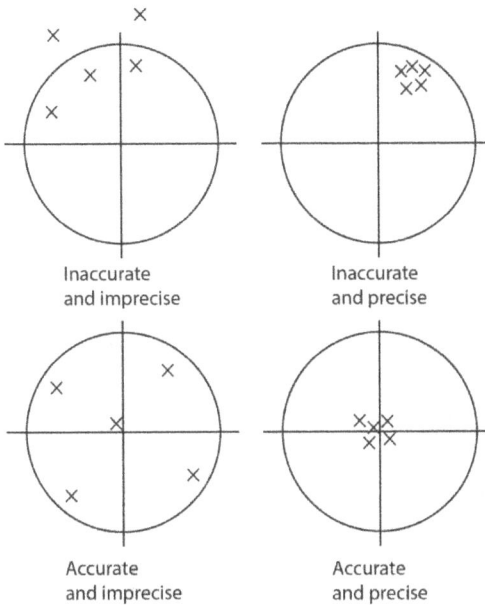

Figure 3.24.  The concepts of accuracy and precision illustrated via arrows hitting a target.

— Resolution is the number of decimal places to which the value is obtained.
— In the top left target in Figure 3.24, the mean of the arrow locations is not around the target centre (*inaccurate*), nor are the arrows well clustered (*imprecise*).
— In the bottom left target, the mean of the arrow locations is close to the target centre (*accurate*), but the arrows are still not well clustered (*imprecise*).
— In the top right target, the mean of the arrow locations is not around the target centre (*inaccurate*), but the arrows are well clustered (*precise*).
— In the bottom right target, the mean of the arrow locations is close to the target centre (*accurate*), and the arrows are well clustered (*precise*).

In the inaccurate cases, the bias is the distance from the centre of gravity of the arrow locations to the centre of the target. Resolution refers to how detailed the measurements are — in the sense of to how many significant figures the measurements can be made.

## Principle 25: Rock mass classification schemes

One of the most useful methods of combining the intact rock and discontinuity properties for assessing rock mass properties is through the use of rock mass classification schemes. The two most popular schemes are the Rock Mass Rating (RMR) developed by Bieniawski and the Q rating developed by Barton. The RMR system uses one property of the intact rock, three properties of the discontinuities, the ground water conditions and the orientation of the discontinuities relative to the engineered structure. The Q system, Figure 3.25, uses four properties of the discontinuities, the water flow and the stress condition. These systems have been successfully and widely used in practice to estimate rock mass properties and to design tunnel supports. As Nick Barton remarked in his 2011 ISRM Müller Lecture, "Discontinuous behaviour provides rich experiences for those who value reality, even when reality has to be simplified by some empiricism."

## Principle 26: Rock engineering systems (RES)

It is important in rock engineering design to establish the objectives of both the project and the supporting analyses. Once the objectives have been established, the physical variables and their interactions can be established. One method, RES, is based on the development of an interaction

$$Q = \frac{RQD}{J_n} \times \frac{J_r}{J_a} \times \frac{J_w}{SRF}$$

(a) The method for obtaining the Q value.

(b) The Q system is ideally suited for tunnels.

Figure 3.25.   The Q system for rock mass classification. (a) The formula for obtaining the Q numerical value for a given location in a rock mass: where RQD is the Rock Quality Designation, see Principle 9; $J_n$ is the joint set number; $J_r$ is the joint roughness number; $J_a$ is the joint alteration number; $J_w$ is the joint water reduction factor; and SRF is the stress reduction factor — see Harrison & Hudson (2000) for for illustrative worked examples using the Q system. (b) The Q system is especially suitable for tunnelling work because long lengths of the rock surface are exposed in a tunnel.

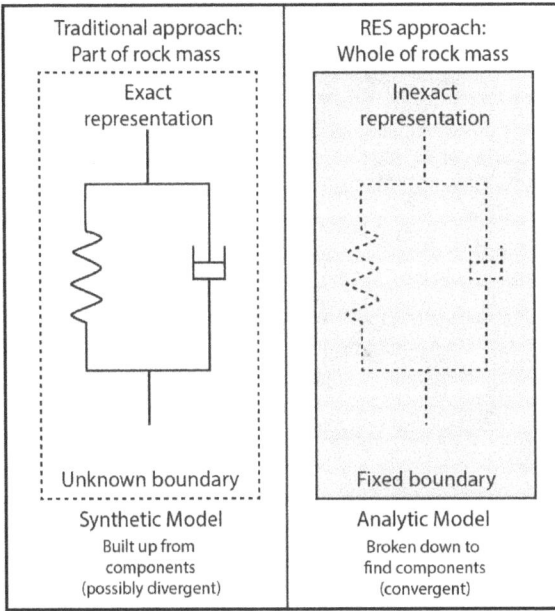

Figure 3.26(a).   Two types of modelling approaches: left, the synthetic model — an exact representation built up by an assumed knowledge of the components; right, the analytical model, an inexact representation in which the components are identified as the work proceeds.

matrix so that the primary variables and their interactions can be identified and coherently presented. The traditional approach is via the left side of Figure 3.26(a) in which the model components are assumed beforehand; the RES approach via the right side of Figure 3.26(a) involves establishing the model components as the work proceeds.

The basic tool of the RES approach is shown in Figure 3.26(b) — the interaction matrix. With this approach, an $n \times n$ matrix is developed with the initially assumed primary variables involved in the rock mechanics/engineering placed along the leading diagonal of the interaction matrix, top left to bottom right. The interaction between successive pairs of the primary variables is then studied in the off-diagonal components of the matrix.

An illustrative example of the analytical RES approach is shown in Figure 3.26(c) where the primary variables are rock structure, rock stress

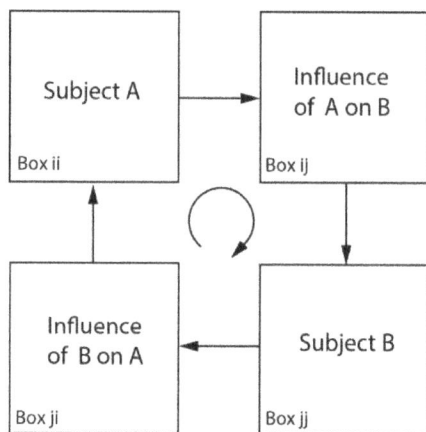

Figure 3.26(b).   The basic tool of the RES approach is the interaction matrix. Note the clockwise convention for studying the respective interactions.

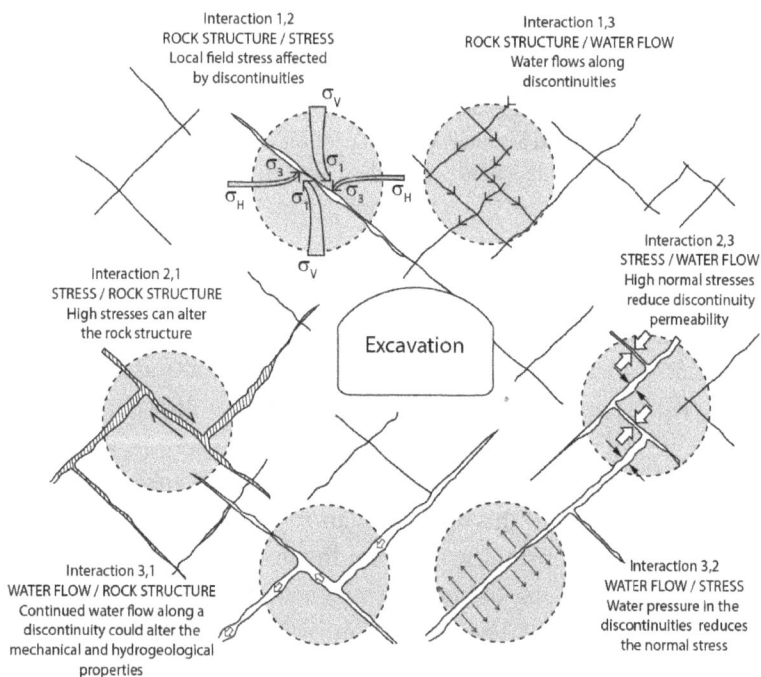

Figure 3.26(c).   An illustrative example of studying the binary interactions involved in a 3 × 3 matrix containing the primary variables, rock structure, rock stress and water flow. In practical applications, the interaction matrix will have between 6 and 12 primary variables and hence between 30 and 132 interactions (i.e., the number of off-diagonal terms in each case).

| | | | | | | |
|---|---|---|---|---|---|---|
| **ROCK STRESS** | Rock fracturing | Normal and shear displacement | No direct effect | No direct effect | Minimal direct effect | No direct effect |
| Stiff rocks attracts stress | **INTACT ROCK** | Fracture type | Primary permeability | No effect | Mineral content | No direct effect |
| Perturbation of stress state | No direct effect | **FRACT-URES** | Secondary permeability | Minimal direct effect | Fracture filling | No direct effect |
| Effective stress | No direct effect | Erosion | **WATER FLOW** | Heating or cooling | Dilution or concentration | No direct effect |
| Coefficient of linear expansion | Increased alteration rate | No direct effect | Convection | **TEMPERATURE** | Solubility | No direct effect |
| No direct effect | Solubility, alteration, replacement | No direct effect | No direct effect | No direct effect | **GROUNDWATER CHEMISTRY** | No direct effect |
| Significant changes | EDZ | EDZ | Alteration of hydraulic head | Heat source or sink | Ionic changes | **PERTURBATIONS** |

Figure 3.26(d). Illustrative example of a rock mechanics/rock engineering interaction matrix with seven leading diagonal terms and hence potentially 42 off-diagonal interactions.

and water flow. The interactions between the primary variables are studied (the interaction matrix usually having from 7 to 10 primary variables) and the significance of these — given the project objective — coded accordingly, see Hudson (1992) and Hudson and Feng (2015). An example $7 \times 7$ interaction matrix is included here in Figure 3.26(d). The identification, coding and establishing the significance of the off-diagonal terms is the key to the RES analytic approach, and the associated modelling of the rock mechanics and engineering is now greatly facilitated by the advent of advanced numerical modelling/simulation. Moreover, the creation of the matrix and hence the identification of the important processes facilitates establishing the required site investigation content, the modelling, the engineering procedures and back analyses, plus the related auditing processes, see Feng and Hudson (2011).

## Principle 27: The rock excavation and support objectives

When rock is excavated for an engineering project, it is necessary to break the rock being removed and simultaneously avoid breaking the remaining rock. The created rock surface (e.g., the slope face or tunnel surface) is thus a critical interface between the excavation and support objectives, Figure 3.27.

## Principle 28: The *in situ* rock block size distribution

The excavation process consists of changing the *in situ* block size distribution in a rock mass to the fragment size distribution after excavation. In Figure 3.28, the x-axis refers to the aperture size in a *hypothetical* mesh large enough to sieve the individual blocks in a rock mass; the y-axis represents the percentage of rock blocks that would pass through the mesh apertures. So, for very large sieve apertures, all the rock blocks will pass through the sieve. On the other hand, for very small sieve apertures, few of the rock blocks would pass through.

Thus, in Figure 3.28, the process of excavation in a fractured rock mass is indicated by the shift from the *in situ* block size distribution to the excavated block size distribution. This is a useful diagram, both conceptually

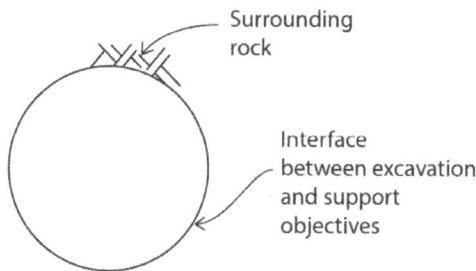

Figure 3.27.   The interface between the excavation and support objectives in a tunnel.

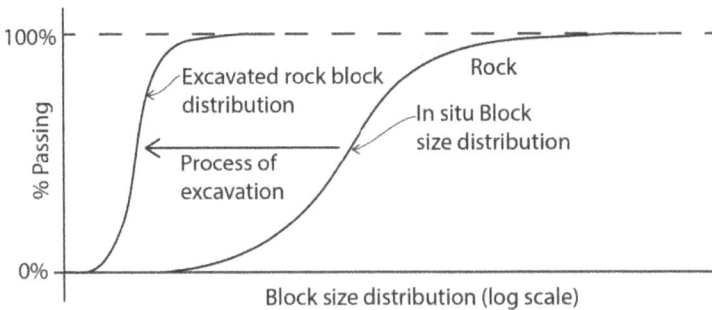

Figure 3.28.   The *in situ* and excavated rock block size distributions.

and for understanding and determining the energy required to reduce the block size distribution to any required level.

## Principle 29:  Rock excavation

There are only two main methods of rock excavation. One is blasting in which large amounts of energy are applied to the rock in seconds with quiescent periods of several hours in between. The other is by mechanised excavation (Figure 3.29(a)) in which a much smaller level of energy is continuously input to the rock (except when the machine is not operating, e.g., for maintenance).

Whereas there is limited potential for increasing tunnel excavation rates using blasting alone, the implication of Figure 3.29(b) is that, there is considerable potential for improvements in tunnel boring machine (TBM) excavation rates — if methods can be developed for increasing the rate of energy input applied to the rock excavation process.

## Principle 30: Optimising rock excavation

Optimising rock breakage by explosives consists of optimising the separate effects of the explosive's stress wave and gas pressure and their interactions with the free face (Figure 3.30). Optimising mechanised excavation consists of optimising the transfer of energy from the TBM cutters to the rock. This involves mechanical engineering considerations, the configuration of the cutters, steering the machine, reducing vibrations, minimising down time, etc.

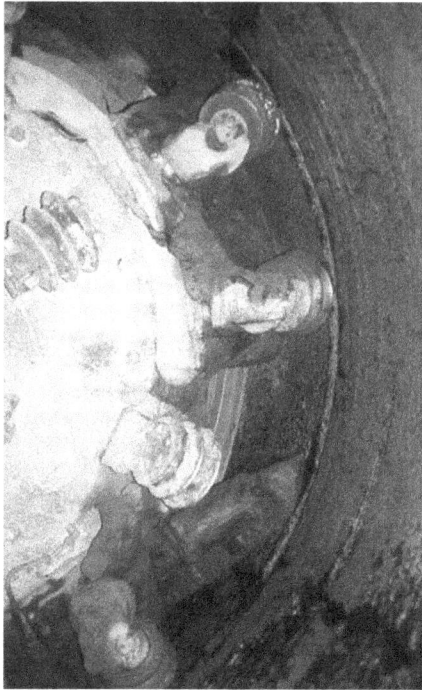

Figure 3.29(a).   TBM with disc cutters.

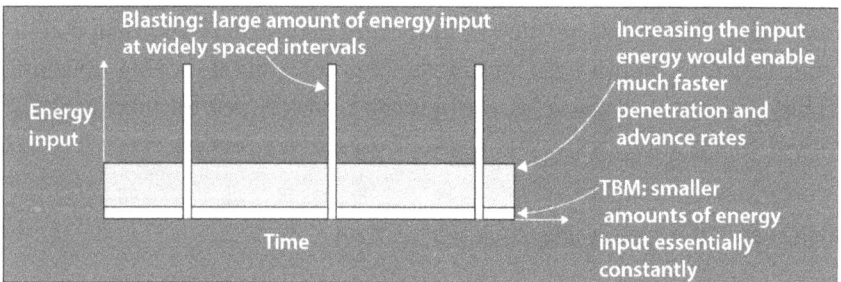

Figure 3.29(b).   The two different energy input modes for blasting and using a TBM.

Figure 3.30. Laboratory tests to simulate pre-splitting using detonators in blocks of perspex (plexiglas) with artificial fractures — in which the purpose is to create a plane face between blastholes in order to demonstrate how compressive stress waves reflect from the fractures as tensile stress waves causing extra unwanted damage; this phenomenon is clearly evident in the central photograph. (a) Stress waves reflected from simulated fractures causing cracking; (b) Stress waves trapped between two inclined fractures causing intense fracturing; (c) All stress wave fracturing trapped between two vertical fractures. See Worsey (1981). Also see the discussion of rock pre-splitting and photographs in Chapter 1, Section 1.3.3.

## Principle 31: The three main effects of rock excavation

After the rock is excavated, the stability of the resultant rock surfaces must be considered. The three primary effects of excavation are shown in Figure 3.31(a) as follows.

1. Displacements occur because rock resistance has been removed.
2. There are no normal and shear stresses on an unsupported excavation surface; and hence, by definition, the surface becomes a principal stress plane involving a local change in the magnitudes and orientations of the components of the pre-existing stress field (see Principle 13 and Figure 3.13(e)).
3. At the boundary of the excavation open to the atmosphere, water pressure is reduced to atmospheric pressure. Thus, if the water pressure in

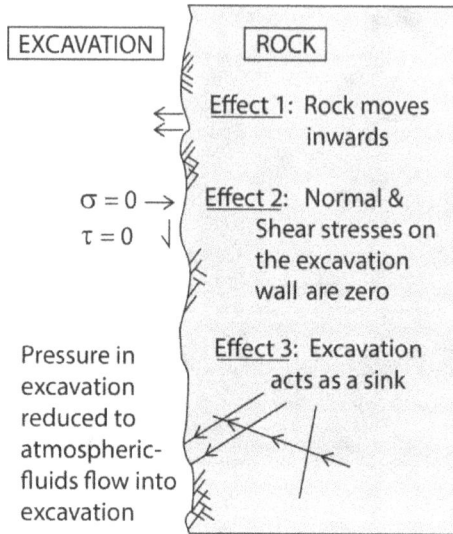

Figure 3.31(a).   The three primary effects of making an excavation in a rock mass.

Figure 3.31(b).   Effect 3 in Figure 3.31(a) — in one of the tunnels excavated for the Jinping II hydroelectric scheme in China.

the rock mass is greater than atmospheric pressure, the excavation will act as a sink with water flowing into it.

## Principle 32: Rock reinforcement and support

To stabilise a rock face after excavation, either no support or rock reinforcement (i.e., rockbolts and/or shotcrete), or direct rock support (e.g., steel arches) may be necessary. The reinforcement strategy is to bolt the rock blocks together so that they behave more like a rock continuum. The support strategy is to maintain the rock displacements to tolerable levels, Figure 3.32.

Figure 3.32.   Shotcrete applied to the foundation of an ancient castle in Spain.

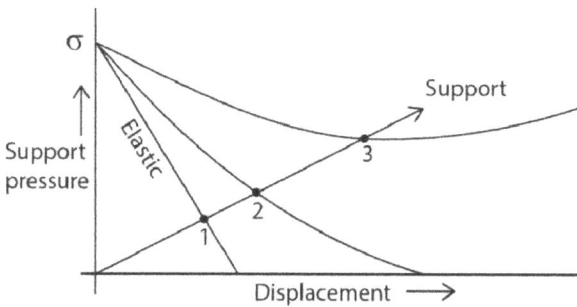

Figure 3.33.   The ground response curve.

## Principle 33: The Ground Response Curve

The Ground Response Curve, Figure 3.33, in which the support pressure is plotted against the boundary displacement, is a useful conceptual framework for considering the stability requirements in continuous and discontinuous rock, and to illustrate the effect of rock damage that might be caused by the excavation process.

Consider the situation where an excavation is being made in a rock mass. To begin, there is a stress state in the rock mass ($\sigma$). As the excavation is created, the stress state normal to the excavation at the excavation boundary will change, as noted in Principle 31. If the rock mass is elastic, the rock will unload along the straight line 1 marked as 'Elastic' in Figure 3.33. If the rock is more fractured or plastic, the rock will unload to a point further along the displacement axis, curve 2 in Figure 3.33. If, however, there is no support installed in a weak, loose, fractured rock mass, the rock mass may continue to displace along the upper curve 3 — eventually requiring more support as it deteriorates. Now, if some form of elastic support is installed during and/or after the excavation process, the elastic support line will intersect these three possibilities at the points 1, 2, and 3 in Figure 3.33. Note, however, that, if the support (lines 1–3) is not sufficiently stiff, it could miss the upper curve entirely and thus have little or no stabilising effect.

The diagram in Figure 3.33 illustrates a series of points.

— A truly elastic rock mass will not need support.
— The stiffer the installed support, the less the rock will displace.

— For a significantly fractured rock mass, an inadequately stiff support will not be sufficient to stabilise the rock mass.

More detailed studies of this ground response concept have been found to be most helpful in understanding reinforcement and support in rock masses.

## Principle 34: Foundations

In Figure 3.32(a), six potential failure modes are shown for a foundation located on anisotropic, discontinuous, rock masses, each containing one fracture set. In the diagram, failure can occur through the intact rock, indicated by the letter R, and/or along a fracture plane, indicated by the letter P.

The possible failure modes for the cases shown in Figure 3.32(a) are

— *Case a*: Failure occurs through rock mass by creation of new failure surface in intact rock. Homogeneous and isotropic failure throughout the rock mass.
— *Case b*: As Case a, but with rock wedge also formed by failure along fractures near the boundary of the outcropping rock.
— *Case c*: As Case a, but with rock wedge formed by fractures below loading area.
— *Case d*: As Case a, but with a rock wedge between failure regions through the intact rock. This is the most general and complex case.
— *Case e*: Plane sliding of single wedge formed by fracture. Failure resulting from the existence of a single sliding wedge along the planes of weakness.
— *Case f*: Failure in intact rock and along fracture. Failure along the family of planes of weakness of the rock mass, with wedge displacement next to foundation surface. In contrast to the previous cases, the rock mass fails at the same time that the family of planes of weakness slides (double failure mechanism).

These six mechanisms can be used to develop criteria for foundation failure, taking into account the possibilities of failure through the intact rock, along the fracture planes, or both.

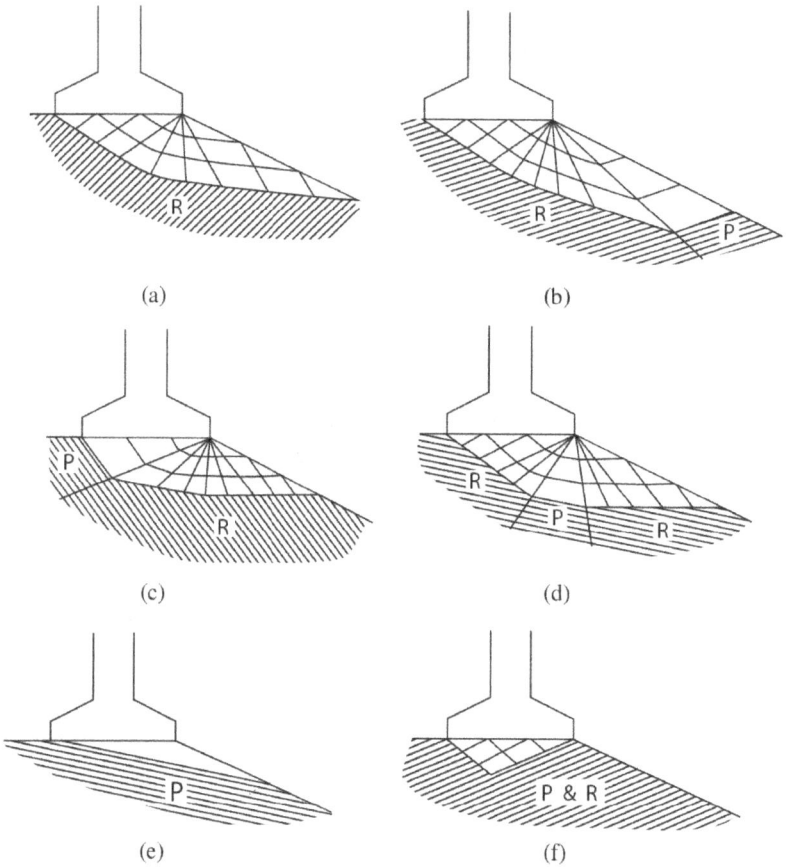

Figure 3.34(a).   Potential for failure of a foundation located on an anisotropic, discontinuous rock mass, from Serrano and Olalla (1998). The letter R refers to failure through the rock mass; P refers to failure along planes of weakness.

Clearly, a knowledge of both structural geology and rock mechanics/ engineering is required to evaluate a given foundation problem along the lines of the cases shown in Figure 3.34(a). So, a study of the abutment foundations of the Clifton Suspension Bridge in the UK (Figure 3.34(b) and the book cover) is presented in Chapter 4 as one of our examples of the synthesis of structural geology and rock mechanics in evaluating rock engineering circumstances.

Figure 3.34(b). An anisotropic, discontinuous rock mass supporting the Clifton Suspension Bridge, UK.

## Principle 35: Slope instability

Slope instability can be caused by failure occurring through weak intact rock or along pre-existing discontinuities in harder rock, Figures 3.35(a), b and c. The four main types of rock slope instability are circular slip, plane sliding, wedge sliding and toppling — all of which have been extensively studied and reported in the rock mechanics literature, see Figure 3.37.

A great deal can be achieved quickly in assessing the potential for instability by using simple solutions for circular slip potential and considering the slope and discontinuity dip and dip directions for failure initiated along pre-existing discontinuities.

## Principle 36: Large-scale rock mass instability

A large-scale rock mass instability is shown in Figure 3.36. In this case, the conjunction of two major fractures or weakness zones formed a wedge, along which slip could occur. The scale of the failure can be seen by the size of the mine roadways. No one was hurt in this wedge slide because the

Figure 3.35(a). Natural rock slope failure in Italy.

Figure 3.35(b). Rock slope in Canada with apparently adversely orientated fractures which could lead to slope instability by plane sliding.

(a) Plane sliding

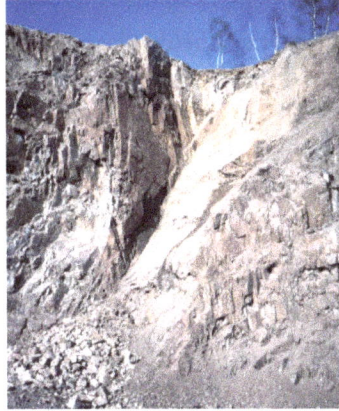

(b) Wedge sliding

Figure 3.35(c).  Types of sliding instability in a hard rock quarry. (a) Plane sliding on a fracture surface; (b) wedge sliding on two fracture surfaces, see Figure 3.37.

Figure 3.36.  Large-scale failure at the Teutonic Bore Mine in Australia. The dust in the picture was formed from frictional effects as the wedge slid down and escaped from the wedge openings and through other fractures in the rock mass.

rock displacements had been continuously monitored at several locations and, when the displacements started to accelerate, the working personnel and equipment were evacuated from the mine.

There are now many large-scale open pit mines in operation across the world and such plane and wedge failures are major risks, especially since the associated rock masses can have complex geological structures. So, this again reinforces the need for a strong link between the structural geologist and the rock engineer.

## Principle 37: Kinematic analyses

When designing surface slopes in rock masses, there should always be an initial kinematic analysis (study of the geometrically possible motions of rock blocks), i.e., given the geometry of the slope and the discontinuities, whether is it physically possible for one or more rock blocks to slide? Because the rock discontinuities tend to occur in sets, there should always be consideration of designing the excavation in harmony with the rock structure. For example, a surface excavation which is elliptical in plan is generally better than one which is circular in plan — because the circular excavation has not taken into account the presence of the fractures and hence not optimised the mine geometry to minimise the possibilities of plane failure, wedge failure and toppling failure Figures 3.37 and 5.39.

## Principle 38: Probabilistic and fuzzy analyses

Because of uncertainty in the input data in design considerations, it is helpful to conduct sensitivity analysis using probabilistic methods, fuzzy maths and other similar techniques.

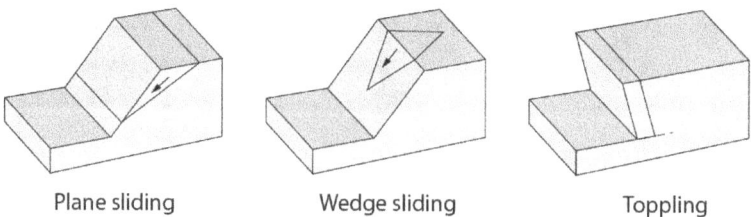

Plane sliding          Wedge sliding          Toppling

Figure 3.37.   The geometries of plane, wedge and toppling failures.

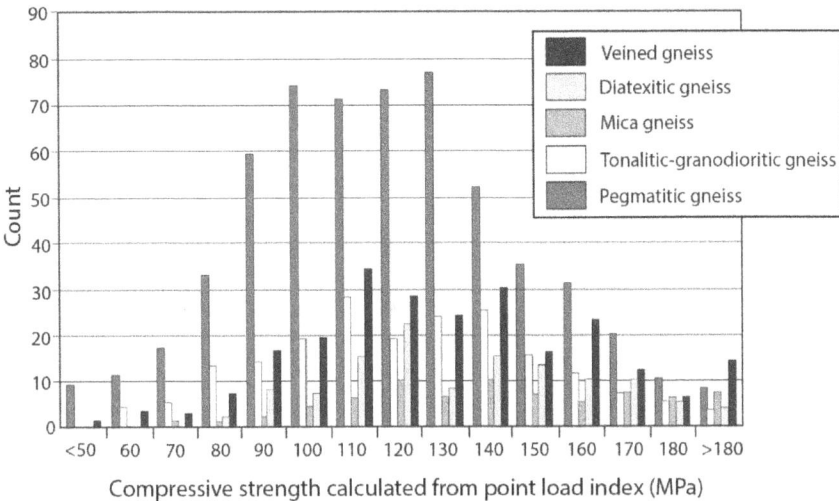

Figure 3.38. Variation of the rock compressive strength of rocks from the nuclear power-plant site at Olkiluoto in Finland, after Hakala *et al.* (2008).

The histogram in Figure 3.38 shows variation in the compressive strength for the five rock types at the nuclear powerplant site at Olkiluoto, western Finland. So in rock mechanics analyses, e.g., the potential for rock spalling where the rock stress values are compared with the rock strength values, not only must this type of rock variability be taken into account but also any natural occurring (as in Figure 2.161) or engineering induced variations in stress, as reported in Hakala *et al.* (2008).

## Principle 39: Shapes and directions of underground openings

The rock fractures do not occur at random orientations and the *in situ* maximum principal stress components will have certain orientations. Thus, where possible, the shapes and directions of underground openings should be selected accordingly to reduce rock damage and subsequent tunnel maintenance, e.g., Figure 3.39 where $\sigma_1$ is the maximum principal stress.

For example, the major transport drifts in a coal mine can be engineered to be parallel to the major principal stress (assuming this is horizontal), thus reducing the concentrated stress around the drifts and hence

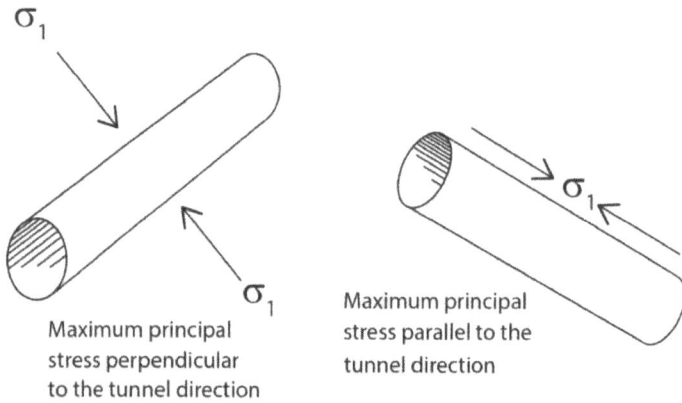

Figure 3.39. Illustrating that the optimal orientation of tunnels to avoid damage caused by stress concentrations is parallel to the maximum principal stress component, i.e., as in the right-hand sketch. The stress concentration around the tunnel in the left-hand sketch will be higher than that in the right-hand sketch.

reducing the tunnel maintenance; use of this principle has been most successful in Chinese coal mines. Of course, in some applications, such as a metro tunnel linking one station to another, there is little or no choice about the tunnel azimuth. However, in other applications, such as coal mining as mentioned above and underground storage/disposal facilities, there can be considerable latitude in the location and orientation of the facilities.

Despite this Principle 39 concept, it should be remembered that the process of excavation not only makes the excavation surface a principal stress plane, see Figure 3.13(e), it also increases the stress component in one direction and decreases it in another. So, it is necessary to assess the regions of high stress and whether failure will occur, and the consequences of such failure. In other words, Principle 39 is a strategic approach but there can be local variations; so a tactical approach is also required. Simple considerations of the stress can be most helpful in this regard: solutions for circular and elliptical excavations can be used to estimate stress concentrations, see Figure 2.16 and the related discussion, and there are many computer programs that can simulate the stresses and displacements around an excavation with a more complex profile geometry.

## Principle 40: Excavation design

There are simple underground engineered structures and there are complicated engineered underground structures. The blasted tunnel in hard rock in Figure 3.40(a) is an example of a simple structure; the large chamber for a hydroelectric project in China in Figure 3.40(b) is an example of a complicated structure.

Because there is a variety in the types of underground rock engineering geometries, it is necessary to provide guidance on the overall design approach. Figure 3.40(c) is a flowchart example of one such approach (from Feng and Hudson, 2011).

In addition, the design flowchart in Figure 3.40(c) is complemented by the risk flowchart in Figure 3.40(d). This flowchart is required because there needs to be a systematic approach to the uncertainty and risk encountered

Figure 3.40(a).   Blasted tunnel in hard rock in the UK. The thin white line is a laser beam used to ensure the correct tunnel alignment.

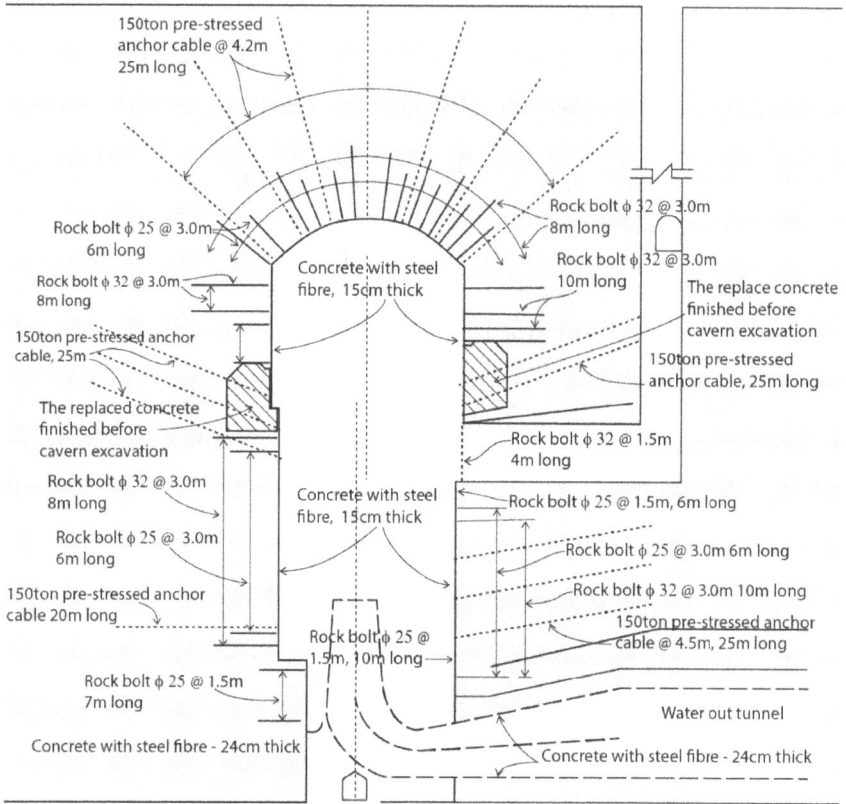

Figure 3.40(b). A variety of stabilising features used in the construction of an underground powerhouse. They include rockbolts of various diameters, ($\phi$ in mm), spacing and lengths, pre-stressed anchor cables and concrete linings strengthened by steel fibres.

in rock engineering. Note that, in the Figure 3.40(d) flowchart, there are two columns: the left-hand one considers risk factors before construction; and the right-hand one considers risk factors during construction. These relate respectively to epistemic uncertainty (i.e., lack of knowledge, such as the rock stress state) and aleatory uncertainty (i.e., essentially unpredictable events, such as at what tunnel chainage a major water bearing fracture may be encountered), see Section 8.3. The former uncertainty is reducible before construction starts through site investigation and computer modelling; the latter is reducible as construction proceeds through observation of the rock characteristics. The work by Bedi (2013) provides a clear

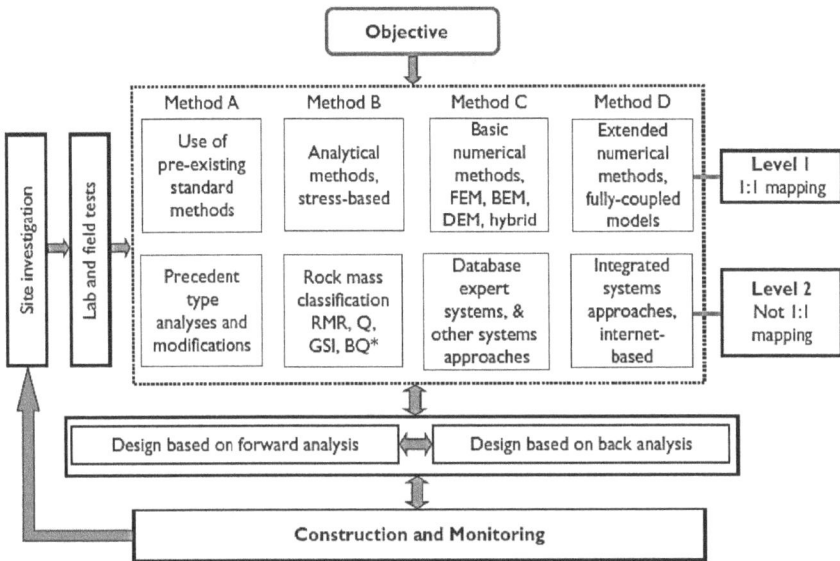

Figure 3.40(c).   Flowchart to guide the design of underground excavations in rock (from Feng and Hudson, 2011). The eight analysis methods within the dotted box are arranged left → right as 'simple' → 'complicated'; the four methods in the upper row are of the 1:1 mapping type — in other words, they directly simulate the excavation and support geometry; the four methods in the lower row are of the non-1:1 mapping type — in other words, they do not directly simulate the excavation and support geometry.

explanation of epistemic and aleatory uncertainties and how rock engineering design can be developed to incorporate the concepts. It is also useful, if not essential, to install some kind of auditing procedure to ensure that the design and construction work proceeds according the requirements (Hudson and Feng, 2015).

## Principle 41:  Computer analyses

Sophisticated numerical codes are now available for assessing the effect of excavation on block movements, stresses, heat and water flow within rock masses. Most of the trends predicted by these programs can be established from the principles already presented. Therefore, the programs come into their own when specific values are required, high speed sensitivity studies are needed, etc. The development of such numerical codes, which has been

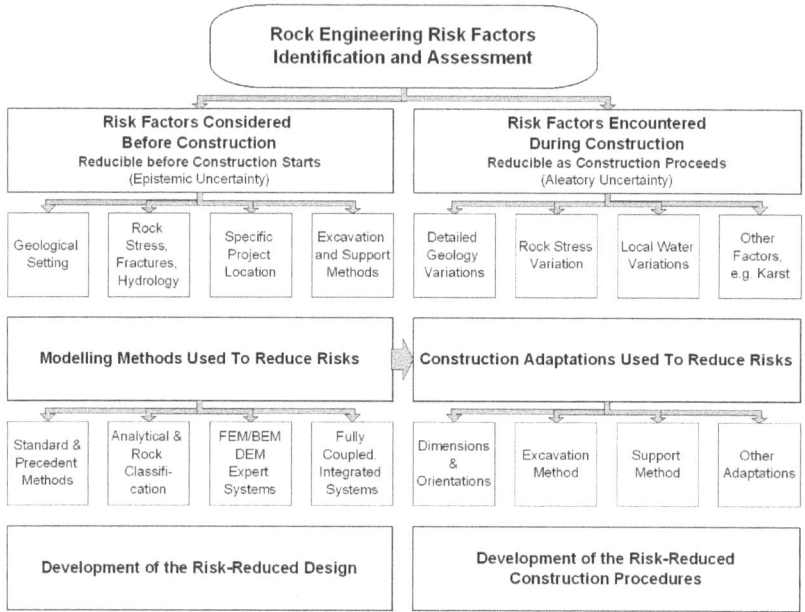

Figure 3.40(d).   Flowchart to guide the assessment of risk for the design of underground excavations in rock, (from Hudson and Feng, 2015).

spearheaded by Dr Peter Cundall of the Itasca Consulting Group based in the USA, has been a revolution in rock mechanics analysis capability.

In particular, it is now possible to use combined geological-thermal-hydrological-mechanical-chemical-engineering modelling to understand the coupled mechanisms that can occur in the rock mass, i.e., *geological*: site geometry, lithology, fractures; *thermal*: heat loads, heat flow; *hydrological*: water pressures, water flow; *mechanical*: rock stress, stiffness, strength; *chemical*: water chemistry, swelling rocks; *engineering*: effects of excavation.

## Principle 42: Excavation Disturbed/Damaged Zone (EDZ) and zone of influence

The EDZ refers to the zone around an excavation in a rock mass where the rock has been damaged, mainly by the introduction of new fractures, e.g., Figure 3.42(a).

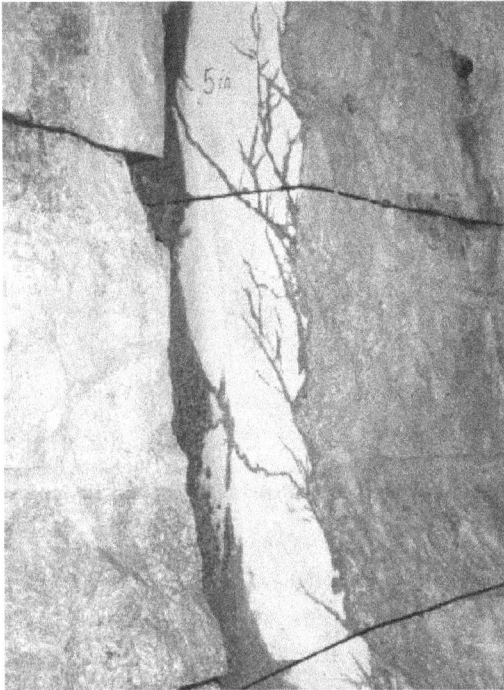

Figure 3.42(a).   Fractures in a tunnel wall induced by blasting and visible via a slot cut in the tunnel wall, Äspö Hard Rock Laboratory, Sweden.

The disturbance caused to the rock mass by excavation is evaluated and then the engineering zone of influence is studied, i.e., the volumetric extent of the rock where the stress (for example) is affected by more than, say, 5%, of its original value. The concept is of help, not only in deciding on excavation spacings, but also on the best excavation sequence.

A useful chart has been developed by Rolf Christiansson of SKB in Sweden (Figure 3.42(b)) which includes the many factors involved and the characterisation methods for assessing the EDZ.

## Principle 43:  Design life and environmental aspects: disorder and entropy

The design life of an engineered project is important because the rock mechanisms are time dependent. The face area in a longwall coal mine has

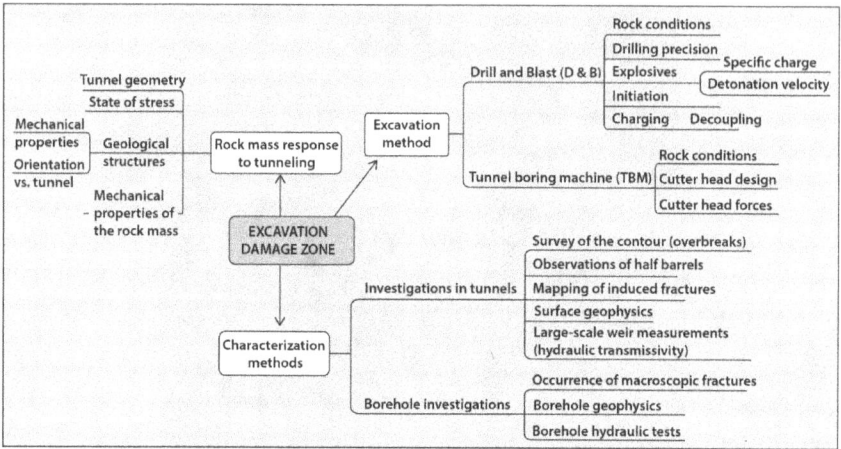

Figure 3.42(b).   Factors affecting the extent of the EDZ. After R. Christiansson, SKB, Sweden.

a life of a few days, an open-pit coal mine can have a life as low as five years, an open pit or underground metal mine can have a life of around 30 years, a civil engineering tunnel 120 years, and the regulators for a radioactive waste repository can ask for a design life of a million years. However, a key point relating to the environmental aspects of all rock engineering construction is that it is not possible to avoid disturbance to the environment: such disturbance can only be minimised to the unavoidable level, Figure 3.43.

In Figure 3.43, the *x*-axis represents a line across, say, an opencast coal mine. The *y*-axis represents the disorder caused by the mining operation. The lower dashed line is the baseline entropy or disorder level before construction; the upper dashed line is the mean entropy or disorder level after construction. The heavy, continuous line represents the order and disorder caused by the mining operation. At the centre of the graph, this line dips because order has been created by the mining operation at this location (in the context of the mining objective). However, outside this area, there must be more than compensatory disorder created. Since all operations cause an increase in entropy, the consequence is that a certain level of disorder is unavoidable.

Therefore, it is not sensible to ask that a mining operation, for example, should not affect the surrounding environment: the consequence of

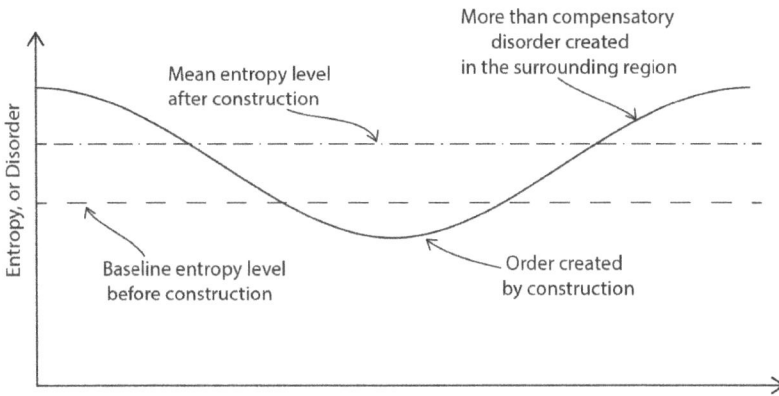

Figure 3.43. A line across a rock engineering site on the *x*-axis and the associated disorder on the *y*-axis. Construction always causes an increase in entropy — which is a measure of the 'disorder' created by using energy. It is not possible to avoid this increase, but it is usually possible to arrange for the disturbance to be reduced to an acceptable level. Note that, whilst the engineering creates order, there is a more than compensatory disorder in the surrounding area.

the entropy principle is that some disturbance is unavoidable. However, through good management, efforts should be made to ensure that there is no additional, *unnecessary* disorder created.

## Principle 44: These principles apply to all rock engineering projects

The principles in this list apply to all rock engineering projects because they are based on the fundamental geometry, properties and mechanics of rock masses. The principles apply, *inter alia*, to foundations, slopes, shafts, tunnels, caverns, mining, radioactive waste disposal, geothermal energy, $CO_2$ sequestration and unusual project objectives such as large scale rock sculptures.

## Principle 45: Stress analysis and stereographic projection

Two subjects which are most useful for rock mechanics and rock engineering design studies are stress analysis and stereographic projection. This is

Figure 3.45. The logo of the rock mechanics group at Imperial College London, composed of the Greek symbol sigma for stress encircling an example stereographic projection containing some fracture plots.

why the logo of the Rock Mechanics Group at Imperial College London in the UK is a composite of the stress tensor, $\sigma_{ij}$, and the outline of some fractures on a stereographic projection, Figure 3.45.

## Principle 46: Continuation of analysis and design via monitoring during construction

The process of analysis, design and site investigation continues during construction when the rock properties are more apparent and the effects of construction clearer. Measured engineering performance parameters, such as tunnel diameter changes and microseismic recording, can provide the best information for continued excavation. Indeed, successful back analysis based on *in situ* monitoring during construction is usually required for a fuller understanding of the observed mechanisms.

Rockbursts are sudden and violent ejections of shattered rock in regions of high natural rock stress as the process of excavation alters the *in situ* stresses and causes a rapid release of strain energy. During the excavation of the Jinping II, 17 km long, headrace tunnels for a hydroelectric project in China, rockbursts were a frequent occurrence because of the extreme depth (up to 2,500 m) where the natural *in situ* rock stresses had high values. When a new section of the tunnel was excavated, the local stresses were altered (as explained in Principles 13 and 31). In Figure 3.46, two tunnel locations are

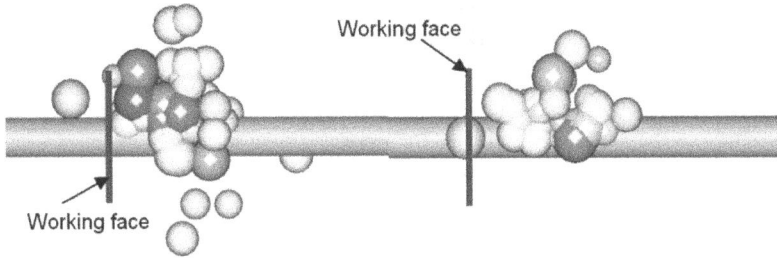

Figure 3.46. Location of microseismicity during excavation of the deep Jinping II hydro-electric headrace tunnels in China. Left working face — where the rockburst risk was not controlled; right working face — where the rockburst risk was mitigated successfully. The spheres represent the location and severity of the rockbursts, the darkness indicating their severity. From Hudson and Feng (2015).

shown where rockbursts occurred during excavation of the tunnels. At the left working face in Figure 3.46, the rockburst risk was not mitigated by engineering measures. At the right working face, it was possible to reduce the severity of the rockbursts by modifying the engineering parameters and the rockburst risk was reduced successfully, from Hudson and Feng (2015).

## Principle 47: The variability of rock engineering projects

Although the principles of rock mechanics remain the same, it should always be remembered in design and risk analyses for construction that each rock mass and each rock engineering project is unique.

Figure 3.47(a) illustrates foundations, slopes, tunnels and caverns, mine stopes (the openings created by metal mining), geothermal energy extraction and radioactive waste disposal, and indicates the rock engineering projects for which the application of rock mechanics is crucial. Moreover, each of these applications has different objectives and design lives. A mine stope's life may be measured in months; civil engineering projects usually have design lives of ~100 years; and the radioactive waste disposal regulator may ask for a design life of hundreds of thousands of years. Some of the rock engineering projects have specific locations, e.g., a metro tunnel from Station A to Station B or a metalliferous mining project; for others, the location may not be pre-determined, e.g., an underground gas storage facility or radioactive waste repository.

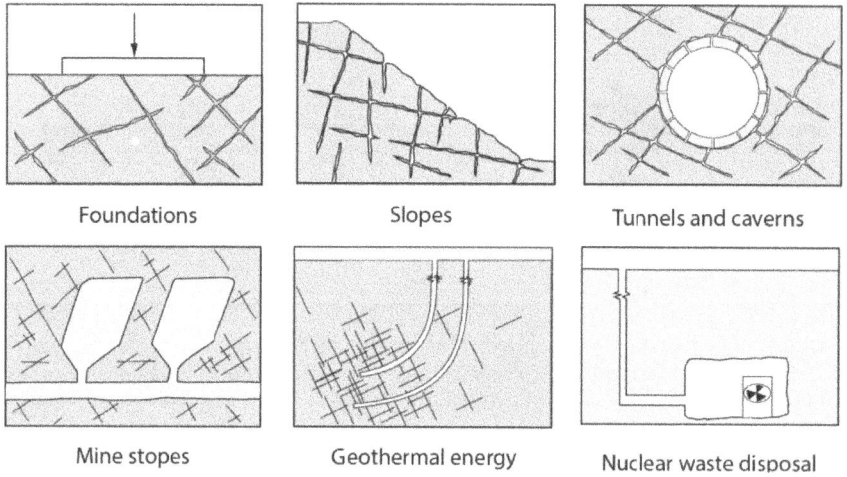

Foundations     Slopes     Tunnels and caverns

Mine stopes     Geothermal energy     Nuclear waste disposal

Figure 3.47(a). The wide variety of rock engineering projects and hence rock engineering objectives.

Figure 3.47(b). Unusual type of problem in Taiwan: assessing the safety of the overhanging rock masses along this road.

For all the projects illustrated in 3.47(a), there is a considerable body of knowledge based on many years of experience, all of it supported by the theory and principles of rock mechanics for engineering. However, each new project is unique in terms of the precise geological conditions and so the geological concepts outlined in Chapter 2 must be used in conjunction with the principles outlined in this Chapter, as we show in Chapters 4–9, the 'synthesis Chapters'. Figure 3.47(b) showing a road in Taiwan is an example of an unusual engineering requirement: estimating the stability of the overhanging rock masses. This is another example where the study needs both geological and engineering expertise.

## Principle 48:  Rock and soil mechanics principles

The principles of rock mechanics have been developed over many years, in particular since the 1960s when the ISRM was formed. Rock mechanics is a specific discipline which takes into account the idiosyncrasies of rock masses. Thus, the companion disciplines of structural mechanics and soil mechanics should be used for rock only with caution, checking that the assumptions on which those principles were developed do indeed apply to rock: in some cases they will; and in most cases they will not.

## Principle 49:  New or unusual rock engineering projects

In the case of a new or unusual project for which there may be no precedent practice (Figure 3.49) the design considerations will rely entirely on the geological and rock mechanics principles plus numerical computer codes, backed by *in situ* feedback at the early stages of construction.

## Principle 50:  Future trends in rock engineering projects

In the future, the following trends are expected in rock mechanics and rock engineering.

— Improved methods of accessing/collating information.
— More emphasis on geophysical methods in site investigation.
— More integration of subjects, see Principle 41 above.

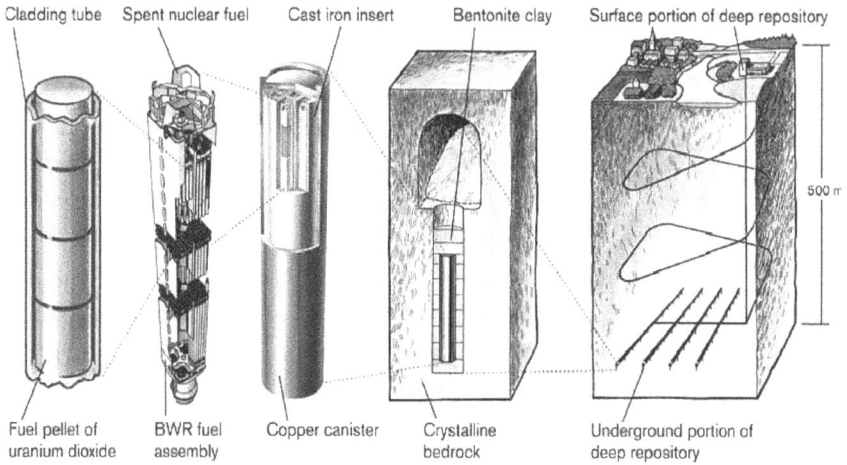

Figure 3.49. The components of radioactive waste disposal from the fuel pellets through to the underground repository, (from SKB, Sweden).

— More use of neural network 'intelligent' computer programs.
— Larger, deeper and longer excavations.
— Increased rates of mechanised excavation.
— More emphasis on 'environmental' aspects.
— More international co-operation, e.g., through organisations such as the International Society for Rock Mechanics, ISRM (www.isrm.net) and the DECOVALEX project (www.decovalex.org).

\* \* \* \* \*

Readers who wish to delve further into the subject of rock mechanics are referred to "Engineering Rock Mechanics: An Introduction to the Principles" by Hudson and Harrison (1997), 444 pages, and "Engineering Rock Mechanics: Part 2, Illustrative Worked Examples" by Harrison and Hudson (2000), 506 pages. More advanced presentations include "Fundamentals of Rock Mechanics" by Jaeger *et al.* (2007), 4th edition revised by R.W. Zimmerman.

\* \* \* \* \*

This concludes our presentations of both the structural geology principles (Chapter 2) and the rock mechanics principles (Chapter 3). The following Chapters 4–11 all discuss and illustrate the synthesis of the two subjects and demonstrate why a knowledge of the geological setting is crucial for the variety of rock engineering projects.

# CHAPTER 4

# ILLUSTRATIVE SYNTHESIS CASE EXAMPLE: THE CLIFTON SUSPENSION BRIDGE, UK

## 4.1 INTRODUCTION: THE CLIFTON SUSPENSION BRIDGE, ITS GEOLOGICAL SETTING AND THE STABILITY OF THE ABUTMENTS

The purpose of this Chapter is to present a case example of the application of structural geology to a rock engineering circumstance, i.e., as an initial case example of the synthesis of structural geology and rock engineering — which is the theme of this book. The case example involves the foundations of the Clifton suspension bridge crossing the river Avon in the south of England, Figure 4.1. The bridge was designed by Isambard Kingdom Brunel in 1830 and it spans the Avon Gorge between Clifton in the east and Leigh Woods in the west, towering 75 m above the river, with a span between the two towers of 215 m. It was completed in 1864, five years after Brunel's death and forms a fitting memorial to this great engineer and an iconic entrance to the city of Bristol. The general course of the river Avon runs from east to west. However, the 2.5 km long Avon Gorge runs approximately N–S, cutting through a resistant limestone ridge. It is

Figure 4.1. The Clifton suspension bridge showing the eastern abutment resting on well-bedded, fractured Carboniferous limestone.

thought that, during the last ice age, ice blocked the river's natural route causing it to cut a gorge through the limestone.

Both of the bridge abutments rest on Carboniferous limestone and, in the following Section, the history of fracturing of the limestone and the build-up of the current fracture network (which controls the bulk properties of the rock mass) is considered. It is the geometry and orientation of this network, together with the orientation of the gorge sides, which determines the stability of the abutments.

Note that the Avon Gorge is located only 20 km from the limestone quarry at Tytherington which we discuss in another case example presented in the next Chapter. Consequently it can be argued that, because the two localities lie within the same rock type and have experienced the same tectonic history, they are likely to have developed the same sets of fractures over geological time and currently contain the same fracture network. The main difference between the two engineering projects is that at Tytherington the layout of the quarry can be designed to exploit the various fracture sets — resulting in a design that enabled stable quarry faces which required minimum effort to extract the rock. But, in contrast, the stability of the bridge abutments is controlled by the fracture orientation in the underlying rock and the orientation of the free faces of the Gorge sides which are pre-determined by the tectonic history of the region and the course of the river. Thus, the Tytherington (mining) project discussed in the next Chapter and the Clifton (civil) project discussed in this Chapter represent the engineering counterpoint of respectively trying to cause rock failure and trying to avoid rock failure.

The geometry, orientation and properties of the fracture network present in the rock mass supporting the Clifton bridge abutments are described in some detail in the next Section to illustrate the methods used by structural geologists to determine the stress regimes operating during the various tectonic events which affect a study area over geological time. This enables them to predict the types and orientations of the fracture sets that might form during these different events and to construct the likely fracture network that would develop in the rock as a result of their superposition. It will be recalled from Section 2.12 that the detailed geometry of the fracture network is sensitive to the order in which the various fracture sets are superimposed. Armed with a model of the orientation and geometry of the fracture network present in the rock mass at an engineering site, the geologist is in a much stronger position to assess its impact on rock stability and to offer guidance on the positioning and orientation of proposed tunnels and surface excavations.

## 4.2 TECTONIC HISTORY AND THE ASSOCIATED DEVELOPMENT OF A FRACTURE NETWORK

An inspection of the geological cross-section through the Avon gorge and the surrounding area (Figure 4.2) shows that the region has been subjected to three major tectonic events. The section, which is oriented north–south, shows the occurrence of a major angular unconformity between the Carboniferous rocks which are folded and thrust and the overlying Permo-Triassic and younger rocks which are relatively undeformed. The folds and thrusts indicate a major compression event oriented approximately N–S. This was the Hercynian orogeny linked to the collision of two major plates, Laurentia in the north and Gondwanaland in the

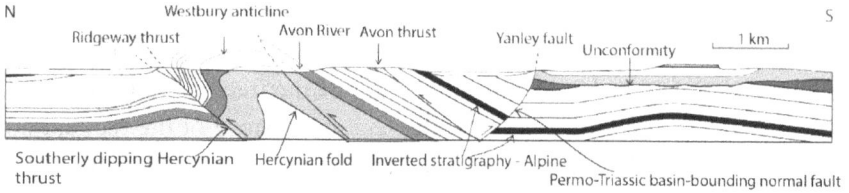

Figure 4.2.   ~N–S geological section through the study area showing the impact of the three major tectonic events that have affected the region, namely the Hercynian folding and thrusting, the Permo-Triassic rifting, and the Alpine inversion. A major angular unconformity separates the folded and thrust Carboniferous rocks from the overlying relatively flat-lying Permo-Triassic rocks.

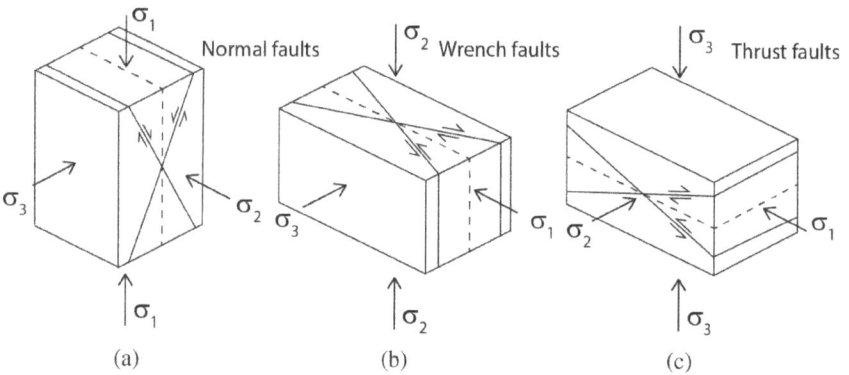

Figure 4.3.   Stress regimes associated with the formation of (a) normal faults, (b) wrench faults, and (c) thrust faults. Each regime can also generate extensional fractures (dashed) which bisect the conjugate shear fractures, see Section 2.3.2 and Figure 2.14(a).

south. This east–west trending mountain belt welded the two plates into a single plate named Pangea. The Bristol Channel is situated on the northern margin of this belt and the Hercynian structures that characterise the region are ~E–W trending folds verging towards, i.e., overturned towards the north, and northward verging, i.e., southward dipping, thrusts. Examples of both these structures can be seen in Figure 4.2, and the stress orientation associated with such a fold-thrust belt is shown in Figure 4.3(c).

The tectonic break-up of Pangea began in the Permo-Triassic and in the region of the Bristol Channel was associated with a major ~N–S extension. This generated a series of E–W oriented normal faults, Figure 4.3(a), which produced a series of basins, including the Bristol Channel

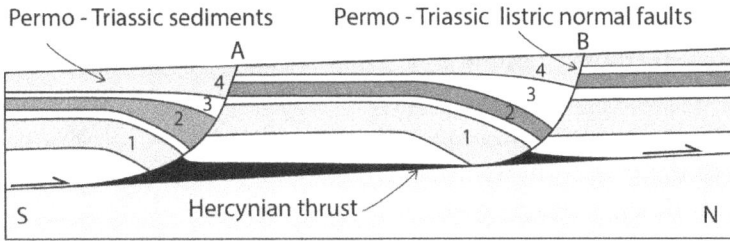

Figure 4.4. A southerly dipping Varsican thrust caused by an approximately N–S compression generated by the Hercynian plate collision. It subsequently acted as a major detachment onto which normal faults detached during their formation during the later Permo-Triassic extension.

and the English Channel Basins. The possible impact of the Hercynian thrusts on the geometry of these later normal faults is illustrated schematically in Figure 4.4. The faults detach onto the low angle thrusts and by doing so generate a curved (listric) profile.

The detailed history of extension linked to the formation of the fault-bounded basins shown in Figure 4.4 can be determined by examining the thickness variations of the different beds infilling the basins. The thickness of some beds remains constant when traced from the hanging wall (the left side of the two listric faults shown) towards the fault, for example, bed 1 as it approaches fault A; and bed 2 as it approaches fault B. This indicates that, during the deposition of these beds, no movement occurred on the faults. Other beds thicken as they approach the fault; bed 2 as it approaches fault A, and beds 1 and 3 as they approach fault B. This thickening reflects movement on the fault during their deposition. An example of a northward dipping, listric, normal fault can be seen on the cross-section of Figure 4.2 where it is labelled as a Permo-Triassic basin bounding normal fault.

The basins continued to open until the Late Cretaceous when opening was arrested by the collision of the African and European plates which generated the Alpine orogeny. This changed the tectonic stress regime from a ~N–S extension to a ~N–S compression. The effect of this on the basin bounding normal faults was to reverse the sense of slip from one of down-dip to up-dip. It can be seen from the cross section in Figure 4.2 that the reverse movement ('inversion') on the Yanley fault caused by the

Alpine orogeny resulted in a net contraction across the fault. For example, the Carboniferous sandstone layer (coloured black) prior to inversion was at a deeper level in the block to the north of the fault (the hanging wall) than the same layer in the southern block (the footwall). However, as a result of the 'reverse' movement on the fault caused by the Alpine inversion, it is now at a higher level.

This brief inspection of the geological cross section across the Avon Gorge and surrounding area has revealed three major tectonic events each linked to episodes of faulting. However, as discussed in Section 2.3 there are two types of brittle failure: namely shear failure (faulting) and extensional failure (jointing), see Figure 4.3. Before discussing detailed field observations relating to the orientation and type of fractures that occur in the vicinity of the bridge abutments, it is useful to consider what regional fractures might be expected to form during the three tectonic events that have affected the rocks making up the Bristol Channel Basin of which the study area, Figure 4.2, is a part.

The first of these gave rise to the Hercynian folding and thrusting. During this time, the least principal stress ($\sigma_3$) was vertical and the maximum principal compression ($\sigma_1$) horizontal and directed approximately N–S, see Figure 4.3(c). It will be recalled that the Bristol Channel Basin is situated on the northern margin of the Hercynian orogenic belt, i.e., away from the central zone where ductile processes dominate the deformation. In the marginal regions, both ductile and brittle deformation occur and a fold-thrust belt is generated. Thus, in addition to forming ~E–W trending folds, this stress regime can generate gently dipping (~30°) thrusts and horizontal and therefore often bedding-parallel, extensional fractures.

The process of forming thrusts often results in the stacking of thrusts one on top of the other and this process leads to an increase in the vertical stress. It is therefore possible to increase the vertical loading to the point when it became the intermediate principal stress ($\sigma_2$). This change in stress regime would generate wrench faults and ~N–S trending vertical extensional fractures, see Figure 4.3(b).

Inspection of the Hercynian faults and folds exposed on the northern edge of the Bristol Channel Basin in Pembrokeshire, South Wales, UK, shows clearly that this transition of stress regimes did occur, Figure 4.5. Several approximately E–W trending folds and thrusts can be seen, but it

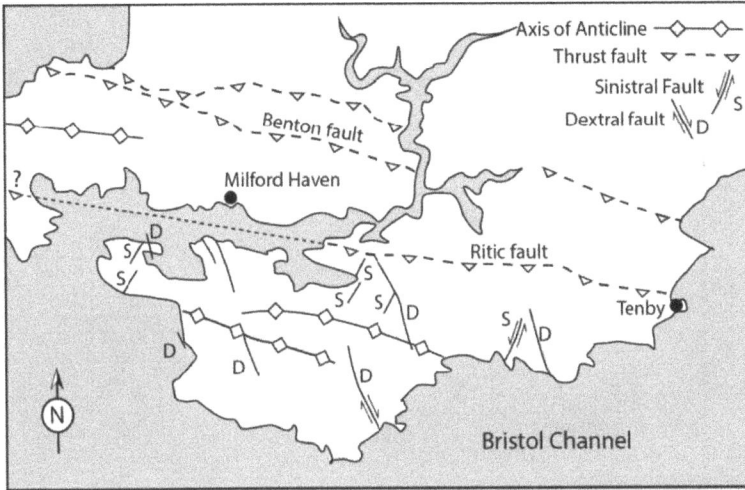

Figure 4.5.   Sketch map of the major Hercynian structures in Pembrokeshire, South West Wales on the northern margin of the Bristol Channel Basin. The structures include the early formed WNW–ESE trending thrusts and folds and the later, cross-cutting, wrench faults (D = dextral, S = sinistral). After Anderson, 1942.

is also apparent that these structures have been cut by later wrench faults trending ~NW–SE and ~NE–SW and displaying a dextral and sinistral sense of shear, respectively, compatible with a ~N–S compression.

The tectonic phase of extension began in the Permian. The maximum principal compression ($\sigma_1$) was vertical and the minimum compression ($\sigma_3$) oriented approximately N–S. The stress regime in a N–S vertical section through the crust at this time is shown in Figure 4.6. It can be seen that both the vertical and horizontal stresses increase with depth and that the difference between them, the differential stress, also increases.

It will be recalled from Section 2.3.4 that it is the differential stress that controls the type of fracture which forms: a low differential stress favours the formation of joints (extensional fractures); and a large differential stress favours the formation of faults (shear fractures). Thus, the fractures that would form in response to the extensional regime shown in Figure 4.6 are normal faults in the basement, i.e., at depths where the differential stress is relatively high. These form the rift basins into which the Permo-Triassic and younger sediments are deposited. The lower differential

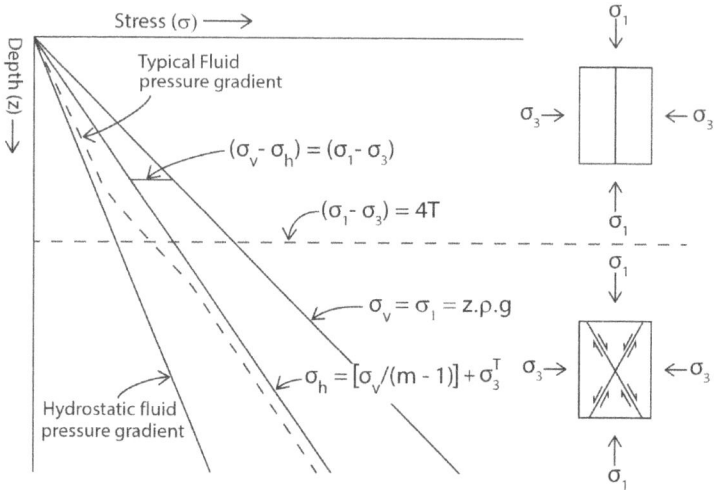

Figure 4.6. Graph showing the increase of the vertical stress ($\sigma_v = \sigma_1$), horizontal stress ($\sigma_h = \sigma_3$) and fluid pressure with depth in the crust when the stress is the result of the overburden and a horizontal, tectonic extension $\sigma_3^T$. As the differential stress increases with depth, so the mode of failure changes from extensional failure (joints) to shear failure (faults). The changeover occurs when the differential stress reaches $4T$, where $T$ is the tensile strength of the rock. See Section 2.3.4.

stresses in these younger rocks would result in extensional fractures forming. These would have the same strike as the basin-bounding faults (~E–W), Figure 4.3(a).

This brief structural analysis of the tectonic evolution of the Bristol Channel Basin and the likely fractures that would form is based on observations of large-scale structures, i.e., the basin-bounding faults of the type shown in Figures 4.2 and 4.4 and the large-scale folds and thrusts shown in Figure 4.5. However, a detailed study of the smaller fractures often produces a more accurate indication of stress field orientation. Such a study of the rocks of the Bristol Channel Basin has been conducted by Nemcok *et al.* (1995). Their work confirms that the Bristol Channel Basin experienced a prolonged rifting event from the Triassic through to the Aptian (Middle Cretaceous).

Working on the South Wales coast in an area midway between Cardiff and Swansea, they observed numerous WNW–ESE trending, normal faults, Figure 4.7, which are small-scale examples of the normal faults that

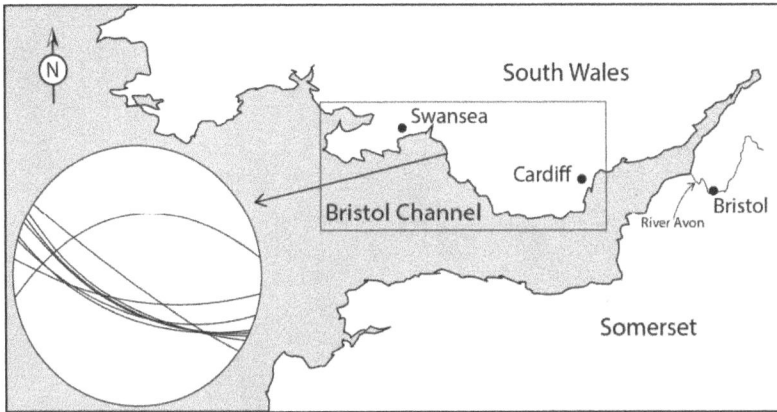

Figure 4.7. A rectangle defining an area on the northern margin of the Bristol Channel Basin in which data on the orientation of rifting-related normal faults in the Permo-Triassic and Jurassic beds have been collected. The faults are plotted onto a lower hemisphere, stereographic projection (bottom left of the Figure). Each arc represents a fault plane. The compass bearing is read clockwise from the top and the ends of the arcs represent the strike of the planes. The strikes can be seen to trend ~WNW–ESE. (After Nemcok *et al.*, 1995.)

bound the Bristol Channel basin. In addition, rift-related extensional veins, also trending WNW–ESE, Figure 4.8, formed in both partially and fully lithified sediments, with evidence of multiple reactivation. This evidence is compatible with the conclusions drawn from the observation represented schematically in Figure 4.4. Here bed thickness changes occur in some beds and not in others — indicating that basin opening occurred in a series of pulses, rather than at a constant, uniform rate. As is discussed later in this Section, further evidence for multiple episodes of extension has also been recorded in the Avon Gorge by the textures of the calcite infillings of the J1 joint set which are interpreted as rift-related extensional veins, (see Figure 4.15 and Table 4.1).

These WNW–ESE trending fractures, observed by Nemcok *et al.* (1995) in the sediments filling the extensional basins, can also be expected in the underlying Carboniferous rocks which are currently exposed in the abutments of the Clifton suspension bridge where, as is discussed later in this Chapter, they are represented by the J1 joint set.

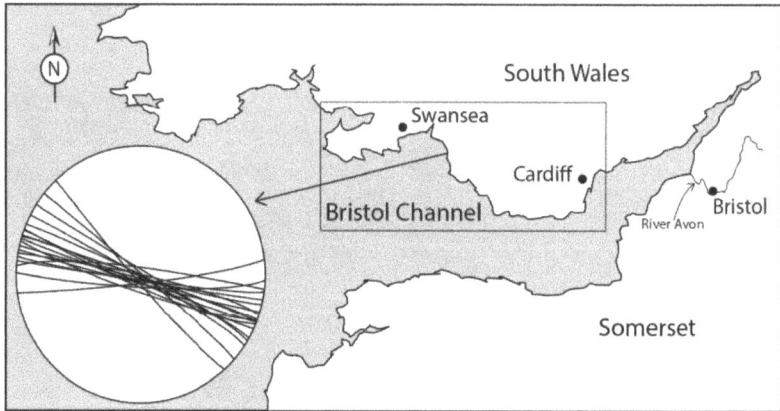

Figure 4.8. A rectangle defining an area on the northern margin of the Bristol Channel Basin in which data on the orientation of rifting-related extensional fractures in the Permo-Triassic and Jurassic beds have been collected. The faults are plotted onto a lower hemisphere, stereographic projection (bottom left of the Figure). The strikes can be seen to trend ~WNW–ESE. (After Nemcok *et al.*, 1995.)

As noted above, the region was affected by a new tectonic regime at the end of the Cretaceous in Aptian times which was linked to the onset of the collision of Africa and Europe and the initiation of the Alpine Orogeny. This event was associated with a new stress configuration in which the minimum principal stress was vertical and the maximum principal compression was orientated approximately N–S. The stress regime was responsible for the reactivation of existing faults and for the formation of new fractures within the basement and the basin sediments. Its effect on the extensional basins was profound. The response to this ~N–S compression was to invert the WNW–ESE striking, basin bounding, normal faults which moved in a reverse, up-dip, direction. This is shown diagrammatically in Figure 4.9(b) and there is considerable evidence for the inversion of these rift bounding faults both on seismic sections and on the constructed cross-sections based on field observations, Figure 4.2.

In addition, the thrust faults and the wrench faults generated in the Hercynian deformation were ideally orientated for reactivation, Figures 4.9(a) and b. In the post-Carboniferous rocks, i.e., rocks deposited after the Hercynian deformation and which therefore do not contain these thrust and wrench faults, the effect of the ~N–S Alpine collision would be

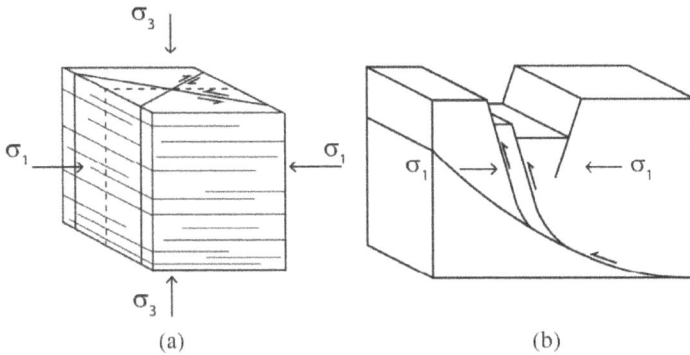

Figure 4.9. (a) The stress regime linked to the Alpine collision which began in the Mid to Late Cretaceous. It generated an approximately N–S compression which closed (inverted) the rift basins shown in (b) and reactivated the Hercynian wrench faults and thrusts shown in (a) and (b) respectively. It also generated new fracture sets both within the basin sediments and in the underlying Carboniferous basement.

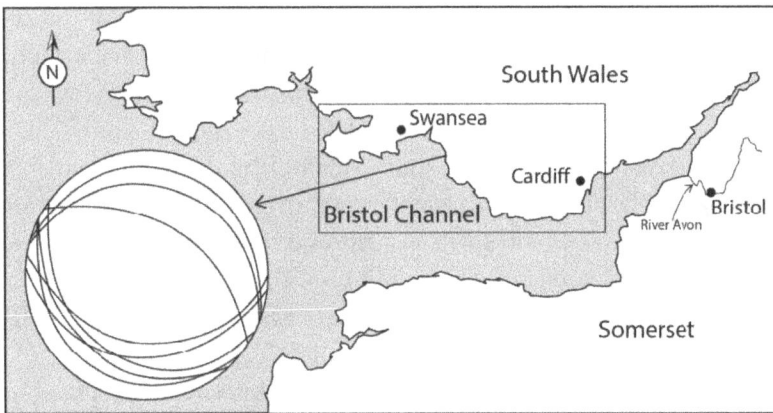

Figure 4.10. A rectangle defining an area on the northern margin of the Bristol Channel Basin in which data have been collected on the orientation of thrusts in the Permo-Triassic and Jurassic beds thought to have formed during the early stages of inversion. The faults are plotted as arcs onto a lower hemisphere, stereographic projection (bottom left of the figure). The strikes can be seen to trend ~WNW–ESE. (After Nemcok *et al.*, 1995.)

to generate thrusts striking ~E–W, see Figure 4.3(c). These can be found in the Permo-Triassic and Lower Jurassic rocks, as shown in Figure 4.10.

It was noted in the discussion of the Hercynian ~N–S compression that, at a later stage in the formation of a fold thrust belt, the vertical stress

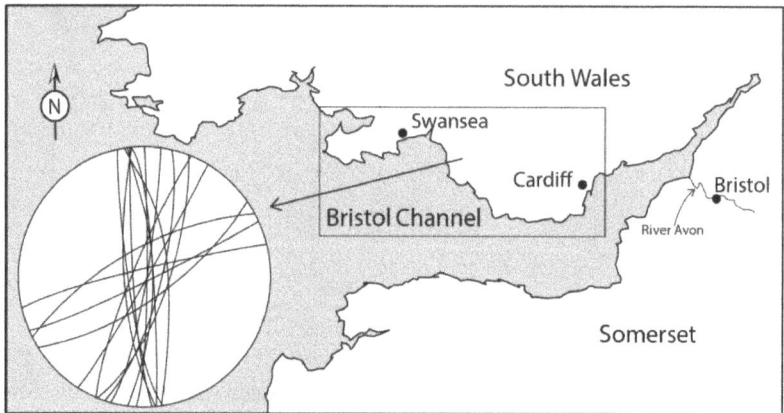

Figure 4.11. Stereographic representation (lower left hand corner) of wrench faults formed in the Permo-Triassic and Lower Jurassic rocks at a late stage in the Alpine inversion of the Bristol Channel Basin. (After Nemcok *et al.*, 1995.)

can change from $\sigma_3$ to $\sigma_2$ with a corresponding change in the faulting regime from thrusting (Figure 4.3(c)) to wrench fault formation (Figure 4.3(b)), see Figure 4.5 and the related discussion. A similar change appears to have occurred in the study area during the Alpine tectonism, and wrench faults linked to this event are shown in Figure 4.11. The extensional fractures that would form in response to the stress configuration that generated the wrench faults would be vertical fractures which bisect the conjugate wrench faults shown in Figure 4.9(a). These have also been recorded in the field and are shown in Figure 4.12.

The above discussion outlines the method by which it is possible to determine the chronology of the tectonic events that have affected the region containing the site of engineering interest, in this example, the footprint of the Clifton Suspension Bridge. Each of these events can produce fracture sets which become superimposed to produce the fracture network currently occupying the rock mass and determining its bulk properties including strength and conductivity. The tectonic events identified are as follows in sequence.

— Thrusting and folding as a result of tectonic compression linked to the Hercynian orogeny which occurred in the Late Carboniferous

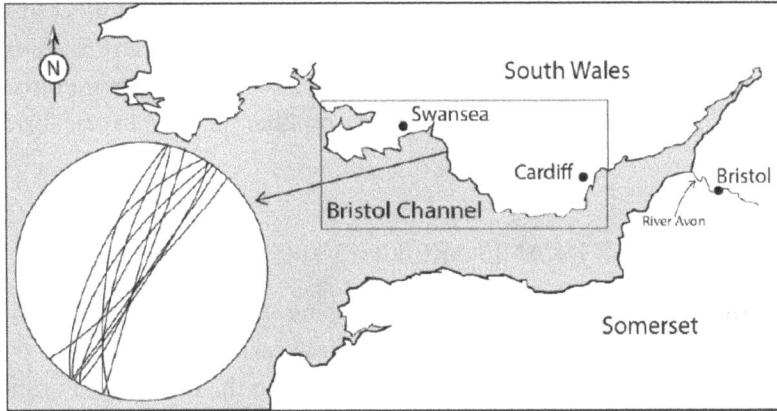

Figure 4.12. Stereographic representation (lower left hand corner) of extensional fractures formed in the Permo-Triassic and Lower Jurassic rocks at a late stage in the Alpine inversion of the Bristol Channel Basin. (After Nemcok *et al.*, 1995.)

(~300 mya). During this time, the rocks were deformed as part of a fold–thrust belt situated on the northern margin of the E–W trending Hercynian mountains.

— Extension (rifting) linked to the break-up of Pangea which began in the Permian and continued into the Cretaceous (~290–100 mya).

— Alpine inversion of the rift basins which began in the Late Cretaceous (~80 mya) and which is continuing today.

This brief study of the tectonic history of the Bristol Channel area allows one to predict the following.

— The types and orientations of the fracture sets likely to have formed in the limestones of the Avon Gorge during the ~300 million years since they were deposited.

— Their relative timing.

— The geometry and orientation of the fracture network generated by the superposition of these fracture sets.

— The likely mechanical properties that this history and the associated fracturing has imparted to the rock mass.

In the following Section, a summary of the fractures observed in the vicinity of the abutments of the Clifton Suspension Bridge is presented, together with a discussion of their likely origin within the context of the tectonic history outlined above. This is followed by a summary Section in which the fracture network observed in the rocks of the study area is discussed in terms of its likely impact on the stability of the bridge abutments.

## 4.3 FRACTURES IN AND AROUND THE ABUTMENT AREAS

The bridge abutments rest on gently dipping, Carboniferous limestone, Figures 4.1 and 4.13. A detailed study of these rocks shows that four fracture sets and the bedding have a major impact on the mechanical properties of the rock mass on which the bridge is founded, Figure 4.14. This latter Figure shows a bedding plane in the vicinity of the Clifton abutment. Four fracture sets, 1–4, can be identified, see Table 4.1. Two are calcite filled and the other two are barren.

Figure 4.13.   View of the Clifton abutment resting on Carboniferous limestone dipping gently to the south and containing both vertical and bedding-normal extensional fractures outcropping on the major J2 joints that form the face of the abutment which is parallel to the gorge.

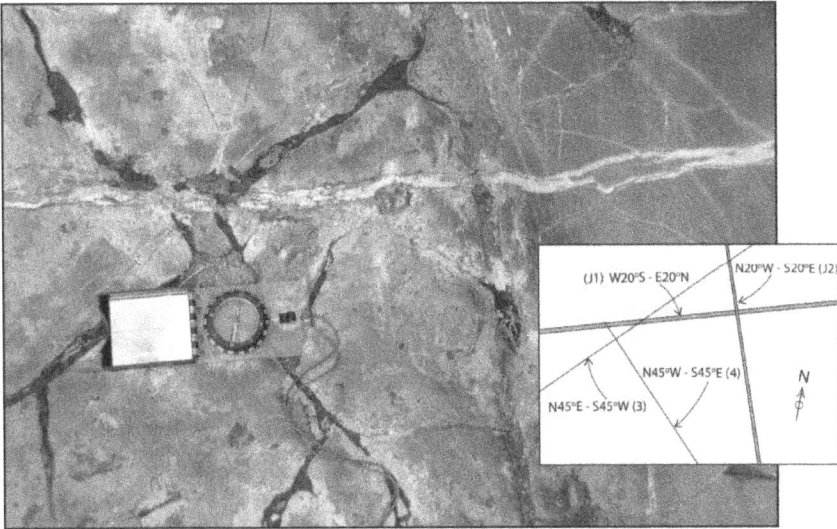

Figure 4.14. A bedding plane close to the Clifton abutment shown in Figure 4.13 cut by four sets of fractures.

Table 4.1. The four major fracture sets which cut the abutment supports of the Clifton suspension bridge.

| Set number | Set description |
| --- | --- |
| Set 1 | — Calcite filled fractures which strike ~W20°S–E20°N (Termed J1) |
| Set 2 | — Calcite filled fractures which strike ~N20°W–S20°E (Termed J2) |
| Set 3 | — Barren fractures which strike ~N45°E–S45°W |
| Set 4 | — Barren fractures which strike ~N45°W–S45°E |

Fracture Set 1, termed here J1 joints, cut vertically through the dipping limestone beds, Figures 4.15(a) and b, which form part of the gently dipping limb of Hercynian folds (the Westbury anticline, see Figure 4.2). One of these major J1 joints forms one face of the limestone columns on which one of the bridge abutments is situated, Figures 4.1 and 4.16. These joints have been superimposed onto this fold and are unaffected by the orientation of the beds. It is argued therefore that they post-date the folding and, in the context of the tectonic evolution of the region described above, correspond to the vertical extensional fracture linked to the Permo-Triassic

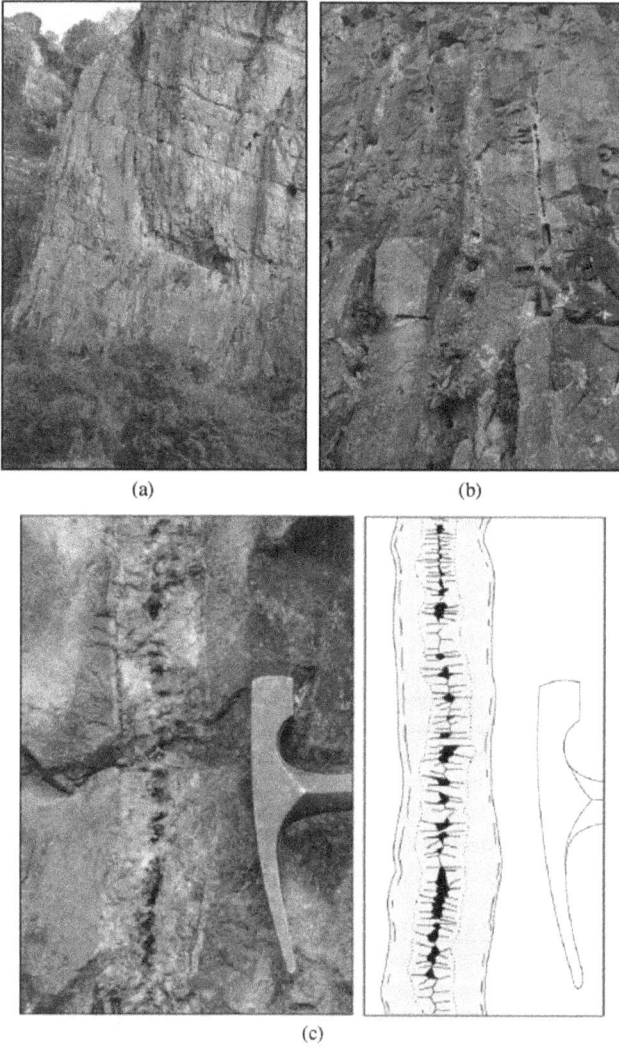

(a)    (b)

(c)

Figure 4.15. (a) Detail of the rock mass beneath the Clifton abutment (see Figure 4.13) showing bedding dipping at ~30° to the south, exposed on a north–south striking, vertical J2 joint. The occurrence of a well-developed, through-going, vertical set of extensional fractures (J1) approximately normal to J2 (~W20°S–E20°N) can be seen cropping out on this face, together with the more strata-bound, bedding-normal fractures. (b) Detail of vertical J1 extensional fractures. The joints are partially infilled with calcite and show textures that indicate several episodes of extension. (c) Vertical extensional fracture, partially infilled with calcite and showing evidence of multiple episodes of opening.

extension (basin opening). Examples of these fractures have been recorded in the younger (Jurassic) sediments of the Bristol Channel basin further west in South Wales, Figure 4.8.

Further evidence for the correlation of these through-going fractures with the extensional fractures linked to the opening of the Bristol Channel Basin is provided by the textures of the minerals that fill or partly infill these fractures, Figures 4.15(b) and 4.15(c). The textures show that the fractures opened in a series of increments, each associated with the precipitation of a layer of calcite on the fracture walls. This observation is compatible with the observations made by Nemcok *et al.* (1995) in their study of the extensional fractures linked to basin opening, see Figure 4.8 and the related discussion.

Each episode of vein development represents an episode of basin extension during which time the resulting extensional fractures acted as transient pathways for the migration of hydrothermal fluids, the cooling of which caused precipitation of the calcite and the eventual infilling of the fracture. Although during most of the extensional events the rate of crystallisation of the calcite infill was equal to the rate of fracture opening with the result that at the end of the period of extension the fractures were filled with a calcite vein and fluids could no longer move along them, it can be seen in the composite vein shown in Figure 4.15(c) that the latest event generated an open fracture. Clearly, the rate of precipitation of the calcite could not keep pace with fracture opening and the calcite crystals which grew into this open space developed characteristic euhedral terminations.

If fractures remain open after an episode of extension (Figure 4.15(c)), then they continue to provide channels for fluid migration even after the period of extension has ceased. Subsequent fluids may not be charged with calcite. For example, rain water (weak carbonic acid as a result of the dissolved of $CO_2$ from the atmosphere) has the opposite effect to the mineralising fluids. Instead of precipitating vein material and filling and strengthening the fractures, it dissolves the limestone along the fractures, resulting in local widening and weakening of the fracture (a process known as karstification). It is noted that open veins of the type shown in Figure 4.15(c)), are found present in both the J1 and J2 fracture sets (see Table 4.1) and local, intense zones of karstification can be found along both in the vicinity of the bridge abutments.

These observations are important when considering the likely properties of the fracture network within the rocks and the resulting integrity of the limestone column supporting the bridge abutments, Figure 4.1. An appreciation of whether vein infill or solution occurs along any of the fracture sets making up the fracture network will help in assessing its bulk properties.

The second set of joints cutting the abutment area trend N20°W–S20°E and has been termed J2 (see Table 4.1). Like the J1 joint set, J2 joints are also through-going and one of them defines the west facing wall of the limestone column on which one of the bridge abutments is situated, Figures 4.13 and 4.15(a)).

Within the context of the tectonic setting of the area discussed above, there are two possible interpretations for the origin of the J2 fracture set, namely that they relate to

(a) Extensional fractures associated with the Late Hercynian wrench fault stress regime, Figure 4.3(b) or
(b) Extensional fractures associated with the Late Alpine inversion, Figure 4.3(b).

Field observations show that, in general, the J2 fracture set cuts the J1 set, indicating that it post-dates the Permo-Triassic extension. Therefore interpretation (b) is favoured. The through-going nature of the J2 fracture set can be seen in Figures 4.16 and 4.17(b) which shows them intersecting a major J1 fracture.

A study of the rocks in the region of the bridge abutments shows that in addition to sets J2 and J1, two other fracture sets occur. These are fracture sets 3 and 4 (Table 4.1) which trend N45°E–S45°W and ~N45°W–S45°E, respectively. Possible interpretations of these two fracture sets are as follows.

(a) They could be conjugate shear joints linked to the Late Hercynian compression, i.e., after the thrust regime had changed to a wrench regime, Figures 4.3(b) and 4.5.
(b) They could be conjugate shear joints linked to the Late Alpine compression when a similar stress regime occurred, Figure 4.3(b).

Because these fractures are cross-cut by the J1 fracture set which is thought to be of Permo-Triassic age, (a) is the favoured interpretation.

As noted, the bridge abutments rest on Carboniferous limestone and one of its most important features from the point of view of its impact on abutment stability is the bedding. The bedding planes represent an important set of continuous fractures, Figures 4.13, 4.15(a), 4.16, and 4.17, which must be added to the four fracture sets listed in Table 4.1 when attempting to determine the geometry and orientation of the fracture network within the rock mass beneath the abutments. The impact of bedding on the properties of the fractured rock mass is at least as important as the two major sets of vertical fractures, sets J2 and J1.

Inspection of Figures 4.15–4.17 shows that the major joints tend to be regular, planar features. In contrast, the regularity of bedding planes in limestone can be highly variable. An important property of limestone is its ability to dissolve, both in response to the weak carbonic acid represented by rain and in response to a stress, see below. During the burial and diagenesis (the latter being the process of chemical and physical changes that occurs in a sediment during its conversion to a rock) of limestone, the overburden

Figure 4.16. The north-facing face of the Clifton abutment support defined by a J1 fracture on which outcrops the sub-vertical J2 fracture set which is parallel to the gorge side.

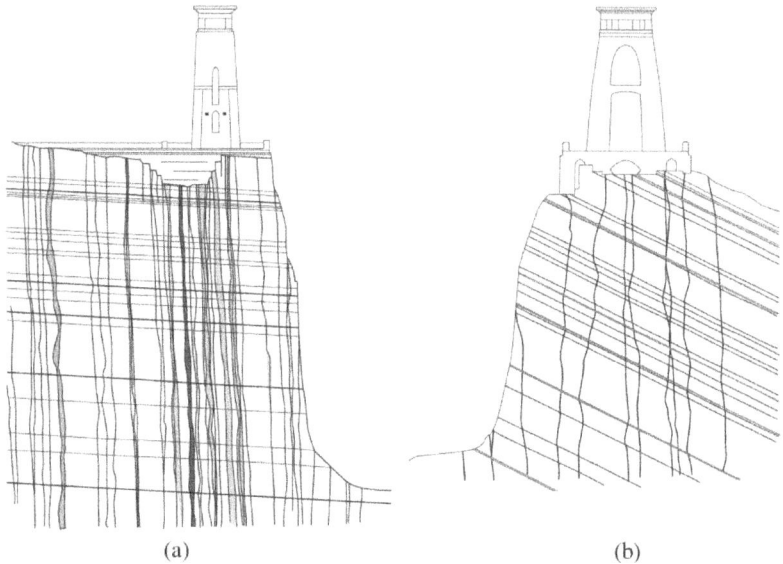

Figure 4.17.   (a) Line drawings of the north-facing wall of the Clifton abutment support. The wall represents a J1 joint face on which the vertical J2 joints and the bedding (which has an apparent dip close to zero) crop out. (b) Line drawing of the west-facing wall of the Clifton abutment support. The wall represents a J2 joint on which the vertical J1 joints and the bedding (which shows its true dip of 30° to the south) crop out. The 'clustering' of the J2 fractures can be seen in a) where they define a fracture corridor.

load can generate pressure solution along the bedding planes when two adjacent beds are in direct contact with each other, i.e., not separated by a clay horizon. This solution of the limestone results in the formation of stylolites, i.e., serrated surfaces along the bedding plane, Figure 4.18. The process of stylolitisation results in the interlocking of the two adjacent beds — with the result that the process of bedding plane slip is severely inhibited.

In contrast, when two adjacent limestone beds are separated by a thin layer of clay, the contact between the two beds is prevented, no stylolitisation occurs and the beds remain flat. In addition, the thin layer of clay between them enhances the possibility of bedding-plane slip. Consequently, these bedding planes are particularly likely to slip and can be exploited during folding when they facilitate slip between the beds as the folds amplify or during thrusting when, as low-cohesion, planar fractures, they can be incorporated into the thrust planes.

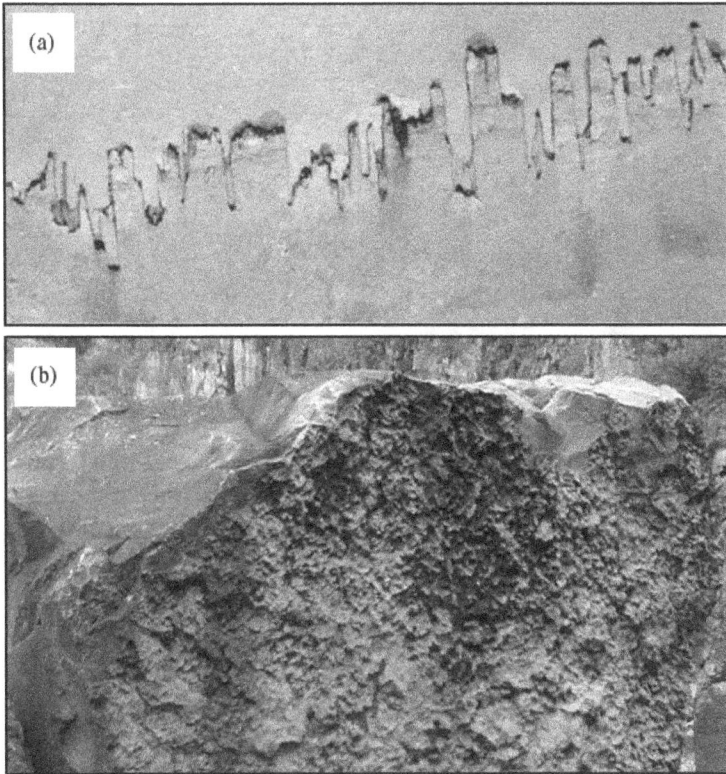

Figure 4.18. (a) A profile through a stylolite formed by pressure solution along a bedding plane in a limestone. (b) Exposed bedding plane along which stylolites have formed during burial and diagenesis as a result of the overburden load. The extremely irregular surface generated prevented bedding plane-slip during the Variscan orogeny when thrusting developed along other bedding surfaces unaffected by solution because of the presence of a thin clay horizon, see Figures 5.19 and 5.20 and the related discussion.

Within the vicinity of the bridge abutments, intense stylolitisation has occurred along some of the bedding planes in the limestone. An example of this is shown in Figure 4.19 where pressure solution has produced an irregular bedding plane, resulting in the beds being locked together. In addition to its impact on movement along bedding during tectonism, the presence or absence of stylolites along a bedding plane will also have a major impact at the present time on the ability of a bedding plane to slip and therefore on the stability of the rock mass containing them.

Figure 4.19.   An irregular bedding plane on the road side just south of the Clifton abutment, showing the effect of pressure solution and the formation of stylolites which are thought to have formed during burial and diagenesis in response to the overburden stress.

It has been argued that fracture sets 3 and 4, Table 4.1, were formed in response to the Late Hercynian compression during the phase when the vertical stress had been increased to $\sigma_2$ by thrusting (see Figure 4.5 and the related discussion). However, it can be seen from the geological cross-section shown in Figure 4.2 that structures related to the early Hercynian tectonics (stress regime Figure 4.3(c)), namely thrusting and folding (see Figure 4.5), are also present in the area. A relatively large-scale thrust has occurred a few hundred metres upstream (i.e., north) of the bridge, and is exposed in the cliffs of the eastern side of the gorge, Figure 4.20. Like the Ridgeway thrust shown in Figure 4.2, it dips steeply to the south. Movement on the thrust has caused folding of the underlying beds and the asymmetry of the minor fold mirrors that of the major fold (the Westbury anticline) shown in Figure 4.2. Fortunately, although the major thrust shown in Figure 4.20 occurs fairly close to the bridge abutments, such structures have not formed in the immediate vicinity of the bridge footprint and therefore do not influence the fracture network of the abutment supports.

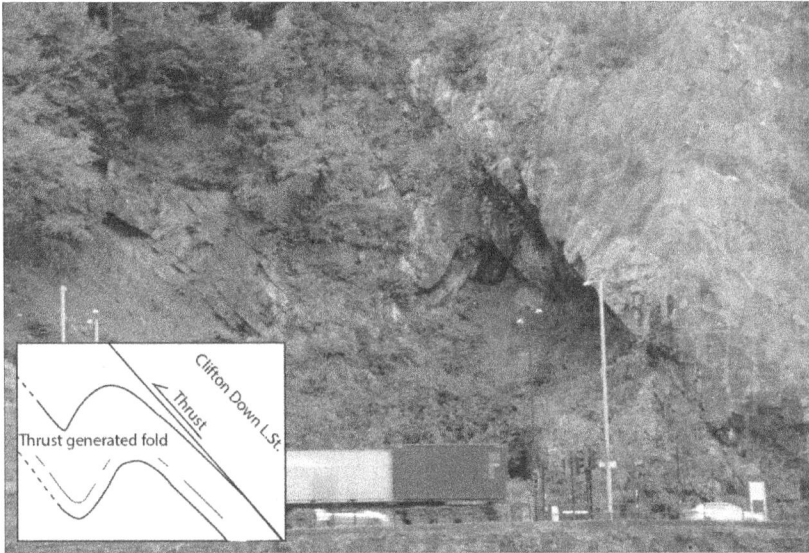

Figure 4.20. View of the eastern side of the Avon Gorge a few 100 m north of the Clifton suspension bridge showing details of the deformation (folding) that occurred below the Clifton Down Limestone as it over-rode the upper Cromhall Sandstone.

Figures 4.16 and 4.17(a) show sub-vertical fractures of joint set 2 cropping out on a J1 joint surface which defines the northern face of the limestone block on which the Clifton abutment rests. This block is cut by three major fracture sets.

— A sub-vertical set of extensional fractures (J1) trending ~W20°S–E20°N which forms the outcrop face shown in the Figures.
— A sub-vertical set of extensional fractures (J2) which cut the outcrop face and which trend ~N20°W–S20°E, Figure 4.17(b). This set is sub-parallel to the gorge side.
— The bedding which has a sub-horizontal apparent dip on the outcrop face of J1 and which dips at ~30° towards ~S20°E as can be seen in Figures 4.13 and 4.17(b).

The important observation here is that all three fracture sets tend to be through-going. Bedding, by its mode of origin, represents a through-going set of fractures. All other fractures are later and may either be constrained by the bedding (i.e., become strata-bound as for example, sets 3 and 4 in Table 4.1) or, like the J2 and J1 fractures shown in Figure 4.16, they may cross-cut the bedding.

Close inspection of Figures 4.16 and 4.17 shows that the J2 fractures are not evenly spaced and tend to cluster into fracture corridors (see Section 2.8.4), a feature that, as discussed later in this Section, has a significant impact on the stability of the rock mass.

## 4.4 EVALUATION OF THE GEOLOGICAL INFORMATION

The regional study of the tectonic evolution of the Bristol Channel Basin presented earlier in this Chapter has been used to analyse and interpret the fractures observed in the rocks immediately surrounding the bridge abutments. The orientations of five main fracture sets have been identified (i.e., those listed in Table 4.1, plus bedding) and a brief study made of their spacing, continuity, distribution and relative timing. It is the fracture network resulting from the superposition of these five sets that will determine the bulk rock mass mechanical properties and therefore the stability of the rock mass on which the bridge is situated. Three of these fracture sets can be observed on the stereographic projection in Figure 4.21, which shows the fractures recorded in the limestone directly beneath the Clifton abutment, Figures 4.15(a), 4.16, and 4.17 (the bedding is not shown).

The chronology of the development of these sets has been discussed earlier in this Section where it was concluded that they developed in the following order:

(a) Bedding — which in the study area dips ~30° to the south.
(b) A barren fracture set, normal to bedding and striking ~N25°E (Fracture set 3, Table 4.1).
(c) A barren fracture set, normal to bedding, and striking ~N45°W, (Fracture set 4, Table 4.1). Fracture sets 3 and 4 are interpreted as a conjugate set of shear fractures which formed during the Late Hercynian compression when the stress regime was appropriate for the formation of wrench faults, i.e., stress regime, Figure 4.3(b).
(d) A calcite filled fracture set striking ~W–E (Fracture set 1, (J1) Table 4.1) linked to the Permo-Triassic extension.

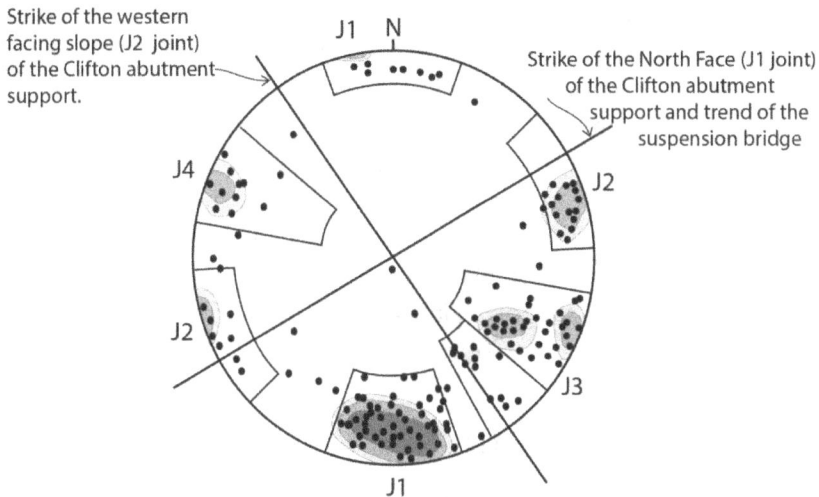

Strike of the western facing slope (J2 joint) of the Clifton abutment support.

Strike of the North Face (J1 joint) of the Clifton abutment support and trend of the suspension bridge

Figure 4.21. A summary stereographic projection showing the orientation of the main joint sets recorded in the rock mass under the Clifton abutment (bedding omitted) see Figure 4.23. Unlike the plotting of the whole joint planes as in Figure 4.7, here just the perpendiculars to the planes (i.e., the poles) have been plotted — as points. As before, the azimuth direction 0°–360° is plotted clockwise from the top and the inclination (i.e., the plunge, the inclination from the horizontal) plotted from 0° at the periphery to 90° at the centre. Examples of the joints J1 and J2 bounding one of the abutment supports are shown in Figure 4.17.

(e) A calcite filled fracture set striking ~N15°W (Fracture set 2, (J2) Table 4.1) linked to the Late Alpine compression.

This chronology is based on knowledge of the tectonic evolution of the area and on a preliminary fracture analysis of the rocks of the study area (see Section 2.12.3).

Before considering the impact of these fracture sets on the stability of the rocks supporting the bridge abutments, it is useful to consider them briefly in turn and determine their spacing, continuity and distribution.

## 4.5 CONTINUITY OF THE FRACTURES

The intensity and continuity of fracture sets 3 and 4 varies considerably in the region. At Tytherington quarry, approximately 20 km to the NW of the abutments, they form major, through-going, vertical fractures (see the Tytherington quarry case example in the next Chapter) with evidence of significant strike–slip movement along them.

In contrast, in the area in the immediate vicinity of the bridge abutments, these fractures are rather poorly developed, are generally normal to bedding and strata-bound, Figure 4.22. However, despite their rather unimpressive expression in the study area, as is discussed below, they are considered to be of great importance in terms of their impact on the stability of the rock mass supporting the abutments and of the cliff faces that define the sides of the Avon Gorge.

## 4.6 DISTRIBUTION OF THE FRACTURES

The J2 joints are not always uniformly spaced, but occur in clusters known as fracture corridors. The tendency for fractures to cluster is thought to be related to the mechanical properties of the rock, discussed in Section 2.8.4, and this tendency can have an important impact on rock stability. It is clear from Figures 4.16 and 4.17(a) that erosion of the rock mass on which the abutment has been built was focused in the vicinity of the fracture cluster. A clear gully formed at this location and the abutment has been constructed within it. However, the rock either side of the corridor is much less fractured with the result that, in the region close to the edge of the rock mass where major J2 fractures define the edge of the block and the Gorge side, the rock is much less fractured and correspondingly more stable.

Recognising the chronology, orientation, distribution and continuity of these five major fracture sets in the vicinity of the bridge abutments provides an understanding of the geometry of the fracture network that currently exists within the rock. This enables the stability of the rock mass to be assessed when it is exposed at free surfaces such as the Gorge walls beneath the abutments. Figure 4.23 shows the orientation of these free surfaces in the vicinity of the bridge. Possible instabilities that might develop at the two abutments are shown in Figure 4.24. It is assumed that the free surface is the Gorge wall which at the location of the bridge is approximately parallel to the J2 joints. The Figure shows two block diagrams containing the five fracture sets. The diagram 'looking east' related to the Clifton abutment shows the bedding dipping south and blocks potentially

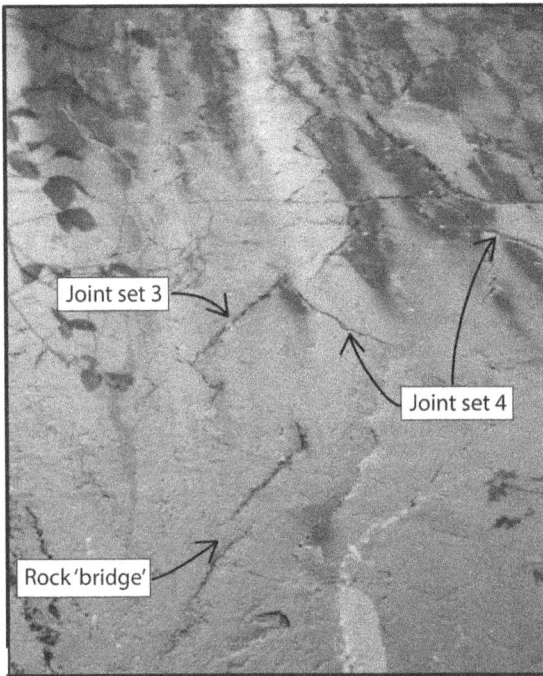

Figure 4.22. A detail of a bedding plane in the Clifton Down limestone downstream of the Leigh Woods abutment (Figure 4.23) showing incipient members of conjugate fractures made up of fracture sets 3 and 4 (Table 4.1) looking north. Note the 'rock bridge' between two offset J3 fractures. Although these bridges tend to inhibit slip of rock blocks on the bedding, they are nevertheless areas of stress concentration and thus prone to failure leading to the release of blocks.

sliding into the Gorge on bedding planes, released by the conjugate fractures striking ~N45°W and ~N45°E. In order to determine whether this mode of failure is likely to occur it is necessary to determine precisely the dip of the beds and the coefficient of sliding friction along the bedding.

In addition to the possibility of plane sliding on the bedding, the existence of the J2 fractures defining the abutment face parallel to the Gorge combined with the bedding planes and the J1 structures which could act as releasing fractures, means that there is also the possibility of toppling instability, see Figures 4.16 and 4.17(a), (see Principle 37, Chapter 3).

The consideration of the rock exposed along the Avon Gorge outlined above illustrates why a knowledge of the geology and the geological

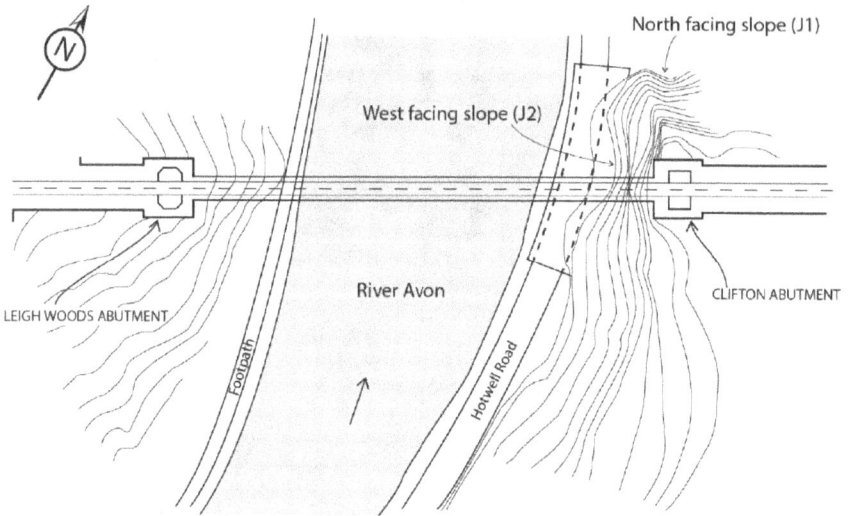

Figure 4.23.   The topography of the Avon Gorge in the vicinity of the Clifton Suspension Bridge. The contours reveal the shape and orientation of the rock mass beneath the bridge abutments defining the free surfaces which in combination with the orientation of the various fracture sets determine the stability of the abutment supports.

evolution of the region is crucial in assessing its stability — because this enables the orientations of the different fracture sets likely to be present in the rock mass to be determined and also reveals the chronological order in which they were superimposed to form the fracture network that now dominates the properties of the rock. Knowing the geometry and orientation of the fracture network and the orientation of the bedding (itself an important set of 'fractures'), it is possible to predict generically the type of instabilities likely to occur in the region. In addition, if a specific site is being considered where the orientations of the free surfaces that define the rock mass at the surface are known, then it is  possible to make specific predictions regarding the type of instability (planar sliding, wedge sliding and toppling) likely to occur and to propose appropriate remedial action. The techniques for doing this are described in Chapter 17 of the 1997 book "Engineering Rock Mechanics" by Hudson and Harrison published by Elsevier Science.

*****

In this case example, the technique has been outlined for obtaining a model of the geometry and orientation of the fracture network within a

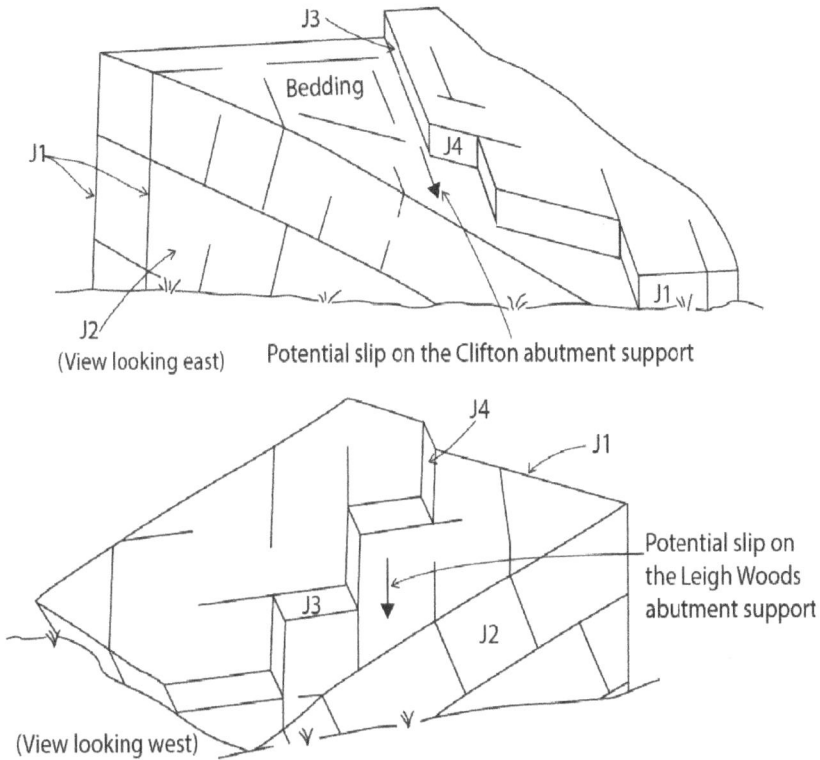

Figure 4.24. Potential sliding instabilities that are predicted at the two bridge abutments. The diagrams show the orientation of the five fracture sets, one being the bedding. The top diagram 'looking east' is related to the Clifton abutment and the lower diagram 'looking west' is related to the Leigh Woods abutment.

rock mass of a study area by combining the regional tectonic evolution of the region with a study of the rock fractures in and around the location of interest. Using the site of the Clifton Suspension Bridge as an example, the resulting model has been used to predict the possible failure modes liable to occur at specific local sites, such as in the vicinity of the bridge abutments and along the Gorge walls. There is no question that this type of failure mode analysis could not have been coherently undertaken without the geological background knowledge presented in this Chapter — and it is one example of why we consider that the synthesis of structural geology and rock engineering is crucial for surface and underground projects constructed on and in rock masses.

# CHAPTER 5

# QUARRIES

In this Chapter, we discuss the roles of structural geology and rock engineering during the design and operation of four quarries: the Grabinex granite quarry in Poland, the Tytherington limestone and Mountsorrel granodiorite quarries in the UK, and the Carrara marble quarry area in Italy.

## 5.1 GRABINEX GRANITE QUARRY IN STRZEGOM, POLAND: "THE MICRO AND THE MACRO"

### 5.1.1 Overview of the quarry and its products

In this case example, we illustrate how microstructural anisotropy, joints and faults govern the geometry of a quarry in Poland. The Grabinex quarry is located near to Strzegom, a medium sized town in south–west Poland, 52 km west of the regional capital Wrocław. The Grabinex quarry has been operating for 18 years and produces 'clear grey' and 'grey yellow' granite mainly for construction purposes, especially setts (rectangular blocks used for paving roads), cobblestones (similar but more rounded) and road base material, amounting to 100,000 tonnes annually. The compressive strength of the granite varies between 70 MPa and 240 MPa. A view of a portion of the quarry is shown in Figure 5.1 and typical granite products from the quarry are shown in Figure 5.2. Note the linear portions of the quarry walls in Figure 5.1.

Figure 5.1.  A portion of the Grabinex granite quarry in Strzegom, Poland. The flat quarry surface around the excavation is an exfoliation plane on which the impression of fractures can be seen, e.g., at the top left of the photograph. The linear nature of the quarrying is created by taking advantage of the microstructural weakness directions of the granite.

(a)                                                                    (b)

Figure 5.2.   Typical quarry products for road construction: (a) Granite blocks and (b) kerb stones.

## 5.1.2 The geological setting and the granite microstructure

The quarry is located in the Strzegom–Sobótka granite massif in Poland. The massif surface is in the form of an extended wedge 45 km long and 10 km wide having a WNW–ESE axis. The overall massif is thought to have been formed during the Upper Carboniferous or the Lower Permian.

The granite outcrop shown in Figure 5.3 shows features recording the flow of the magma that have been retained after it has solidified. A feature of cooled and crystallised granite which is most helpful for quarrying is that it exhibits a strength anisotropy — due to the original magma flow pattern and the fact that the granite shrinks on cooling. However, the type of flow features in Figure 5.3 at Strzegom are an exception: generally the magma flow characteristics are more consistent, exhibiting an orthotropic anisotropy, see Principle 16 in Chapter 3, which is consistent across the quarry site.

An additional feature, clearly evident in the quarry, is natural exfoliation jointing sub-parallel to the ground surface. This occurs when the pluton is exhumed, the weight of the overlying rock is removed, the vertical component of the stress field is reduced to effectively zero, and fracturing develops, see Figure 5.4 and Section 2.14.4. The exfoliation fractures tend to decease in intensity with depth, see Figures 5.4–5.6.

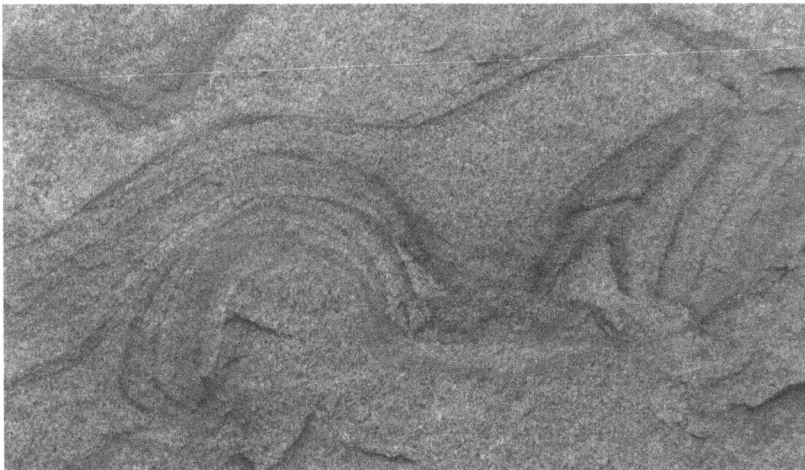

Figure 5.3.    Flow features in the Strzegom–Sobótka granite massif, image ~2 m across.

Figure 5.4.   Sub-horizontal exfoliation jointing in the Strzegom–Grabinex quarry in SW Poland.

Figure 5.5.   Decrease in the intensity of the exfoliation fractures with depth.

Figure 5.6. Background: Exfoliation fracture frequency decreasing with depth; foreground: anthropogenic exfoliation fracture (below the two observers) caused by release of overburden during quarrying.

These exfoliation surfaces can be exploited during quarrying to provide the working surfaces, e.g., as in Figure 5.5.

The granite also contains joints and faults, some of which can cause problems during quarrying, see Figure 5.7. Other structural geology features explained in Chapter 2 are evident in the quarried walls, e.g., the curvature and termination of one joint at another, see Figure 5.8 and Section 2.12.2.

The quarrying procedure, which has been in operation in the area since the Middle Ages, is significantly helped by the anisotropy of the intact granite — which has three characteristic orthogonal splitting directions: the 'rift', the 'grain' and the 'hardway'. These three planes of weakness are found in many granites, see for example Figure 5.9 which shows the orientation of these planes in the Inada granite in Japan, cored to determine the strengths of these three planes. In the Strzegom–Grabinex quarry in SW Poland these three planes are differently oriented, see Figure 5.10;

Figure 5.7.   Fault in the quarry and evidence of a pegmatitic intrusion into the fault in the lower part of the picture.

Figure 5.8.   Joint curving and then terminating at another joint. The vertical lines are the 'half-barrels' of the drillholes that were blasted with detonating cord and are ~0.5 m apart.

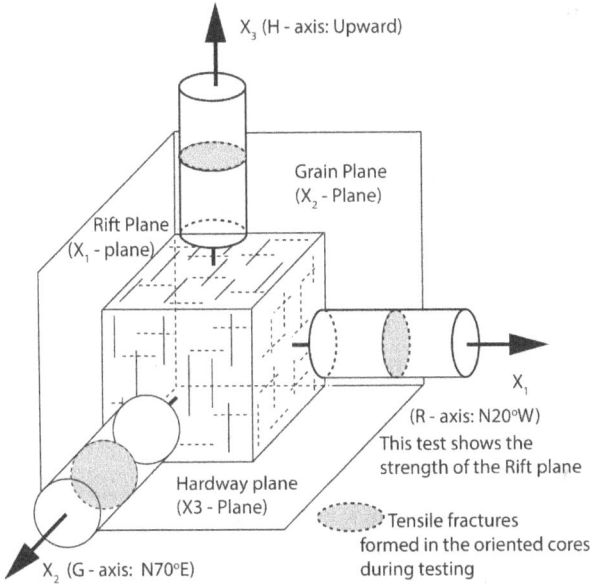

Figure 5.9. Illustration of the rift, grain and hardway planes in Inada granite. This diagram is from Fujii *et al.* (2007) and also indicates orthogonal cores taken for testing. The hatched planes within the three samples are parallel to the three key planes within the rock.

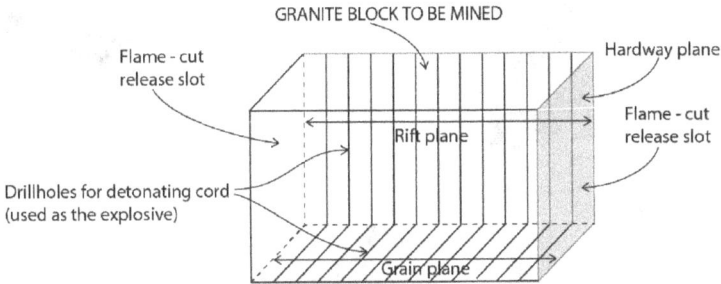

Figure 5.10. The mining technique involving blasting boreholes drilled in the vertical rift and horizontal grain planes and flame-cut release slots in the vertical hardway plane at each end of the block. Note the similarity with pre-split blasting described in Chapter 1.

the granite splits most easily in the rift direction, here a vertical plane, and also splits in the grain direction, the horizontal plane. In the other vertical direction, the hardway, as the name implies, it is more difficult to split the granite.

### 5.1.3 The mining technique

The quarry takes advantage of the preferential splitting planes in the mining technique by releasing the granite blocks along the rift, grain and hardway planes, as shown in the sketch in Figure 5.10. Flame-cut slots are created parallel to the hardway plane at each end of the block, then closely spaced, small diameter boreholes are drilled vertically in the rift plane and horizontally in the grain plane. Detonating cord is used to create essentially simultaneous explosions in these boreholes, thus releasing the block and causing it to move forwards slightly. The high velocity detonating cord is a thin, flexible plastic tube filled with PETN, i.e., pentaerythritol tetranitrate.

The large block thus released can then be reduced in size by hand using the wedge and feathering sledgehammer system illustrated in Figure 5.11. This further spitting of the block is achieved by applying repeated blow to a line of wedges that cause a tensile stress perpendicular to the rift plane, Figure 5.11, see also Figure 9.8(b).

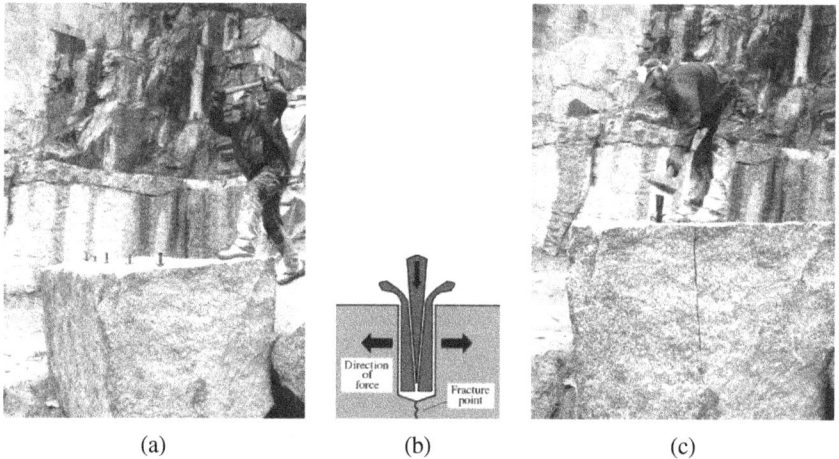

(a)  (b)  (c)

Figure 5.11. The 'wedge and feathers' method of splitting the granite. (a) Striking the wedges in the line of wedge and feather sets, (b) the arrangement of the central wedge and the two feathers in the vertical rift plane, (c) the crack being developed in the rift plane.

Figure 5.12. The orthogonal nature of the quarry geometry which directly reflects the orthotropic microstructural granite anisotropy.

Thus, by repeatedly taking advantage of the preferential splitting planes in the granite during extraction of the blocks, the macro geometry of the quarry takes on an orthogonal form (see Figures 5.1 and 5.12). Some of the quarry surfaces will exhibit the half-barrels of the boreholes used in the rift and grain planes; the other surfaces will exhibit the flame-cut effect, e.g., Figure 5.13.

The other factor governing the overall quarry layout is the presence of faults in the granite, Figure 5.14.

\*\*\*\*\*

The authors are grateful to the Grabinex company and especially the Mine Manager Mr Stanislaw Paluch and the translator Mr Janusz Kurc for their kindness in explaining the quarry operation.

Figure 5.13.   The orthogonal geometry of the quarry walls. Note the surface texture of the flame-cut granite.

Figure 5.14.   Fault in the Grabinex quarry, exposed as a face above the block displaying the half-barrels and as a fracture cutting this block.

## 5.2 THE IMPACT OF THE TECTONIC HISTORY OF THE REGION ON THE QUARRYING OF THE CARBONIFEROUS LIMESTONE AT TYTHERINGTON, SOUTH GLOUCESTERSHIRE, ENGLAND

Tytherington quarry is situated 40 km NE of Bristol in the UK in Carboniferous limestone. The excavation is occurring in the SE dipping limb of a syncline with a wavelength of ~15 km. Within the quarry, the beds dip constantly at 30° and consist of a series of relatively massive limestone beds ranging in thickness between 0.5 m and 3 m, sometimes separated by thin beds of clay a few centimetres thick, Figure 5.15.

The Quarries Regulations (1999) require every excavation to have a design based on sound geotechnical principles. The following discussion shows how an understanding of the geological evolution of a site can help

Figure 5.15. Limestone dipping at 30° exposed in Tytherington Quarry. The strata are situated on the limb of a large fold formed as a result of Hercynian deformation which occurred in the Late Carboniferous period.

to facilitate this requirement. A brief description of the geological evolution of the area since the deposition of the limestones is outlined and it is demonstrated that an understanding of this evolution can be used to account for the majority of the fractures developed in the quarry, thus enabling its optimum orientation to be determined and to provide input into the most efficient methods of extraction.

### 5.2.1 Conclusions derived from the regional and local studies of fracturing in and around the quarry

The tectonic evolution of the Bristol Channel Basin has been discussed in some detail in Chapter 4 to which the interested reader is referred. Only a brief summary is given below.

— The Carboniferous rocks of the Bristol Channel basin were profoundly affected in Late Carboniferous times by the Hercynian orogeny which generated approximately E–W trending folds and thrusts, the latter dipping gently to the south. Later in this tectonic event, major wrench faults striking NW–SE (dextral) and NE–SW (sinistral) developed. This collision welded together two major plates (Laurentia in the north and Gondwanaland in the south) and generated a single tectonic plate: Pangea. The post-Carboniferous evolution of the region (and the world) was dominated by the break-up of this plate which began in the Permian.

— The Bristol Channel region experienced a prolonged rifting event from the Triassic (300 mya) through to the Aptian (125 mya), which generated dominantly WNW–ESE trending normal faults (which formed the Bristol Channel Basin) and extensional veins which show the same trend in both partially and fully lithified sediments, with evidence of multiple reactivation, see Figure 4.15.

— The basin was inverted (began to close) during the Lower Cretaceous and Tertiary by a ~ NW–SE orientated maximum principal compression ($\sigma_1$), the result of the collision between the African and European plates which resulted in the formation of the Alps. Evidence from the neighbouring Wessex basin suggests inversion from the end of the Cretaceous (~70 million years ago) through to the Late Oligocene and

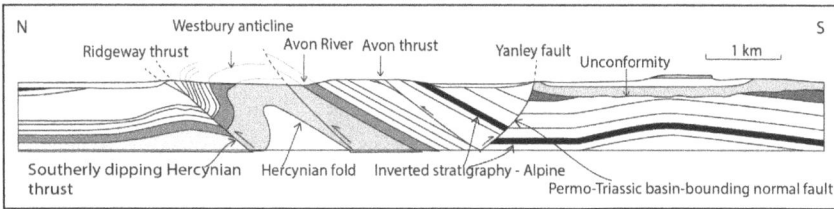

Figure 5.16. ~N–S geological section a few kilometres west of the quarry showing the impact of the three major tectonic events to affect the region, namely the Hercynian folding and thrusting, the Permo-Triassic rifting and the Alpine inversion.

Early Miocene (~20–25 million years ago). By the time of this inversion, the sediments of the basin had become fully lithified. The earliest stages of this Alpine deformation resulted in the reactivation of the thrusts and strike-slip faults in the Carboniferous rocks generated as a result of the Hercynian orogeny, (see Figure 4.5 and the related discussion in the Clifton Suspension Bridge case study).

— This tectonic evolution of the Bristol Chanel can be clearly seen in the structures displayed in the north–south orientated geological cross-section drawn a few kilometres to the west of the study area at Tytherington quarry, Figure 5.16. It shows Hercynian thrusting and folding, Permo-Triassic extension (rifting), and Alpine inversion of the rift basins and the resulting reversal of stratigraphy.

This brief study of the tectonic history of the Bristol Channel area allows one to predict the following: the types of fractures that are likely to have formed in the limestones in the quarry during the ~300 million years since they were deposited, their relative timing, their orientation and their likely current mechanical properties.

In the following Section, a summary of the fractures observed in the vicinity of the quarry is presented, together with a discussion of their likely origin within the context of the tectonic history outlined above. As noted, in the study area the onset of the plate collision that resulted in the Hercynian orogeny produced a series of southerly dipping thrusts and northerly verging folds, Figure 5.16. These structures are the result of a stress field in which the minimum principal compression ($\sigma_3$) was vertical, Figure 5.17(c). In addition to the relatively large-scale folds and thrusts

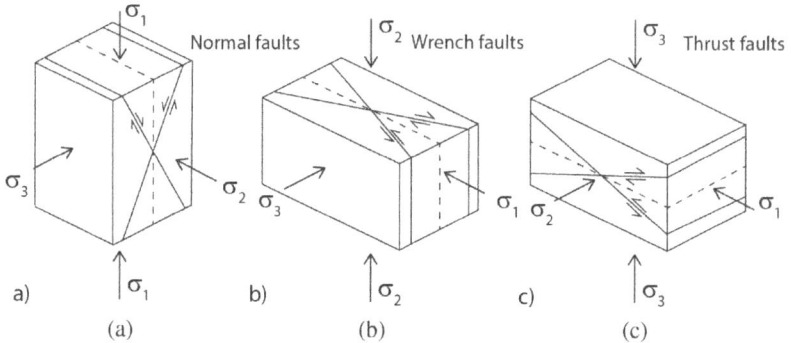

Figure 5.17. The three stress configurations linked to the three major types of faults, normal ($\sigma_1$ vertical), wrench or strike-slip ($\sigma_2$ vertical) and thrusts ($\sigma_3$ vertical). The orientations of the extensional fractures that would form under these stress configurations are indicated by dashed lines.

shown in Figure 5.16, small-scale examples of these thrusts also developed and these are apparent in the beds exposed in the quarry.

Two examples are illustrated in Figure 5.18 which show strata-bound, embryonic thrusts cutting across the bedding at ~30°. The bedding dips at ~30° (see Figure 5.15) and so the thrusts are currently horizontal. It is argued that the thrusts pre-dated the folding which caused the beds to dip and formed originally with the typical dip of 30° as shown in Figure 5.17(c). There has been almost no movement along the thrust shown in Figure 5.18(a) and the bedding is therefore not displaced. In Figure 5.18(b) approximately 100 mm of movement has occurred along the thrust and a small fold has formed at the thrust tip. Two stages in the evolution of a thrust are shown in Figures 5.18(c) and 5.18(d). As movement occurs along the thrust, a small fold begins to develop — as can be seen in the example shown in Figures 5.18(b) and 5.18(c). Further movement along the thrust leads on to the formation of a fault–bend fold, Figure 5.18(d).

These two diagrams of thrust development show how the thrust exploits the bedding. Although theoretically they should form with a dip of 30°, (see Figure 5.17(c)), along parts of the thrust they exploit the planes of easy slip offered by some of the bedding planes and, as a result, are made up of a series of 'flats' and 'ramps' which gives them a staircase profile. The flats occur where the slip is along the bedding and the ramps

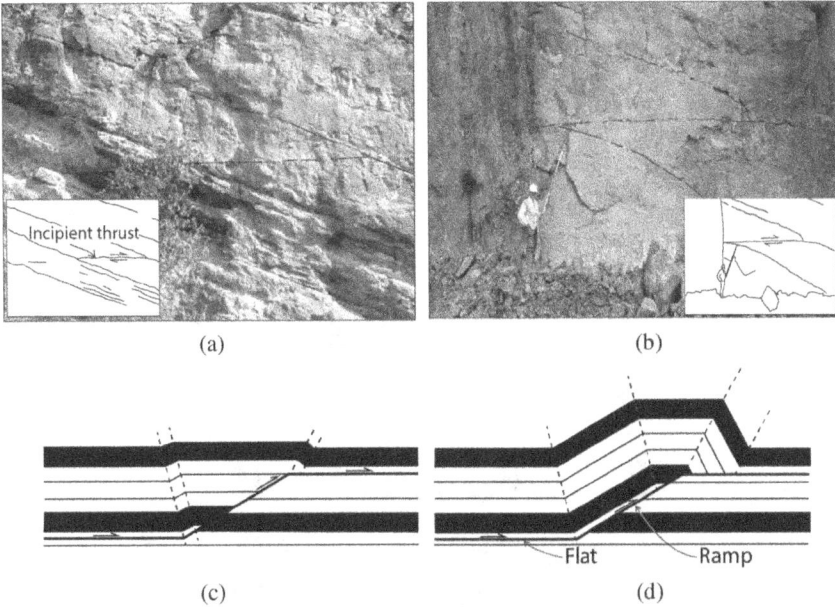

Figure 5.18. Examples of small-scale thrusts cutting across bedding at 30°. In (a) the thrust has almost no movement along it, whereas in (b) approximately 100 mm of movement has occurred forming a small fold at the thrust tip. Two stages in the evolution of a thrust are shown in (c) and (d).

where the thrusts climb across the strata. Movement of the thrust over the ramp produces a fold (a fault-bend fold or ramp anticline).

It is observed that, within the quarry, the mechanical properties of the bedding planes which separate the beds are varied. When two adjacent beds are not separated by a thin clay horizon and are in direct contact with each other, pressure solution occurs along the bedding plane in response to the vertical loading that occurs during burial and diagenesis (the processes linked to the changing of a sediment into a rock). This leads to the formation of stylolites — serrated bedding surfaces within the rock mass, Figure 5.19. The process of stylolytisation results in the interlocking of the two adjacent beds with the result that the process of bedding plane slip is severely inhibited. (See also Figure 4.19 in the Clifton case study in Chapter 4).

In contrast, when two adjacent limestone beds are separated by a thin layer of clay, contact between the two beds is prevented, no stylolytisation

Figure 5.19. (a) Top photograph — Profile through a stylolite which has formed by pressure solution along a bedding plane in a limestone. (b) Bottom photograph — Exposed bedding plane along which stylolytisation has occurred during burial and diagenesis as a result of the overburden load. The extremely irregular surface generated prevented bedding plane-slip during the Hercynian orogeny when thrusting developed along other bedding surfaces unaffected by solution because of the presence of a thin clay horizon.

occurs and the beds remain flat. Furthermore, the thin layer of clay between them facilitates bedding-plane slip and, as a result, these bedding planes are utilised by the thrusts. Two adjacent bedding planes are shown in Figure 5.20, one along which stylolytisation has occurred and the other, along which there is a thin inter-bed of clay, where it has not.

In addition to the rocks responding in a brittle manner to the Hercynian stress field which resulted in the formation of the thrust faults

Figure 5.20. Top photograph — A 2 m thick limestone bed bounded by two different types of bedding planes, an upper irregular stylolite plane and a lower, smooth plane defined by a thin layer of clay. Bottom photograph — Another example where a bedding plane has remained planar because the presence of a thin layer of clay prevented the formation of a stylolite.

described above, they also responded in a ductile manner by buckle folding, an example of which, the Westbury anticline, can be seen in Figure 5.16. The fold axes are parallel to the strike of the thrusts. These folds produce local strain variations as illustrated in Figure 5.21(a). Layer-parallel stretching occurs at the hinge of the fold in the outer arc and layer-parallel compression at the inner arc. A surface of no strain (the neutral

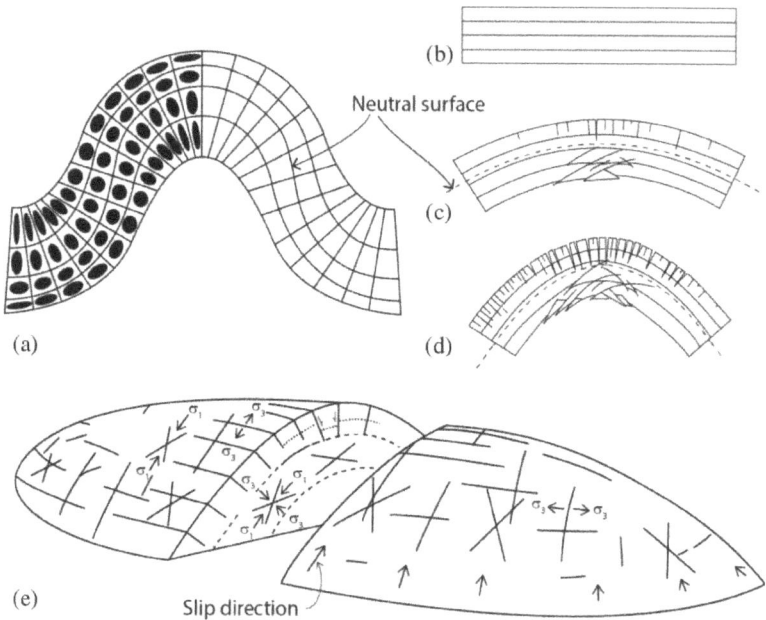

Figure 5.21.    (a) The strain distribution in a buckled homogeneous isotropic layer. Stretching occurs in the hinge region above the neutral surface and compression below. The development of the fractures that might result from this local deformation is shown in (b)–(d). After Ramsay (1967). (e) The different sets of fractures that can be generated in a pericline as it amplifies. After Stearns (1964).

surface) runs through the fold and separates these zones of extension and compression. The local stress fields linked to these deformations can give rise to fracturing, see Figures 5.21(c) and 5.21(d).

The section through the fold shown in Figure 5.21(a) is taken at right angles to the fold hinge and is known as the profile section. It can be seen that the greater the curvature, i.e., the tighter the fold, the more intense the fold related fracturing will be, Figures 5.21(c) and 5.21(d). Field observations of buckle folds show that their three dimensional geometry is that of an elongate dome known as a pericline, Figure 5.21(e). Because there is also a curvature along the fold hinge, there is a commensurate stretching that can lead to fracturing. These fractures will form normal to the fold hinge and are termed cross-axial. The various fractures found in association with

buckle folds are summarised in Figure 5.21(e). A knowledge of this association is useful when working on a site where folding has occurred.

In addition to fault–bend folds and buckle folds discussed above,
Figures 5.18 and 5.21, there is a variety of other folds found in nature.
All have a characteristic set of fractures associated with them. Thus, if the
type of folding affecting a potential site is known, then it is possible
to predict the folding-induced fractures that are likely to be present.
A review of the fractures linked to the various types of folds can be found
in Cosgrove (2015).

As discussed above, Figure 5.17(c), thrusts form when the least principal compression is vertical. However, as thrusting continues, the thrusts
stack one on top of the other with the result that the vertical stress
increases and may become $\sigma_2$. This change in stress regime results in a
change over from thrusting and folding to wrench faulting, Figure 5.17(b).
This is clearly illustrated in Figure 5.22 which shows the effect of the
Hercynian orogeny on the rocks of Pembrokeshire, South Wales, i.e., a

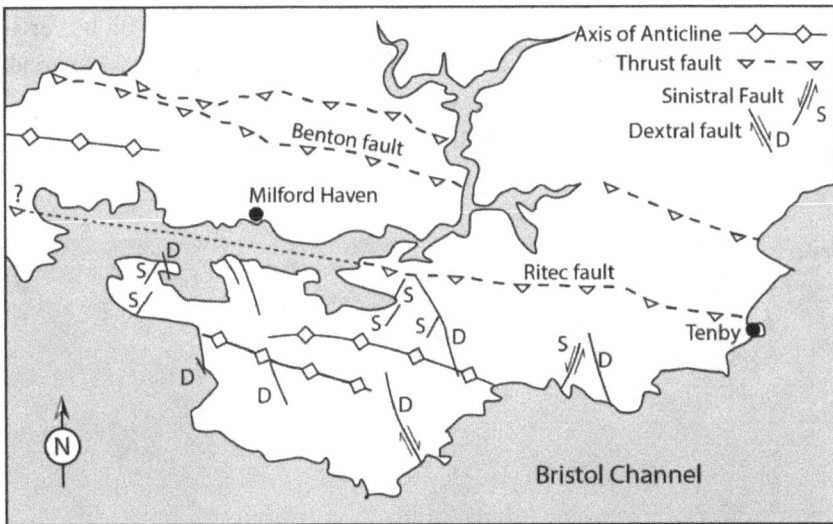

Figure 5.22.  The approximately E–W trending folds and thrusts associated with the stress
regime acting during the early stages of the Hercynian orogeny, cut by the dextral and
sinistral wrench faults associated with the stress regime that characterised the late stages.
After Anderson (1942).

region to the north and to the west of the quarry which has experienced the same tectonic history. Here, the approximately E–W trending thrusts and folds have been displaced by conjugate wrench faults striking ~NW–SE and ~NE–SW.

Evidence for these late-stage Hercynian wrench faults can be found in the quarry. Unlike the earlier formed thrust faults, which tend to be strata bound and which formed when the overburden stress was small, the wrench faults, Figures 5.23 and 5.24, are through-going probably as a result of them forming under a higher layer normal stress, see Figure 2.136 and Section 2.8.5. It will be recalled that the ability of a fracture to propagate across an interface between two beds is sensitive to the normal stress acting across the interface. The greater the stress the more likely it will be that a fracture will propagate from one bed to an adjacent bed and therefore for the fractures to be through-going.

It will be recalled from Section 2.3 that a particular stress regime can generate both extensional and shear fractures depending on the magnitude of the differential stress. The extensional fractures form normal to the least principal compression. Thus, in the quarry at Tytherington the extensional fractures linked to the wrench fault stress regime will be vertical and will bisect the conjugate set of wrench faults that have formed, Figures 5.17(b) and 5.25. These fractures are observed at Tytherington and a summary diagram showing all the Hercynian structures observed at the

Figure 5.23. Two examples of wrench faults cutting the limestones of the quarry. They are vertical, through-going fractures and on some, e.g., (a), slickensides, parallel to the dotted lines, provide clear evidence of sub-horizontal movement.

Figure 5.24.   View of a quarry bench face showing the closely spaced vertical fractures linked to the formation of wrench faults. Note the 'zig-zag' plan of the quarry face as it follows alternatively the two conjugate wrench faults, see Figure 5.25.

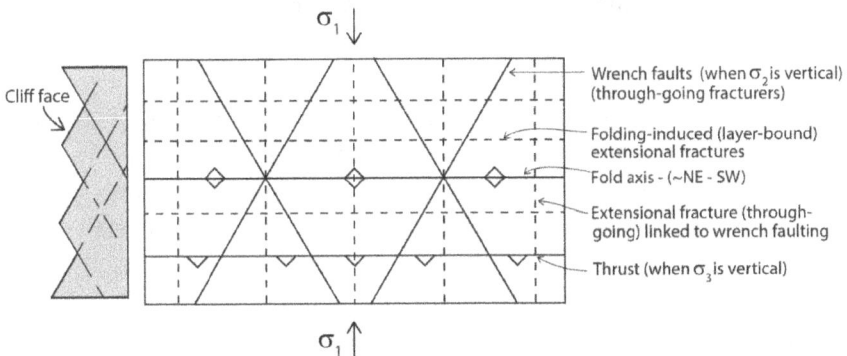

Figure 5.25.   Summary diagram of the structures formed in the quarry related to the Hercynian orogeny. They include; thrusts, folds, fold-related, strata-bound, extensional fractures, through-going wrench faults and wrench-related through-going extensional fractures. See text for a discussion of the chronology of their formation. The shaded diagram is a plan of a cliff/bench face whose orientation is controlled by the conjugate wrench faults, see Figures 5.23 and 5.24, after Cosgrove and Sedman (2005).

quarry is given in Figure 5.25. It shows the early formed thrusts, the later formed folds and the axis-parallel fractures associated with them (see Figures 5.21(a)–5.21(d)) and the latest fractures linked to the wrench fault stress regime, namely the wrench faults and associated vertical extensional fractures.

It is this Hercynian fracture network that dominates the fracturing of the quarry despite the fact that, as noted above in the Section discussing the conclusions drawn from the regional and local studies of fracturing in and around the quarry, two episodes of deformation affected the rocks of the region after the Hercynian orogeny. These are the crustal extension which started in the Permo-Triassic and which opened the Bristol Channel Basin during the Mesozoic and the compression associated with the Alpine collision which closed the Basin in the Late Cretaceous and Tertiary. The effect of these two events on the fracturing of the rocks in SW England can be clearly seen in Figures 4.7–4.12 in Chapter 4 which show the effect of this extension and compression on the Mesozoic rocks of the region, i.e., rocks that post-date the Hercynian deformation. In contrast to these younger rocks, the Carboniferous rocks of the quarry were affected by the Hercynian deformation and, as noted above and summarised in Figure 5.25, contain the fracture system linked to that deformation. Because the Alpine compression direction (~NNW–SSE) is approximately co-axial with the Hercynian compression the effect of this later collision is to **reactivate** the Hercynian fractures rather than generate new ones. In addition, the locally developed outer arc fractures formed during the Hercynian folding (Figures 4.2 and 5.22) may have propagated and become less localised in response to the Mesozoic extension.

The fracture network that dominates in the quarry, Figure 5.25, determines the ideal orientation of the quarry. The main quarry walls follow the main vertical fractures (i.e., the extensional fractures linked to the wrench phase of the Hercynian deformation) and the vertical extensional fractures formed parallel to the fold axis formed during the earlier phase of this deformation (Figure 5.26). In places in the quarry, the NW–SE trending faces are defined by the through-going wrench faults rather than the associated extensional fractures. When this occurs, the face is made up of alternating sections of each of the conjugate sets. This gives the face a characteristic serrated profile, see Figures 5.24 and 5.25.

Figure 5.26. Aerial photograph of the ~NW and NE orthogonal working faces following the two dominant vertical fracture sets at the Tytherington quarry in the UK, cf. the rock microstructure dominating, as in Section 5.1. The scale can be judged by the width of the M5 motorway crossing the bottom right-hand corner of the Google Earth photograph.

## 5.3 GRANODIORITE QUARRY, CENTRAL ENGLAND

In Section 5.1 we presented the case of the Grabinex granite quarry in Poland where the microstructural anisotropy governed the overall layout of the quarry. In Section 5.2, the Tytherington Quarry in England was described in the context of how the orientations of the quarry benches were governed by the structural geology. We now describe a granodiorite quarry in England where the suite of fracture systems exposed by quarry- ing causes all three modes of surface rock failure: plane sliding, wedge sliding and toppling, as presented in Principle 37 in Chapter 3, and repro- duced below in Figure 5.27.

A cross-section of the overall geological setting is shown in Figure 5.28. The granodiorite intrusion is of Lower Palaeozoic age. This large granitic boss intrudes discordantly into the Swithland slates and Stockford slates, of late Pre-Cambrian to Cambrian age. The Mercia mudstone overlaid a significant part of the site indicating the intrusion is pre-Triassic. Mineralogical dating suggests the intrusion is of mid-Devonian age (395–345 mya) associated with

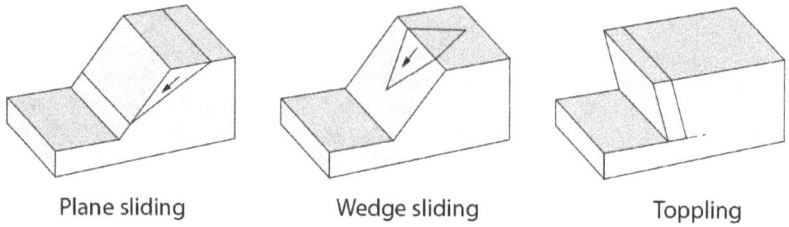

Plane sliding          Wedge sliding          Toppling

Figure 5.27.   The three main modes of surface rock failure induced by rock fractures.

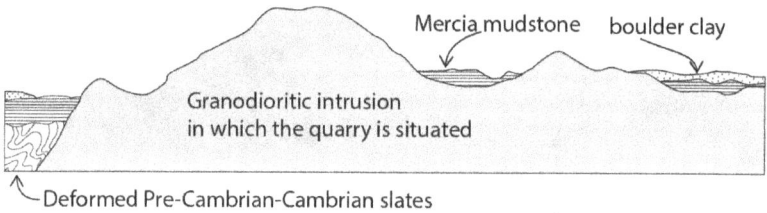

Mercia mudstone    boulder clay

Granodioritic intrusion
in which the quarry is situated

Deformed Pre-Cambrian-Cambrian slates

Figure 5.28.   Geological cross-section at the granodiorite quarry.

the Caledonian orogeny. The unconfined compressive strength of the rock is of the order of 211 MPa, so any quarry failure will be via the fractures. The quarry layout, as a plan and section is shown in Figures 5.29(a) and 5.29(b) , together with the traces of the main fracture sets.

As a result of the fracturing present in the rock mass, all three of the Figure 5.27 slope failure modes can potentially occur: plane failure, wedge failure and toppling/fallout. Examples of these are shown in Figures 5.30–5.37.

From the photographs, it is evident that the fracturing in this rock mass can lead to all modes of hard rock slope failure. Moreover, the rock is inhomogeneous across the site as evidenced by the failures in Figure 5.35 and the unfractured core in Figure 5.38. However, if there are recurring dominant fractures that can cause severe quarry bench problems, such as the one shown in Figure 5.33, then the quarry shape can be adjusted to reduce such failures, as indicated in Figure 5.39 and by variations on this theme.

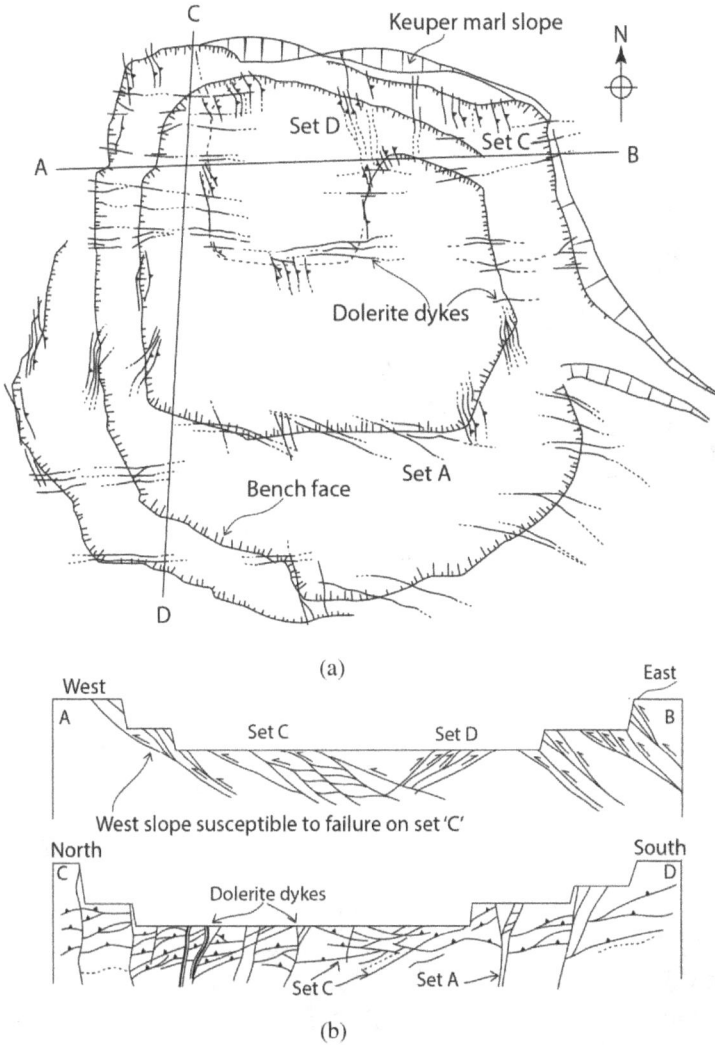

Figure 5.29. (a) Plan view of the granodiorite quarry with the main fracture systems indicated. The important E–W trending fracture set A is clearly evident, together with the dolerite dykes, which form sub-parallel to these features. (b) Sectional views of the grano-diorite quarry with the main fracture systems indicated. Note that Fracture Sets C and D are conjugate thrusts linked to the same tectonic event. Vertical fractures, which are apparent in the N–S section are the ~W–E trending Fracture Set A — which is sub-parallel to a suite of dolerite dykes.

Figure 5.30. A general view looking North showing Fracture Set C dipping ~35°–40° to the East. These are conjugate shear fractures linked to Fracture Set D, see Figure 5.29(b), top section. All three types of slope failure can be detected in this photograph.

Figure 5.31. A detail showing the eastern dipping shear fractures, which in combination with a joint set normal to these planes gives rise to the blocky nature of the rock mass. The white line is a scanline tape. Plane and wedge block failure are evident.

Figure 5.32. Evidence for shear displacement which has produced a wide shear zone by slip on a thrust linked to Fracture Set C (looking north, white horizontal line is a measuring tape).

Figure 5.33. Shear zone resulting from movement on a thrust (Fracture set C). This type of shear zone can cause plane failure problems during rock excavation.

Figure 5.34. Water seepage along a brecciated fracture zone belonging to Fracture Set C.

Figure 5.35. Set A fractures looking East. Note the persistence of this sub-vertical set as it cuts the face, approx. 50 m high. All modes of slope failure are evident here.

Figure 5.36.   Set A, looking West, bench ~12 m high. The through-going nature and size of these fractures causes bench instability.

Figure 5.37.   Highly fractured and planar bench faces.

Figure 5.38. One box of core from a vertical drillhole. Some deeper zones in the rock mass can contain few fractures, as evidenced by these core sticks.

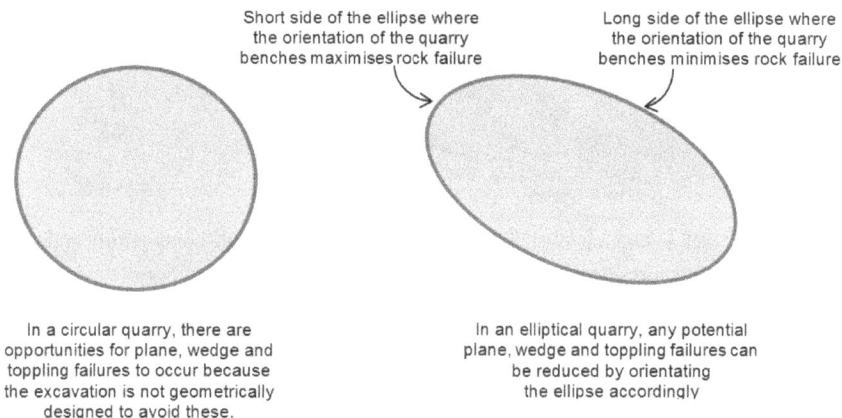

Short side of the ellipse where the orientation of the quarry benches maximises rock failure

Long side of the ellipse where the orientation of the quarry benches minimises rock failure

In a circular quarry, there are opportunities for plane, wedge and toppling failures to occur because the excavation is not geometrically designed to avoid these.

In an elliptical quarry, any potential plane, wedge and toppling failures can be reduced by orientating the ellipse accordingly

Figure 5.39. Illustrating that the occurrence of bench failures can be reduced by using a suitably orientated, elliptical quarry plan, rather than a circular plan.

*****

The authors are grateful to A. P. Crowe for assistance with the photographs in Section 5.3.

## 5.4 THE CARRARA MARBLE QUARRY, ITALY

In the Frontispiece to this book, we show photographs of a Carrara marble quarry and the Marble Arch monument in London which is faced with Carrara marble. These two photographs represent the 'raw rock' and an 'engineered product' — and thus the synthesis of structural geology and rock engineering which is the theme of this book. So, we now finish this Chapter 5 on the subject of quarries by explaining some features of the Carrara marble quarries in Italy (both above and below ground) with the

Figure 5.40.   The location of the Carrara marble district (black dot) in Tuscany, Italy, from Coli *et al.* (2010).

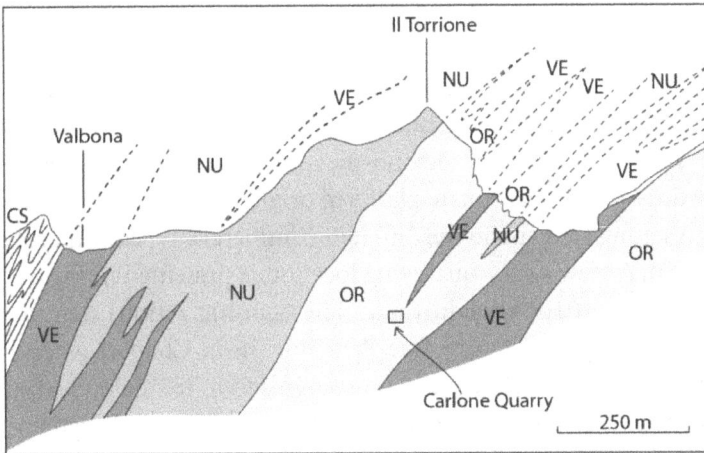

Figure 5.41.   The geostructural setting of the Carlone marble quarry area: Cross-section — marble types: NU = Nuvolato; OR = ordinary white; VE = veined.

emphasis on structural geology aspects and quarrying the marble. It was in this area that Michelangelo obtained the marble for his masterpieces.

The location in Italy and the structural geology setting of the Carrara marble is shown in Figures 5.40 and 5.41. The marble derives from the

Figure 5.42.    A view of the Carrara marble quarrying operation.

tectono-metamorphic deformation of a Hettangian (about 200 mya) carbonate platform. The Carrara marble is the result of three tertiary (27–12 mya) overprinted tectono-metamorphic deformations onto a massive limestone of carbonate platform origin. Different lithofacies in the platform setting gave rise to the different commercial types of marble (Coli *et al.*, 2010). A view of the quarrying location is presented in Figure 5.42.

Coli *et al.* (2010) explain that, "The tectonic actions determined a global, orthotropic structure [see Principle 16 in Chapter 3] and three principal planes of weakness can be recognised in the field. Such planes, at right angles relative to each other and called by quarrymen *verso*, *secondo* and *contro*, control the exploitation and excavation of the Carrara marble via planes along which blocks are cut. They sub-divide the rock mass into prismatic blocks, the sizes and shapes of which determine the commercial grade." Note that the *verso*, *secondo* and *contro* planes are directly analogous to the rift, grain and hardway of the granite described earlier in this Chapter in Section 5.1.2 and illustrated in Figure 5.9. In the granite quarry, the rock is extracted via flame cutting and 'gentle'

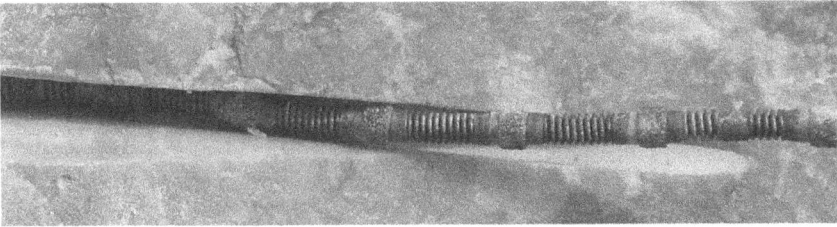

Figure 5.43.    Cutting the Carrara marble with diamond encrusted wire inserts.

blasting; the Carrara marble is cut out using a wire cutter which has diamond coated inserts, Figure 5.43.

Because the fractures are mainly distributed into bands, it was possible to categorise the Carrara marble rock mass into four typologies: intact, scattered fractured, systematically fractured (*finimento*), intensely fractured (intersection of two *finimento*) (Coli *et al.*, 2010). The rock mass also contains high stresses which can lead to rockbursts, see Principles 13 and 46 in Chapter 3. Similarly, the rockbursts encountered at the Jinping II site in China, Principle 46, occurred in a marble rock mass.

The two views of one of the Carrara marble quarries are shown in Figures 5.44(a) and 5.44(b). These show the style of surface quarrying and the fractures traversing the marble rock mass. The quarry is located in a zone of intense ductile folding deformation which involves many of the Carrara marble lithotypes. The Carlone quarry, Figure 5.41, which is underground was opened in ordinary white marble, between a syncline and an anticline, marked respectively by flames of Nuvolato (cloudy) marble and veined marble. Towards the NW, the quarry shaft is bounded by intense deformation bands of the *contro* type (Coli *et al.*, 2010). Because of the *in situ* high rock stresses and the increase of these stresses immediately around the underground openings, rockbursts have occurred. Figure 5.45 shows an underground room where a rockburst has occurred. As with tunnels, the intensity of the rockbursts can be mitigated by orientating the long axes of the underground rooms to be parallel with the major principal stress (Principle 39 in Chapter 3).

In fact, the Carrara marble can also retain built-in stresses as a result of its tectonic formation and these may result in bowing of the marble

(a)

(b)

Figure 5.44.   (a) View of Carrara marble quarrying (see Figure 5.41 for the geostructural location of the quarry). (b) Additional view of the Carrara marble quarrying (to the right of the view in Figure 5.44(a)).

Figure 5.45.    An underground room in the Carlone marble quarry where a rockburst has occurred (note the irregular nature of the far wall).

cladding on prestige buildings, see the bending of the slabs on part of Finlandia Hall in Helsinki, Finland, shown in Figure 5.46.

*****

The four quarry examples described in this Chapter all illustrate the importance of the link between structural geology and rock engineering — and in particular that the engineering operations should be in harmony with the geological aspects. The method of extracting granite from the Polish quarry explained in Section 5.1 is optimised by orientating the excavation geometry in line with the known microstructural anisotropy, which in turn then determines the macroscopic geometry of the quarry. The benches in the UK limestone quarry described in Section 5.2 are determined by the jointing pattern. The photographs taken in the UK granodiorite quarry presented in Section 5.3 illustrate the range of plane, wedge and toppling failures that can occur during excavation in a heavily fractured rock mass; however, and if required, these can be reduced by

Figure 5.46. Convex bowing of the Carrara marble slab cladding on Finlandia Hall, Helsinki.

adopting an appropriately orientated elliptical rather than circular plan shape for the quarry. Finally, in Section 5.4, it was noted that the effect of underground excavation in a highly tectonised marble for quarry operations can lead to rockbursts, but the intensity of these can be reduced by suitable orientation of the quarry operations.

# CHAPTER 6

# DAMS

Following the previous two Chapters containing studies of the Clifton suspension bridge and four quarries, we now present three case studies of dams to demonstrate further the need for a synthesis between structural geology and rock engineering. The dams are in Brazil and Scotland.

## 6.1 THE ITAPEBI HYDROELECTRIC DEVELOPMENT ON THE JEQUITINHONHA RIVER, BRAZIL

This case example concerns a rock mass beneath a dam abutment and is an example where a combination of primary fractures (bedding) and secondary fractures (tectonically-induced joints) controls stability. The Itapebi Hydroelectric Power Plant, Figure 6.1, is located in the basin of the Jequitinhonha River, in the municipality of Itapebi, in the southern region of the state of Bahia, Brazil.

The rock on which this structure has been constructed is in an ancient crystalline shield of Pre-Cambrian age. It is made up of metamorphic rocks of the Espinhaço Super Group and the outcrops in the vicinity of the dam consist of layered, granitic gneisses intercalated with sub-horizontal lenses and layers of a biotite-amphibolite schist. The layering (foliation) and the interbedded schists strike NW and dip gently towards the NE, Figures 6.2(a) and 6.2(b).

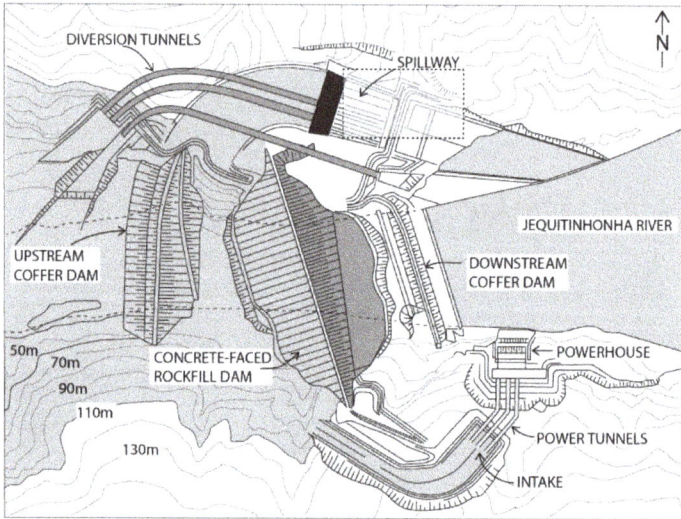

Figure 6.1. General layout of the dam and power plant at Itapebi. The river flows from west to east (left to right), after Albertoni (2009). The dotted rectangle in the vicinity of the spillway indicates the approximate site of the rockslide.

Figure 6.2(a) Gently dipping, dark, biotite and amphibolite-rich layers (see the half-arrows) in the granite gneiss that characterise the bedrock of the dam site.

Figure 6.2(b).   A thick biotite–amphibolite layer (dark) within the granite gneiss now beneath the right abutment of the dam. From Albertoni (2009).

The fracture network in the rock comprises sub-vertical joints in the following directions: NS/85°E, N85°W/80°S, N50°W/75°SW and N42°E/40°SE. Shear fractures striking N–S and E–W, and sub-vertical faults striking N–S also occur, Figure 6.3. The valley, which cuts more than 100 m into the bedrock, trends approximately E–W and the river flows from west to east, Figure 6.1. A study of the dam site shows that the formation of the valley caused the local opening of the N–S and E–W trending joint sets. These openings permitted the infiltration of water to depths of more than 50 m, which produced an alteration in the schist layers, principally at their contacts with the granite gneiss.

Inspection of the orientation of the fracture sets and the foliation within the granite gneiss indicates clearly the potential for downstream (~eastwards) slipping of blocks on the weathered schist layers if support is removed. The E–W trending and N–S trending vertical fracture sets are ideally orientated to act as release joints and, as noted above, are already open.

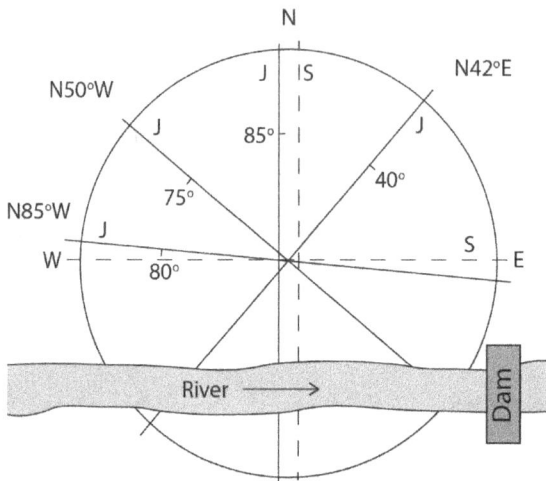

Figure 6.3. The orientation of the major fracture sets cutting the bedrock at the dam site. J = extensional fractures (joints) and S = shear fractures (faults).

In the left hillside in the region of the left abutment and the spillway, i.e., on the northern bank of the river, Figure 6.1, the erosive relief of the valley promoted the release of large blocks of rock with the potential for slipping downstream on one of the numerous schist horizons, Figure 6.4(a).

The commencement of the work on the dam abutments produced excellent exposures that permitted access to the schist layers, Figures 6.2(a) and b. The diverse sub-horizontal lenses and layers of the schist were mapped on each bank in the vicinity of the abutments. A lack of correlation occurs across the valley as a result of the major discontinuity which the river follows and the anastomosing nature of the layers which bifurcate and join (see Figure 6.7).

The layers and lenses of the schist present undulations on the scales of decametres, metres and centimetres, and attain lengths greater than 500 m. Their thickness varies between 0.20 m and 3 m, and the granite gneiss package between the layers varies in thickness from 5 m to 20 m, Figures 6.2(a) and 6.2(b). The layers and lenses of schist dip on average between 10° and 20° to the NE (~downstream), but inflections of the layering can generate local dips of up to 30°.

Figure 6.4(a).  View (looking south) of the rock mass in the region of the left abutment of the dam. The foliation and the eroded schist layer which dip gently to the east (left) can be clearly seen.  The lack of a retaining buttress of rock at the eastern end of the rock mass results in a high probability of slip occurring on the schist. Vertical ~N–S trending, releasing joints intersect the exposure face, itself an E–W trending vertical fracture.

Figure 6.4(b).  The slip of the rock mass shown in Figure 6.4(a) on a schist layer, which occurred in July 2001.

Figure 6.5. Plan view of the slide on the left abutment of the dam. The large block shown in Figure 6.4 can be clearly seen on the high terrace, cut by two N–S trending release fractures and a fracture trending E–W. Smaller blocks form a coarse talus in front of the main 'intact' slipped block. From Albertoni (2009).

In July of 2001, during the excavation for the spillway, a large rock slide occurred in the left abutment. The site of this slide is shown in Figures 6.1 and 6.4(a) and its actual occurrence is shown in Figure 6.4(b). The slide took place along a schist layer with a mean inclination of about 13° and along a length of approximately 210 m, involving a rock volume of about 170,000 m³. Figure 6.5 is a photograph of the plan view of the left abutment taken immediately after the rockslide. The large, slipped block shown in Figure 6.4(b) can be seen on the high terrace. It is cut by two secondary N–S trending vertical release fractures and a single fracture trending E–W.

The uncommon and unforeseeable nature of a rock slide of this magnitude led to a detailed evaluation of its causes, requiring a re-evaluation of the shear-resistance parameters of the weathered layers of the schist horizons, whose assigned values had been obtained by triaxial and direct shear tests on blocks of undisturbed samples taken from the excavations at the beginning of the job. A re-evaluation was also made of the geomechanical model in order to determine the type and quantity of the reinforcements to be utilised in the foundations of the structures, without affecting the contractual commitments for the start of power generation. Different intervention approaches were initially considered for treating the schist layers present in the foundations of the main structures of the project. Meanwhile, the underground interventions were given priority since they permitted the simultaneous execution of the reinforcements to the foundation and of the superstructures themselves, with a minimum of interference between these jobs, thus allowing the time schedule of the contracted services to be maintained.

The most appropriate solutions from both a construction point of view and as regarding site safety were:

— Reinforcements of the foundations through the construction of concrete shear keys implanted along the tunnels and trenches excavated in the rock, in order to intercept the layers of schist, Figures 6.6 and 6.7;
— excavation of complementary drainage tunnels in the abutments to assist in relieving the uplift pressures imposed by the reservoir; and
— placement of tendons through the shallow schist layers permitting safe excavation and concrete placement of shear-keys and structures — subsequently, these tendons were incorporated into the final containment elements and considered in the global safety analysis of the structures.

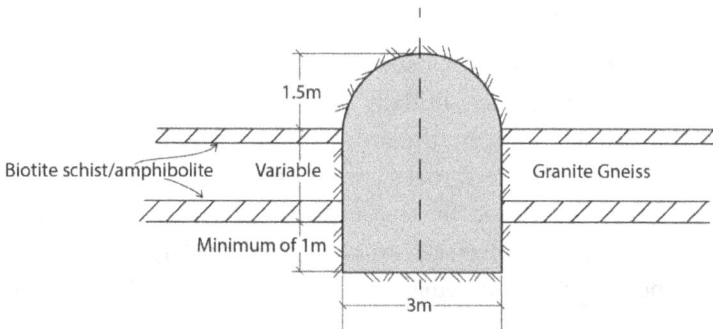

Figure 6.6.   Diagram showing a typical shear-key tunnel positioned to stabilise the schist horizons beneath the concrete face. After Albertoni (2009).

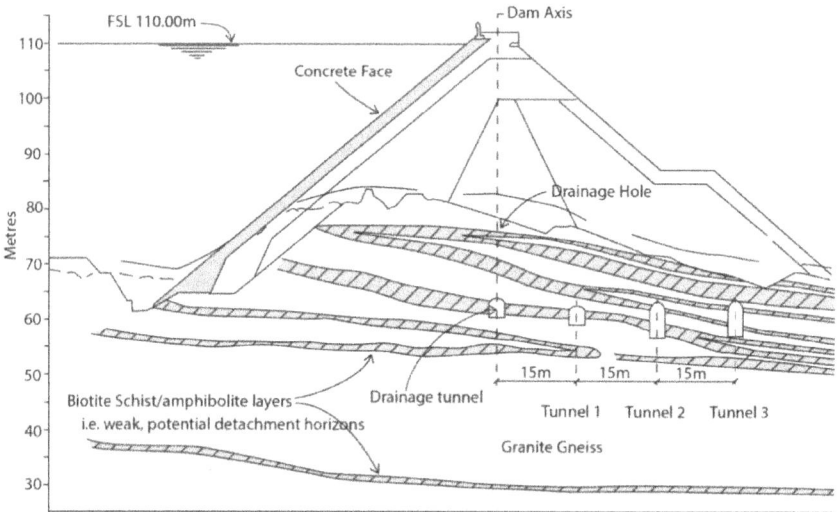

Figure 6.7. Diagram showing the anastomosing schist horizons beneath the dam and the position and dimensions of the shear-key tunnels designed to stabilise the foundation. After Albertoni (2009).

Figures 6.6 and 6.7 show the typical shear-key section and the foundation reinforcement of the dam in the area of the abutments.

In Section 6.2 below, we provide another but similar plane sliding case study.

## 6.2 PLANE SLIDING ALONG SECONDARY PLANES

The planes on which the rock mass at the Itapebi dam slid were primary planes, i.e., bedding planes, initially in shales which were subsequently metamorphosed to form a biotite-amphibolite schist. In addition to these primary planes of weakness, rocks contain secondary planes of weakness (joints and faults) induced by tectonism which, if appropriately orientated, can also impact adversely on rock stability. An example is briefly described below where remedial action was necessary because of such structures.

The setting is one in which thrust faults cut at a low angle through a granite which is acting as the foundation of a hydroelectric scheme. It will be recalled (see Section 2.3.2, Figure 2.14(c) and the related discussion) that thrusts form in a stress regime where the maximum principal compression is horizontal and the least principal stress is vertical. Although it can be seen from that Figure that in theory thrusts are predicted to dip at around 30°, for a variety of reasons their dip is often much less. For example, and as discussed in Section 5.2 in relation to the Tytherington Quarry, when thrusts propagate through well-bedded, horizontal strata they frequently exploit the pre-existing planes of weakness represented by the horizontal bedding planes. When they do 'climb' through the strata, they do so at the predicted dip of ~30° until they reach a higher, weak, bedding plane which they then follow. The resulting profile of the thrust is that of a staircase made up of 'flats', where the thrust tracks the bedding, and 'ramps', where it climbs the stratigraphy. In such thrusts, there are large portions of the fault that are sub-horizontal. Low angle thrusts are also common in unbedded rocks such as the granite making up the foundation of the project under consideration. The crust contains a variety of large-scale low angle fractures onto which smaller scale faults detach. In doing so, they need to change their dip from its original value to that of the dip of the detachment plane. This gives rise to curved faults termed 'listric' (as described in Chapter 4, Figure 4.4).

It was noted in Sections 2.10.6.1 and 2.10.6.2 that, during the formation of a shear fracture and during its subsequent evolution, there is a progressive degradation of the surrounding rock and the generation of a zone of deformation, a fault zone. The rocks adjacent to the fault become highly fractured and the shear movement along the fault planes generates fault breccias and fault gouges. Shearing on the thrusts in the study area resulted in the formation of breccias along the fault plane and the fracturing of the surrounding granite. Consequently, compared to the intact granite, the faults and fault zones are highly permeable and provide conduits along which hydrothermal fluids can migrate and into which weathering agents can penetrate.

Both these effects can produce a further weakening of the fault zone. In the upper portion of the crust where the rocks tend to behave in a brittle manner, shear failure results in faults which do not move at a constant rate but in a series of pulses. This is linked to the build-up of shear stress along the fault until it is of sufficient magnitude to initiate slip. This slip releases the stress and movement ceases until the stress again builds up to the critical value needed for slip. This process of stick-slip movement is accompanied by the release of seismic energy during each slip event and by the propagation of a pulse of fluid under high pressure along the fault, Sibson (1973) and (1995).

The interaction of these high pressure, high temperature fluids with the fault and the surrounding rock is of two types: namely mechanical and chemical. The mechanical effect is to produce a reduction in the effective stress, $(\sigma - p)$ see Section 2.5.3, which can give rise to intense hydraulic fracturing along the fault further degrading the intact granite. The chemical effects depend on the nature of the hydrothermal fluid and the rock type through which it is passing. The example under consideration is granite and, as will be seen, the chemical interaction between it and the fluids has a major impact on the strength of the fault zone.

Granite is composed mainly of quartz and feldspar with minor amounts of mica, amphiboles and other minerals, Figure 6.8. The two main components of the rock react very differently to hydrothermal fluids. The quartz ($SiO_2$) is essentially stable. A small amount may go into solution but it does not alter to other minerals. It is resistant to alteration and remains as an intact strong grain (it ranks as level 7 in the 10 levels of hardness of the Mohs scale of mineral hardness). In contrast, the feldspar ($KAlSi_3O_8$) is particularly susceptible to alteration by hot, acidic fluids and readily breaks down to kaolin (china clay: $Al_2Si_2O_5(OH)_4$):

$$2KAlSi_3O_8 + 2\,H^+ + H_2O = Al_2Si_2O_5(OH)_4 + 2\,K^+ + 4SiO_2\ (aq).$$

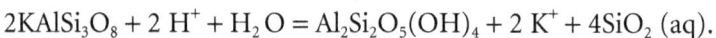

The result is a that the quartz grains, originally held together in a matrix of feldspar (see Figure 6.8) are freed and form a coarse quartz sand along the fault planes. Initially, the kaolin is disseminated evenly throughout the grains.

Inspection of the fault planes at the site shows that the kaolin has now concentrated into discrete, fault-parallel seams within the quartz sand, Figure 6.9. It is thought that this concentration was caused as a result of fluid migration along the fault planes. This redistribution of the kaolin by the fluids into discrete layers makes the faults particularly prone to slipping.

The final geological event to have affected the strength of these fault zones is that of weathering. As the fault approaches the surface during the

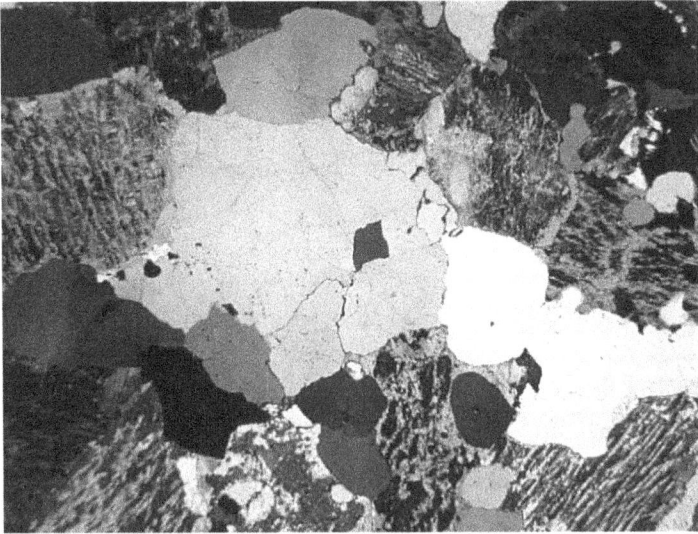

Figure 6.8. A thin section of a granite shown under polarised light. The plain crystals, which range in tone from white to black are quartz and the textured crystals are the feldspars. (×20 magnification.)

Figure 6.9. The appearance of sub-horizontal thrusts zones on a tunnel wall. The thrusts have experienced intense mechanical and chemical alteration. The thin horizons of kaolin (white) within the weathered and altered layers can be seen.

process of uplift and exhumation, the overburden and lateral confining stresses are reduced and the fault zone is able to dilate allowing surface weathering fluids to gain access and interact physically and chemically with the rock to weaken it further.

Thus, the sub-horizontal thrust planes cutting through the dam foundations have been weakened by three processes.

1. The original generation of the thrust which resulted in the granite being mechanically deformed along and adjacent to the thrust.
2. The movement of hydrothermal fluids which were pumped along the fault as it continued to develop in a series of pulses and which attacked the fault zone both mechanically and chemically.
3. The ingress of weathering agents which occurred as a result of dilation of the fault zone in response to the removal of the overburden and lateral confinement during the last stages of exhumation.

So the granite, which in its undeformed and unaltered state would form a strong foundation, is significantly weakened by these low dipping shear zones along which extensive layers of kaolin occur.

$$*****$$

Mechanically the problem is almost identical to that associated with the low dipping planes of schist at Itapebi (Section 6.1). There the planes of weakness that threatened the integrity of the dam foundation were **primary**, i.e., bedding planes now metamorphosed to layers of mica schists. The intrinsic weakness of the mica, the alignment of the mica crystals within the schist sub-parallel to the bedding and the low dip of the planes result in a high probability of 'plane sliding failure' see Principle 37 in Chapter 3. At the present site, the planes of weakness are **secondary**, i.e., thrusts generated during an episode of tectonism. The planes have been weakened by mechanical and chemical processes linked to the thrust formation and evolution. As a result of their gentle dip, low cohesion and the thin layers of kaolin which formed along the faults, they have a high potential to fail by 'plane sliding'.

Remedial action similar to that proposed for the Itapebi dam described in Section 6.1 would also be appropriate in this case to counter the threat to stability posed by the thrusts cutting the granite. This would include the

Figure 6.10. Tunnel intersecting the sub-horizontal planes of weakness. Infilling the tunnel with cement provides a key which ties the intact rock layers either side of the planes, together.

reinforcement of the foundations through the construction of concrete shear keys implanted along the tunnels and trenches excavated in the rock in order to intercept the thrusts. Such a tunnel is shown in Figure 6.10 which shows clearly the sub-horizontal thrusts and the concentrations of kaolin along them.

## 6.3 MULLARDOCH DAM, SCOTLAND

This section describes a dam problem where the engineer had to make a decision in the absence of sufficient information.

### 6.3.1 The Mullardoch Dam

The Mullardoch Dam in Scotland, 48 m high and 727 m long, is a mass concrete gravity dam located 48 km south–west of Inverness and was completed in 1952. The dam has a counter-intuitive geometry consisting of two arms meeting at an angle of 140° on a rocky knoll, with the apex pointing downstream, Figures 6.11 and 6.12. A detailed description of the construction of the dam is given in Roberts (1953). The topography and fractured nature of the Pre-Cambrian rocks on which the dam is built are shown in Figures 6.13 and 6.14.

The relatively massive rock mass of the Moine Series shown in Figure 6.14 is of metamorphosed Pre-Cambrian origin having a glaciated surface and containing sets of vertical and horizontal orthogonal joints. There is evidence that the horizontal fractures are not through-going and abut against the vertical fractures. This indicates that they formed later, supporting their interpretation as exfoliation joints. There is also evidence of possible glacial plucking.

## 6.3.2 The problem

The problem arose when there was a sudden increase in leakage associated with the widening and extension of horizontal cracks between the inspection tunnel inside the dam and the upstream face of the dam, and vertical

Figure 6.11.   The Mullardoch concrete gravity dam in Scotland which consists of two arms meeting at an angle of 140° on a rocky knoll, with the apex pointing downstream.

Figure 6.12.   The rocky knoll at the apex of the two arms of the Mullardoch dam.

Figure 6.13.   View upstream from the dam showing the low relief, glacial topography that characterises the site.

Figure 6.14.   View downstream from the dam showing the orthogonally fractured, crystalline rock, the sub-horizontal glaciated rock surface, and possible exfoliation fractures (Section 2.14.4 and Section 5.1, Figures 5.1 and 5.4).

cracking was also observed in the concrete within the inspection tunnel (Figure 6.15).

The difficulty in considering remedial action hinged on the fact that the exact cause of the cracking was unknown. Different suggestions were put forward as follows.

— The cracking in the concrete was being induced by the overall repeated daily temperature fluctuations.
— Or by differential concrete expansion and contraction caused by the summer sun.
— Or by chemical effects in the concrete.
— Or that the dam was being compressed by increasing geological stress acting across the reservoir and hence along the dam foundation — with the dam's bilinear geometry causing cracking to occur at the vicinity of the apex.

None of these could be definitively established as the cause.

Figure 6.15.   Vertical cracking in the concrete around the inspection gallery tunnel within the Mullardoch dam. The width of the image is 0.5 m.

### 6.3.3 The two potential solutions

Two solutions, Figure 6.16, were considered to alleviate the stresses within the dam and hence reduce any further cracking:

Option 1:  Cut a vertical slot through the dam at the apex to relieve the longitudinal force acting at the apex and then waterproof the slot;

Option 2:  Install post-tensioned anchors to strengthen the dam.

The objective of the first option is in line with Figure 3.13(d) in Chapter 3: the stress component perpendicular to a free surface is zero. The cutting of the slot would therefore directly arrest any further cracking. The objective of the second option was to strengthen the dam so that it could resist the additional stresses. One can immediately see the problems with both these options. Cutting the slot is an elegant solution for relieving the stresses, but it would not stand up to scrutiny should the dam collapse. Reinforcing the dam with post-tensioned anchors would only increase the stiffness of the dam and hence attract further stresses and possibly further

**Option 1**: Cut a slot through the dam to relieve the longitudinal force acting at the apex

**Option 2**: Install post-tensioned anchors to improve and strengthen the dam

Figure 6.16.   The two options for counteracting the stresses which may be the cause of fractures in the Mullardoch Dam.

cracking. It is well known in tunnelling that significantly increasing tunnel support can be detrimental to the tunnel stability — because the increased support stiffness attracts further load.

So which option would be the best engineering solution? The choice was difficult because the reason for the dam cracking had not been established. Cutting a vertical slot at the dam apex is an elegant solution and did have a precedent in the United States, as described by Abraham and Sloan (1978): "In 1972, cracking in the concrete dam at TVA's Fontana project embarked TVA on a four-year program of investigation, analysis and repair. Thermal expansion in the downstream face of the dam caused a longitudinal thrust that cracked the curved blocks near the left abutment. To retard cracking while a permanent solution could be devised, TVA sprayed reservoir water on the dam face to cool the concrete, pressure-grouted the crack and installed post-tensioned cable tendons to keep the crack from opening further. The final solution was to cut a vertical slot 100-ft deep across the dam to interrupt the longitudinal thrust. The slot was cut by drilling overlapping holes into the top and downstream face of the dam. Two flexible rubber seals prevented the water from entering the slot. Frequent monitoring of the dam shows the slot has solved the cracking problem."

### 6.3.4 The adopted solution

Because of uncertainty concerning the source of the dam cracking and the viability of the slot cutting solution, it was decided after much discussion to post-tension the central four blocks of the dam using 26 vertical tendons which were successfully installed despite the restricted and congested working area on the crest of the dam, Figure 6.17. Underslung helicopter lifts were used for lifting and placing the tendons in position.

Transverse tendons were also installed to counteract induced tensions above the roof of the inspection gallery. The tendons were provided with double corrosion protection and were re-stressable and fully detensionable. Performance of the dam during the initial post-tensioning period was

Figure 6.17.   Anchor post-tensioning in progress (from Hayward, 1990).

satisfactory except that the horizontal cracks did not close as much as anticipated when the anchorage cables were stressed. The innovation displayed by all those involved in the problem and its solution was recognised when the British Construction Industry Small Project Award was conferred for the project.

*****

Following Chapter 4 describing the geology beneath the Clifton suspension bridge in the UK and Chapter 5 describing the features of four quarries in Poland, the UK and Italy, in this Chapter we have discussed the importance of understanding the structural geology relating to dam foundations. In the case of the foundations of the Brazilian dams, the emphasis has been on the need to understand the geology in detail in order to establish appropriate engineering protective measures. In the case of the Scottish dam, it was explained that an engineering decision had to be made in the absence of a full understanding of the site conditions. Both these examples again support our contention that the structural geology analyses and rock engineering design should go hand-in-hand, even if the site circumstances cannot be fully understood, as in the case of the Scottish dam.

# CHAPTER 7

# OPENCAST COAL MINING

## 7.1 INTRODUCTION

Following Chapters 4, 5, and 6 which demonstrated the links between structural geology and rock engineering in bridge abutments, quarries and dams, we now move to opencast coal mining with an interesting case example site in central England. This case example demonstrates that the reasons for the engineering problems on site could not have been understood without a combination of the structural geologist and rock engineer working together.

In the case of opencast coal mining, the procedure is to expose a portion of a relatively near-surface coal seam by removing the overburden, excavating the coal, and then replacing the overburden. This is done in stages, so that newly removed overburden can be used to fill in the previously mined regions as the mining progresses across the site. Although the procedure can temporarily create a local environmental disturbance (Figures 7.1 and 7.2), at the end of the process the land is returned to its original green field condition. It is within this surface mining context that geological processes, which have occurred since the coal was originally formed, can create problems, as we explain in the next Section.

Figure 7.1. View of the opencast coal mining site. Note the scale of the site from the size of the dump trucks.

Figure 7.2. View of the opencast coal mining site showing the benches that are developed as the mining proceeds.

## 7.2 PROBLEMS ENCOUNTERED AT THE SITE

At the site in question, four inter-related geological and engineering problems were encountered which significantly affected the site operations:

— coal losses/difficulty in recovering coal, see for example, Figure 7.3;
— instability of excavated slopes;
— absence of suitable road making material; and
— dump trucks using more fuel than expected.

The site study involved establishing the underlying geological reasons for the problems experienced, and explaining how these directly and adversely impacted on the problems listed above.

The content of the study began by developing an understanding of the geological evolution of the site, both by outlining the regional context and

Figure 7.3.   The effect on a coal seam of reverse movement (inversion) on a normal fault. The coal seam (centre of picture) is displaced from the usual horizontal orientation to vertical.

describing the exact geological circumstances encountered on site. The reasons for the problems experienced could then be established and their controlling factors discussed. There was focus on aspects of the geological evolution of the region that impact directly on the geometry of the coal seams, the descriptions being concerned primarily with the tectonic evolution of the region, its subsequent exhumation, and complex interaction with the Pleistocene ice-cap.

## 7.3 REGIONAL AND LOCAL GEOLOGICAL SETTING

The site was situated in the Middle Coal Measures sequence of south Staffordshire on the south-eastern limb of the Stafford Basin. The Stafford Basin is a broad open structure with a half wavelength of approximately 30 km and limb dips of a few degrees, plunging gently to the north–east. There exists a series of major faults and smaller folds, with wavelengths of a few kilometres trending northeast–southwest, which parallel this geological structure.

Since the deposition of the Coal Measures rocks, the region has not been subjected to any major tectonic event — as it lies to the north of the main east–west trending Hercynian deformation belt of Devon and Cornwall which formed in the Late Carboniferous. Although the site is even further north of the Tertiary Alpine orogeny (an orogeny being a mountain-building event) this deformation did impact on the mine.

The geological structures observed at the site were the result of four main events:

(a) **Normal Faulting:** a dip–slip feature, Figure 2.14, dipping at approximately 60° and associated with an approximately north–south Mesozoic extension which occurred approximately 200 million years ago.

(b) **Fault Inversion:** reactivation of some of the faults developed in (a) above by imparting a reverse dip–slip component. This inversion was associated with a north–south compression which occurred approximately 60 million years ago linked to the collision between the African and European tectonic plates which gave rise to the Alps.

(c) **Exhumation:** uplift and erosion which result in a geological reduction of overburden and development of extensional features. This occurred between 50 and 10 million years ago.

(d) **Glaciation:** generated complex compressional and shear stresses approximately 10,000 years ago and earlier during previous ice ages. These features are part of a widespread deformation episode affecting the excavation.

On the scale of the opencast operation at the site, the large-scale folding of the Stafford Basin was not apparent. The beds dipped by only a few degrees and the dominant tectonic structures were faults. The main faults to affect the site were normal faults (i.e., faults where the slip occurs down the dip of the fault plane) trending WNW–ESE. They were probably formed during the north–south Mesozoic extension, which generated the east–west trending basins of the Bristol and the English Channels, as discussed in Chapters 4 and 5, Section 5.2. The site was also cut by normal faults trending NNE–SSW. These faults were relatively small-scale compared with the WNW–ESE faults. Faults with a WNW–ESE trend had displacements measured in tens of metres and those trending NNE–SSW had displacements measured in metres.

It was clear from the exposures at the site that some normal faults had been reactivated as reverse dip–slip faults (i.e., pushed back up the original fault surface). This 'inversion' probably represents minor tectonism in a region marginal to the Alpine belt and is the result of relatively weak north–south compression that occurred in the Tertiary. The effect of this reactivation had a dramatic, albeit local effect on the orientation of some of the coal seams at the site, Figure 7.3.

The strata were also cut by numerous extensional fractures (joints). In the coal, these joints were sub-vertical. The trends of the fracture sets in both the coal and the sandstone layers (Figure 7.4) were approximately the same, but they were much more closely spaced in the coals. They were formed during exhumation of the rocks because of the removal of the overburden by geological processes and the release of stored strain energy. Their orientation is therefore related to the stress fields associated with tectonism, i.e., most likely the stress field in place at the time of the removal of the overburden.

## 7.4 GLACIAL HISTORY AND ITS EXPRESSION ON SITE

The site in central England is situated well north of the southern limit of the Pleistocene glaciation (Figure 7.5) and has been subjected to the

Figure 7.4.  Vertical joint sets in a sandstone stratum at the site above a coal seam (photo shows an ~1.5 m horizontal extent).

repeated advances and retreats of a major ice cap.  Examples of the variety of glacial features that have been preserved in the area are eskers and kames: these are mounds of glacial debris — the eskers being along the direction of glacial flow and the kames being across the flow. These have enabled the ice stands associated with the final retreat of the ice cap to be mapped out. They trend approximately east–west and illustrate that the retreat was from south to north and that it occurred in a series of pulses. This is compatible with the regional direction of ice flow shown in Figure 7.5.

One of the remarkable and unexpected features found at the site was the development of structures that are incompatible with the accepted tectonic history of the region. These phenomena were mainly restricted to an 'upper zone', which extended as far down as 25–45 m below rockhead. In comparison with the relatively simple tensional system required to form the macro-geological structure (the normal faults and joints), a totally different mode of origin needed to be invoked to account for these features. Their nature and position relative to rockhead were consistent with, and symptomatic of, a complex and intense compressional stress regime. Such

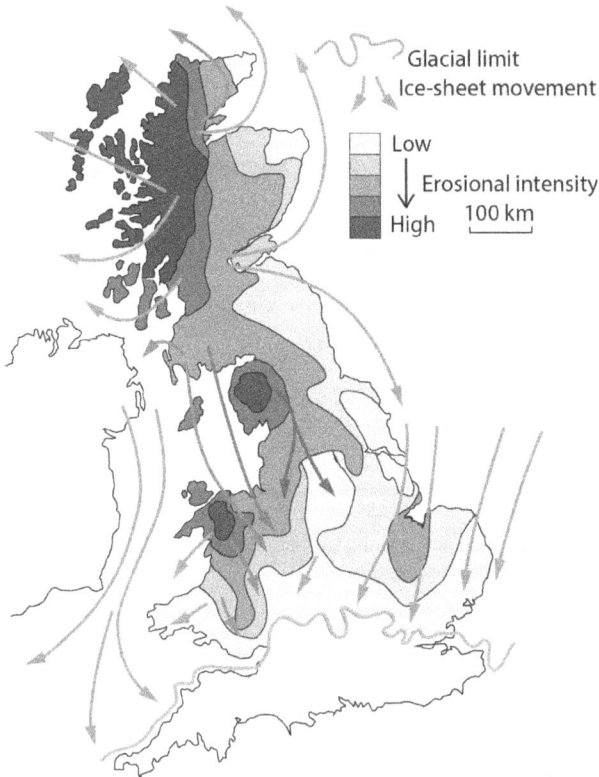

Figure 7.5. Centres of ice formation and directions of movement in Great Britain during the most recent Ice Age, after Palmer and Yates (2005). Note the centre of ice formation in the Lake District to the northwest of the site in central England and the direction of movement to the south.

features can only form in response to severe glacio-tectonic deformation and loading.

It is worth detailing the broad array of features observed during excavation (their nature, spatial extent, and dimensions) in order to appreciate the heterogeneity and unpredictability of glacio-tectonic deformation.

## 7.4.1 Linked imbricate thrust stacks

In this case, the rock is deformed in a manner so that it resembles tiles on a roof, and typically affected large areas at the site. As a consequence of the

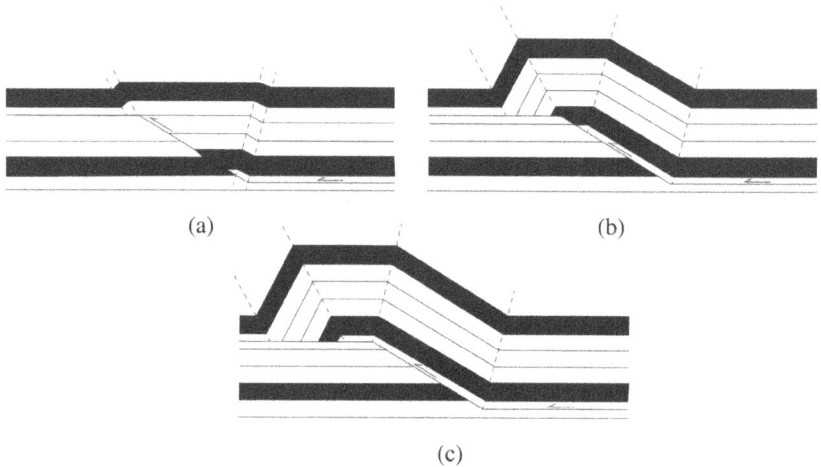

Figure 7.6.   Sketch showing three stages in the formation of a ramp anticline, (a)–(c), by movement on a thrust which slides along the bedding plane, climbs across a bed on a ramp, and continues along a higher bedding plane.

lubricating qualities at the coal/rock interface, shear deformation would tend to be localised along coal seams, which would become cut by thrusts as the deformation advanced through the coal. Movements on these thrusts (ramps) produce 'ramp anticlines', Figure 7.6.

Continued movement on this thrust becomes more difficult as it is 'locked' by the proximate strata. It is eventually abandoned and a new thrust initiated in the footwall, i.e., the rock beneath the ramp anticline. This process of footwall collapse may repeat, leading to the formation of an imbricate stack of thrust slices, known as a duplex. Dimensions and scales of such a feature observed at the site are shown in Figure 7.7. There was also the disruption to the Coal Measures strata below the coal seam.

### 7.4.2 Individual thrusts

Individual thrust sheets were often observed affecting both the coal seams and the Coal Measures. A large-scale example was initially exposed in the form of an elongated asymmetrical dome (termed a pericline). On excavation, the pericline was found to be a single thrust-generated, ramp anticline, Figure 7.8.

Figure 7.7. Imbrication in the Charles F seam caused by footwall collapse, see text. The lower diagram shows the thrusts more clearly; also Figure 7.10. The consistent orientation of the thrusts gives the structure an asymmetry that declares the direction of the ice-flow (from right to left). This brittle response to layer-parallel compression should be compared with the ductile response shown in Figure 7.11. (Truck for scale.)

The asymmetry of the thrust/pericline with a steep limb dipping south could be used to indicate the direction of ice flow. The trend of the pericline axis (125°) is compatible with that of a structure generated in response to a principal compression from the 35° direction. The plunge of the axis exposed in the excavation was towards 125°. It is interesting to note that the strike of the dominant vertical cleat in the seam affected by the thrust/pericline, and therefore, the principal stress responsible for cleat formation, was north–south. This orientation is incompatible with it being related to the deformation event forming the thrust/pericline and provides definitive evidence that the exhumation/cleat structures formed at a different time, Figure 7.8(d). Detailed analysis of the structures associated with the thrust/pericline showed it to be overturned at the (visible) apex of the fold close to rockhead. Perhaps the most exceptional feature of

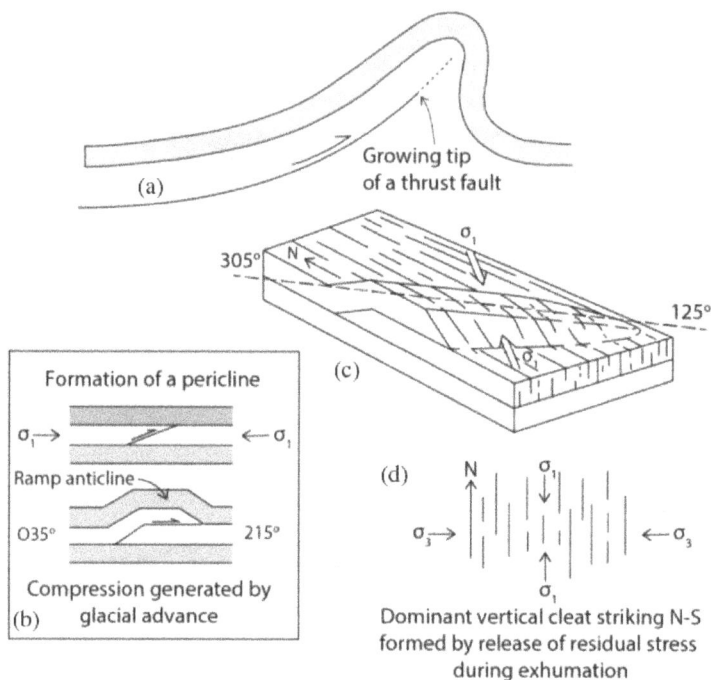

Figure 7.8. (a) The link between a thrust and an associated fold. The fold's asymmetry is determined by movement on the thrust. (b) Schematic representation of the formation of a ramp anticline. (c) A sketch of the pericline (ramp anticline) formed at the site in response to a horizontal compression caused by the advance of a glacier. The length of the exposed part of the pericline was ~10 m. (d) Formation of dominantly vertical cleats in the coal at the base of the excavated face.

the thrust/pericline examined was the extent of its vertical continuity. The axial plane could be traced from rockhead through to a coal seam 45 m below.

### 7.4.3 Listric-shaped thrusts

Thrusts are low angle faults which typically have a dip of ~30° (see Figure 2.14(c)). However, they are often seen to link with sub-horizontal planes of low strength such as salt layers and over pressured shales which act as detachment horizons. As they do so, their dip changes from ~30° to zero generating a curved or listric geometry (see the normal faults, Figure 4.4).

Figure 7.9.  Listric thrusts, from upper right to lower left, curving into the top surface of a coal seam exposed at the bottom of the cliff face. The coal is a horizon of low shear strength.

Coal is predominantly graphite, one of nature's most efficient lubricants, and it is therefore not surprising to find that the coal seams have acted as horizons onto which the thrusts detach. A series of listric shaped thrusts detaching onto a coal seam is shown in Figure 7.9.

### 7.4.4 Tectonised clay within fissures

As a result of glacial loading, highly disturbed clay is often observed forced into/along the pre-existing cleats (fractures in the coal seam) in near surface seams. Accommodation of the clay within these discontinuities highlights the cross-cutting orientations of pre-existing fracture patterns. At the site, the scale of clay infill varied from impersistent 'smears' up to a width of 1.5 m and running for several metres along strike; in fact, the entire vertical extent of a fracture through the coal seam may be infilled. If there is any subsequent movement on these fractures, the coal is effectively disrupted into a series of discrete blocks bound by clay-filled joints. Moreover, there were also observed larger and more discordant clay-filled fissures which were independent of any pre-existing structural control, some of which extended for many tens of metres along strike and were up to 2 m wide.

### 7.4.5 Complex, chaotic, and severe geometrical disturbances

Features regularly observed within disturbed zones were complex folds, thrusts, rapid seam thinning, bed attenuation/seam loss, bed rotation and contortion. This irregularity reflects various structural detachments activated simultaneously to produce the overall complexity, e.g., Figure 7.10.

### 7.4.6 Folding associated with the above features

At the site, a continuum existed from open to tight fold geometry. The folds, the result of glaciation, were occasionally overturned and isoclinal, and

Figure 7.10.  Severe disturbance to a coal seam outcropping along the top of the exposed face. The seam has been cut by numerous imbricate thrusts (all dipping at ~30° to the right), movement along which has resulted in the formation of an imbricate stack, see Figure 7.7.

(a)

(b)

Figure 7.11. (a) Typical fold patterns in a thin coal seam within the glacially disturbed Coal Measures. Note the vertical imprints of the excavator teeth. (b) Sketch of the fold patterns in the photograph above.

their amplitude and wavelength varied in scale from millimetres up to a maximum of 2 m. The persistence and style of the folding reflects the widespread occurrence and severity of the glacial overprint. Typical small scale folds in a thin coal seam are shown in Figure 7.11.

## 7.4.7 Development of foliation/fabric

It became increasingly apparent in those regions exposed to the most severe glacial stress that an irregularly orientated foliation can be superimposed on strata. This foliation cut across primary bedding features, often rendering them unrecognisable, and in the most severe cases obliterating them completely.

## 7.4.8 Summary of the near-surface geological conditions at the site

The area affected by glacial deformation was widespread, both laterally and vertically, effectively 'blanketing' a major part of the excavation. The extraordinary dimensions of the thrust/pericline, which had a total length of ~20 m, Figure 7.8(c), provided an indication of magnitude, scale, and severity of stresses applied to the strata. The majority of features within this zone of deformation were mostly localised in extent, which reflects the irregularity, unpredictability and complexity of the forces involved in their formation.

Implicit in the features outlined is the assumption that they were formed in response to complex and intense shear stresses generated within a 'deformation zone'. This zone probably coincides with the freeze/thaw zone that formed beneath and in front of the ice cap. During periods of glacial advance, ice movement would have been from north to south and it would have occurred as a series of surges. Each surge of the ice sheet as it advanced would impart a shear stress on the underlying rocks. These surges of glacial advance are thought to be responsible for severely disrupting the Coal Measures at the site. The base of the zone of glacial deformation would closely follow the topography; thus it is likely that gently dipping seams would move in and out of the influence of glacial deformation as the overlying topography rises and falls. The importance of coal seams as decoupling low-friction horizons, i.e., weakness zones along which slip can occur (see Figure 7.9), was apparent in many faces excavated at the site.

It is to be expected that the magnitudes and orientations of the forces induced by a moving ice sheet above the strata would depend significantly on the exact surface topography, local rises being more affected than local dips in the surface. The idiosyncratic nature of the glacial action was also illustrated by the fact that another opencast coal mining site some kilometres away had not experienced such disturbances to the strata.

In summary, the evidence indicated that there had been an all-pervading, glacially induced deformation of strata. A vertical section through the site strata for rock engineering purposes indicated three zones.

— ZONE 1: An upper zone, showing abundant evidence of severe glacial deformation in addition to 'normal' tectonic structures,
— ZONE 2: An intermediate zone where the effects of glaciation were less marked, and
— ZONE 3: A lower zone which contained tectonic structures, such as normal faults and joints, but which was otherwise relatively undisturbed.

## 7.5 ENGINEERING PROBLEMS ENCOUNTERED ON SITE

As mentioned earlier, the three main problems relating to the geological and geotechnical conditions that were experienced at the site were:

— Coal losses/difficulty in recovering coal.
— Instability of excavated slopes.
— Absence of suitable road making material.

These are discussed in turn in the following sub-Sections.

### 7.5.1 Coal extraction and recovery problems

In terms of coal extraction and recovery, it was the unforeseen and unpredictable nature of the glacial disturbance that had the most profound impact on coaling operations. The susceptibility of coal to external stress has been highlighted as critical in the formation of decoupling horizons and failures. It was the expression and nature of these failures in the coal that were paramount in determining the efficiency of coal recovery in this case. The transformation of originally planar seams into a discrete and complex array of blocks by normal faulting was disruptive enough but, when compounded by abrupt changes in level, irregular dip, bed contortion, rapid thinning and areas of total seam loss, the operation became even more difficult, as evident in the Figures already presented.

Furthermore, the appearance of clay-filled fissures required additional cleaning to remove adhered clay. In summary, the effects of unforeseen glacial disturbance reduced the potential coal tonnage by a combination of reducing coal area, reducing thickness, and increasing cleaning losses.

### 7.5.2 Instability of excavated slopes

Slope failures experienced on site originated in the rock strata below rockhead and propagated backwards and upwards into the superficials. Such failures occurring in the drift above were localised and did not threaten the overall integrity of excavated slopes. The initiation and propagation of slope failures proceeded in response to the criteria discussed below.

#### 7.5.2.1 *Rock weakening by glacial action*

There was a 'blanketing' nature of glacial action and effects on strata to variable depths. In addition to the formation of decoupled horizons, the overall effect on the Coal Measures within this zone was for them to be severely weakened and disturbed.

#### 7.5.2.2 *Rock weakening by water percolation and exposure*

Water movement in the disturbed upper zone weakened the Coal Measures, a process exacerbated still further by exposure after excavation. This was investigated using the slake durability testing apparatus. See Section 7.6.

#### 7.5.2.3 *Presence of weak mudstone–coal surfaces (creep mechanism)*

Failure initiation was expected at all mudstone/coal interfaces within the upper two glacially disturbed zones. At these interfaces, there was a high permeability zone with increased water movement lubricating the interface and low frictional resistance. Once excavation had removed the confining stress by excavating below rockhead, the thoroughly weakened Coal Measures could then creep, initiating a failure surface above the coal. As a

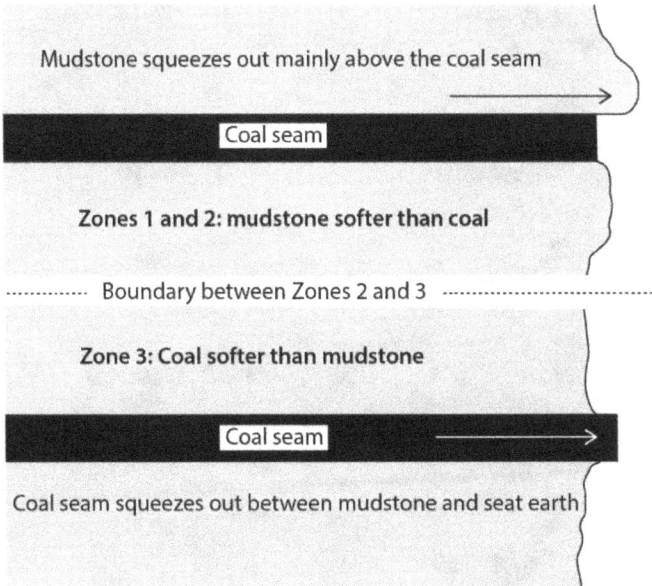

Figure 7.12. Indicating that creep of the mudstone or coal is a function of the relative strengths of each. Note: the seat earth is the layer beneath a coal seam.

consequence of overburden load on the weak mudstone once the failure initiated, it was essentially self-perpetuating, the failure surface becoming more vertical as it travelled back upwards through the weak mudstone. Conversely, in Zone 3 which has only experienced tectonic deformation, the coal was relatively weaker than the original higher strength mudstone and so the coal seam could creep out (Figure 7.12).

Suitably orientated, pre-existing discontinuities (faults and joints) within the upper Coal Measures also facilitate movement by acting as release surfaces for the creep mechanism.

### 7.5.3 Deterioration of the mudstone

In addition to the problems relating to unpredictable coal seam geometry and slope failure, the weak nature of the mudstone (Figure 7.13) caused additional problems. The glacially disturbed and weakened mudstone had broken down further through a combination of exposure, water and

Figure 7.13.   The fractured and weak nature of the mudstone (note the imprints of the excavator teeth).

mechanical action so that the rock in Zones 1 and 2 was too weak to provide adequate haul road material.

## 7.6 SLAKE DURABILITY OF THE MUDSTONE

For both the slope instability issues and the haul road degradation, it is important to have an objective measure of the susceptibility of the mudstone to degrade under the combined action of exposure, water and mechanical action. There is a specific Suggested Method recommended by the International Society for Rock Mechanics for this purpose (Ulusay and Hudson, 2007).  In the introduction to the test, it is mentioned that "An abundant class of rock materials, notably those with a high clay content, are prone to swelling, weakening or disintegration when exposed to short term weathering processes of a wetting and drying nature.  Special tests are necessary to predict this aspect of mechanical performance. These tests are index tests; they are best used in classifying and comparing one rock with another." Accordingly, the mudstones in the three zones were evaluated by this test procedure.

The results were in accordance with the zoning described earlier, Section 7.4.8. For the near surface seams, Zone 1, five of the six results were classed as 'very low durability' and the sixth of 'low durability'. The samples corresponding to Zone 3 also showed a consistent result with five of the six samples in the 'medium durability' class. These tests show that the near surface mudstone had a significantly lower slake durability index than the mudstone well below and provided objective evidence justifying the classification of the rocks at the site into three zones of decreasing strength based on the amount of deformation they had experienced. This result corresponded with the occurrence of the slope instability and the haul road degradation problems.

The overall trend of the slake durability results showed an increase in the mudstone quality with depth. The slake durability of these mudstone samples, as measured by the ISRM standardised test, has shown that the mudstones at the site are extremely susceptible to the effects of stress release, moisture and temperature changes, and mechanical loading on excavation.

## 7.7 CONSIDERATION OF OTHER FACTORS

The main reason for the three problems experienced on site was the damage caused to the strata by glacio-tectonic disturbance. However, other potential factors considered are discussed below.

### 7.7.1 Residual stress

It was found that bedding plane surfaces often showed evidence of having a polished, striated nature. In addition to shear along the bedding planes, it will be recalled that the site is cut by numerous thrusts and normal faults and it is possible that locally high stresses were locked into the rock close to these discontinuities. The same phenomenon occurred on a micro-scale within the mudstone itself as the material was deformed. However, it is unlikely that these local residual stresses were responsible for the major deformation of the coal seams observed on site and it was concluded that this factor did not need to be considered separately: any stress locked into major shear zones would only exacerbate problems in the vicinity of the shear zones, whereas the problems on site were all pervasive. It is likely that there were residual stress effects within the mudstone material itself, but these had already been taken into account through an understanding of the damaged zone near surface and through the results of the slake durability tests.

### 7.7.2 Previous underground mining

There had been significant underground mining at the site; drawings indicated that some of the seams below the site had been extensively and irregularly mined. The subsidence associated with underground mining occurs immediately above mined out areas and on either side — following an 'angle of draw' line of approximately 35° outwards and upwards from the edges of the mined out panels to the surface. This subsidence occurs within the strata, as well as being manifested by surface subsidence. Naturally, rocks closer to the mined out areas will be more disturbed by these mining subsidence effects than those further away.

Thus, any effect of underground mining at the site should increase with depth below the surface. As has been described, the opposite was the case, with the problems following the three zones already described and the slake durability results. The problems at the site were associated with near surface conditions: in the vicinity of the seam at the lowest part of the site, adverse mining conditions would be at an optimum but there was little evidence of any disturbance. So, although there were mined out areas beneath the site, these did not contribute to the three problems described earlier.

### 7.7.3 Water conditions

The presence of water is always significant in slope instability and the associated slake durability analyses. At the site, the water table and the location of high permeability zones were irregular because of the stratified and faulted nature of the rocks. The rock strata have widely different permeabilities and the faults can act as both water flow conduits and water flow inhibitors, depending on the idiosyncrasies of the local circumstances.

However, the slake durability test encompasses the effect of water because the test involves rotating a copper mesh drum containing the mudstone samples underwater, and thus simultaneously takes into account the combined effect of water and mechanical action. The main problem is not the water *per se,* but the fact that the mudstone is so susceptible to degradation in the presence of water. Water merely compounds a problem already initiated during glacial action by reducing still further the intact rock strength.

### 7.7.4 Stratal and structural inhomogeneity

The basic inhomogeneity of the Coal Measures strata can cause problems by virtue of adjacent rocks possessing different mechanical properties. However, this is not a problem if the constituent rocks themselves are strong. The Measures in Zone 3, the strongest zone, although containing 'normal' Coal Measure lithologies, were relatively stronger than those in Zones 1 and 2 and correspondingly suffered less failure. Thus, whilst the contrasting properties of coal and mudstone have been highlighted as significant in manifesting failures in the upper Coal Measures, it is not these that were fundamental to failure, but the unexpectedly very weak nature of the rocks as a consequence of intense glacio-tectonic deformation. Structural inhomogeneity (faults, joints, etc.) can cause problems, especially where unpredicted. However, these tectonic types of inhomogeneities were relatively well predicted and by themselves did not contribute to the site problems.

## 7.8 CONCLUSIONS

It is clear from the earlier Sections of this Chapter that the Coal Measures strata at the site have been severely disrupted and weakened as a result of unexpected local glacio-tectonic deformation.

The reasons for the site problems can be summarised as follows:

— **Coal recovery and extraction:** The Coal Measures have been disturbed *en masse* by glacio-tectonic processes on a variety of scales and to varying intensities causing deformation and dislocation to the original tabular geometry of the seams. The unforeseen scale, nature, depth and severity of disruption resulted in difficulties in recovering all the coal expected to be present.

— **Instability of excavated slopes:** Slope instabilities were generated in the Coal Measures rather than in the drift above. As a result of glacio-tectonic processes, the underlying strata were significantly weakened, allowing instabilities to occur in Zones 1 and 2. These slips are initiated by lubricated creep movement at the decoupling horizon between the mudstone/coal and propagate backwards and upwards into the superficials under the weight of the overlying Coal Measures. Failures are controlled by the inherent weakness of the strata and this can be independent of the dip and strike of bedding, jointing and faulting. Similarly, instabilities can occur regardless of the geometry and orientation of excavated slopes.

— **Absence of suitable road making material:** Because of glacio-tectonic processes, strata in Zones 1 and 2 had been fundamentally weakened and this was compounded by water percolation and exposure following excavation. An objective measurement of rock strength was provided by slake durability tests. The majority of the Coal Measures exposed in one site area had very weak rock mass characteristics and were unsuited to mechanical loading from heavy earth moving plant.

*****

This Chapter 7 case example on opencast coal mining has further illustrated just how important a knowledge of structural geology is to rock engineering. In this opencast coal mining application, the emphasis has been not so much on the *a priori* understanding of the geological setting but the structural geologist's ability to understand the nature of the site within the context of and during the opencast mining process itself. The reasons for the engineering problems and the necessary remedial actions could not have been understood and implemented without a combination of structural geology and rock engineering.

# CHAPTER 8

# UNDERGROUND ROCK ENGINEERING AND RISK

## 8.1 INTRODUCTION

Chapters 4–7 have been concerned with surface rock engineering examples. We now consider underground rock engineering projects and the associated aspects of risk. In order to be able to coherently design an underground rock engineering project, one has to be able to predict the future. For example, what will happen when a tunnel is constructed of this diameter, at this depth, excavated in this direction, in this rock mass? Will rock blocks fall into the tunnel? Will the concentrated rock stress damage the tunnel? Will the tunnel encounter any major faults? If one is unable to answer questions such as these, the design of the project is subject to some level of uncertainty and risk. In fact, we cannot obtain full information about any rock mass before engineering begins — so all rock engineering projects are subject to some level of uncertainty and hence risk. Reduction of the risks is directly related to reduction of the uncertainties and so, for improved rock engineering design, we must concentrate on the uncertainties involved. Lack of attention to this subject could endanger lives and/or prejudice the functionality of the completed project.

The Chapter begins with an explanation of epistemic and aleatory uncertainty, then follows with further notes on risk analysis techniques and 'black swan' events. Three illustrative rock engineering application examples are discussed: shale gas extraction, $CO_2$ storage and radioactive waste disposal — with the emphasis being on understanding the structural geology successfully and achieving the engineering project objectives. The Chapter concludes with 10 recommendations, concentrating on the further development of the subjects of uncertainty and risk as they apply to geology, rock mechanics and rock engineering. Especially important is the need to validate coupled numerical modelling and to make the programs more accessible to practitioners. This, together with the wider application of the techniques used for the design of radioactive waste repositories, will significantly reduce the impact of uncertainties and risks inherent in all rock engineering modelling, design and construction.

## 8.2 BOOKS RELATING TO ROCK ENGINEERING RISK

There are many books on the subject of risk, dealing with both the subject of risk in general and risk directly connected with rock engineering. The following useful general books on the subject of risk are highlighted: "Risk" by Adams (2001); "A Practical Guide to Risk Management" by Coleman (2011); and "Fundamentals of Risk Management" by Hopkin (2012). There are also books directly or indirectly relating to rock engineering/geotechnical risk: "Risk and Reliability in Ground Engineering" edited by Skipp (1994); "Managing Geotechnical Risk" by Clayton (2001); "Geotechnical Risk in Rock Tunnels" edited by Campos *et al.* (2006); "Geotechnical Baseline Reports for Construction" by Essex (2007); and "Risk Management for Design and Construction" by Cretu *et al.* (2011).

In his book "Risk", Adams (2001) notes in the wider risk context that, "Rarely are risk decisions made with information that can be reduced to quantifiable probabilities." This is also true in site investigation and rock engineering because of the difficulties of both the epistemic and aleatory uncertainties (see next Section) inherent in understanding and quantifying the variability of rock masses and their response to engineering perturbations. Adams refers to the Royal Society reports (1983, 1992) on risk assessment highlighting the differences of opinion between physical and social scientists concerning the nature and meaning of risk. These reports define risk as the probability that a particular adverse event occurs during a stated period of time, and distinguishes between objective and perceived risk.

## 8.3 TWO TYPES OF UNCERTAINTY: EPISTEMIC AND ALEATORY

The understanding of uncertainty is clarified through consideration of the two types of uncertainty, epistemic and aleatory, which are characterised by their distinct features, as listed in Tables 8.1(a) and 8.1(b).

Thus, epistemic uncertainty, e.g., relating to lack of fundamental knowledge about the rock conditions, can be reduced through systematic structural geology studies, site investigation and during the construction itself. On the other hand, aleatory uncertainty, e.g., when the next water inrush will occur, relates more to specific difficult-to-predict factors.

Table 8.1(a).   The characteristics of epistemic uncertainty.

| Epistemic Uncertainty |
| --- |
| — *Relating to knowledge, from the Greek,* episteme, *for knowledge* |
| — Subjective uncertainty, reducible uncertainty |
| — Due to lack of knowledge of processes or quantities in the rock engineering system |
| — Reducible through further investigation |
| — Reducible through expert information |
| — Generally has one value, but this is not known |
| — Conceptually resolvable |
| — Can be expressed as a 'degree of belief' probability distribution (Bayesian approach) |
| — Sensitivity analysis can reduce epistemic uncertainty |
| — Relates to assumptions made about rock mechanics processes in, e.g., the computer model. For example, the inclusion of fractures |
| — *Example:* Lack of understanding of coupled phenomena, lack of data |
| — *Example:* What type of geological structure will be encountered by the tunnel? |
| — *Example:* Will an earthquake damage the tunnel? |
| — *Problem:* Knowing all the factors involved |

Table 8.1(b).   The characteristics of aleatory uncertainty.

| Aleatory Uncertainty |
| --- |
| — *Dependent on chance, an uncertain outcome, from the Latin,* alea, *game of chance, dice* |
| — Random uncertainty, irreducible uncertainty |
| — Due to chance, intrinsic randomness in the rock engineering system |
| — Modelled via probability, statistics |
| — Not reducible through expert information |
| — Generally has more than one value |
| — Conceptually not resolvable |
| — Usually expressed as a probability density or cumulative distribution function |
| — Sensitivity analysis can reduce aleatory uncertainty using probability density functions |
| — Relates to varying the material parameters in the computer model |
| — *Example:* Variability in fracture density, variation in life of tunnel boring machine cutters |
| — *Example:* At exactly what chainages will water-bearing fractures be encountered by the tunnel? |
| — *Example:* When will the next earthquake occur? |
| — *Problem:* Appropriately representing variability |

Both epistemic and aleatory uncertainties are unavoidable and ubiquitous in rock engineering projects, but both of them reduce as construction proceeds.

Kahyaoglu (2012) provides an intriguing tunnelling example — which has been slightly modified by the current authors — of reduction in epistemic uncertainty through improved structural geology understanding as construction proceeds. Consider that a tunnel is being driven from right to left in Figure 8.1. At chainage (distance along the tunnel) point A, all is going well with steeply inclined strata being intersected. The same conditions are still encountered at point B, but at point C conditions worsen and there are roof falls. This continues until point D when steeply inclined strata are again encountered — but these are at a different orientation. The geological advice is now updated with the information *that a box fold is being traversed*, indicating that future tunnelling conditions will now

Figure 8.1. Example of reduction in epistemic uncertainty as a tunnel is being driven, with the full structural geology context becoming apparent at chainage point D.

probably revert to be similar to those encountered in the A–B region, albeit in reverse order. At point D, knowledge of the rock structural conditions has been significantly enhanced, epistemic uncertainty has been reduced, and the rest of the tunnel length becomes a risk that can be identified and quantified from the knowledge gained. This was not the case at the earlier points A and B. The example illustrates the point in the sentence below Table 8.1(b) — that both epistemic and aleatory uncertainty reduce as construction proceeds.

## 8.4  RISK ANALYSIS TECHNIQUES

The subjects of risk and the risk analysis techniques available have recently been highlighted in the paper by Brown (2012b) on "Risk assessment and management in underground rock engineering — an overview". Brown provides "brief overviews for some of the risk analysis and evaluation tools that are commonly used in geotechnical engineering and tunnelling, and may be suitable for use in large underground construction projects", specifically fault tree analysis (FTA), event tree analysis (ETA), consequence or cause–consequence analysis, bowtie diagrams, probabilistic risk analysis

(PRA), decision analysis, multi-risk analysis, analytical hierarchy process (AHP), Bayesian networks, and fuzzy logic and other artificial intelligence (AI) methods. Brown (2012b) notes that, "It is important to recognise that in geotechnical engineering, including in underground rock engineering, many of the risk sources or hazards arise from geotechnical uncertainty or error". The risk subject has continued to be studied in the context of different applications, e.g., the paper by Barton (2012) on "Reducing risk in long deep tunnels by using TBM and drill-and-blast methods in the same project — the hybrid solution".

In the overview paper, "Rock engineering, uncertainty and Eurocode 7: implications for rock mass characterisation", by Harrison (2012), the subjects of epistemic and aleatory uncertainties, interval analysis, fuzzy numbers, probability boxes, Bayesian statistics, selecting the uncertainty model, EUROCODE 7 and limit state design, and rock mass characterisation are described. Harrison and Hadjigeorgiou (2012) further discuss the challenges in selecting appropriate input parameters for numerical models and conclude that, "…not only does the challenge of obtaining appropriate input parameters remain, it may even be growing due to the increased sophistication of numerical techniques."

## 8.5 'BLACK SWAN' EVENTS

Another aspect to consider is the occurrence of either highly unlikely and/or completely unexpected conditions and/or events. In his book, "The Black Swan: The Impact of the Highly Improbable", Taleb (2007, 2010) points out that there can be outliers which are not included in probability distributions — the Black Swan events that lie "outside the realm of regular expectations". They have low predictability together with large impact and represent our epistemic knowledge limitation.

One such event is illustrated in Figure 8.2. A soft-ground tunnelling machine with pressurised bentonite at the face excavating through sand in the UK unexpectedly encountered a large granite boulder, resulting in considerable machine damage and difficulty. The tunnel was just above an old land surface on which the erratic boulder had come to rest after being transported there in earlier geological times by glacial action.

Another such Black Swan event, perhaps more related to regulations but still completely unexpected, occurred in January 2013 when 27 tonnes

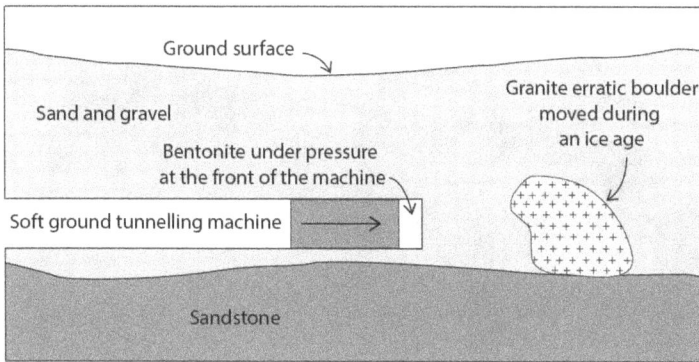

Figure 8.2. 'Black Swan' event: a large granite boulder unexpectedly encountered by a shallow, soft ground tunnelling machine in the UK.

of brown cheese caught fire when being transported by truck through the Brattli Tunnel in Tysfjord, Norway (Tunnels and Tunnelling, 2013). Brown cheese is made from whey containing up to 30% fat which caused the fire to last for five days and toxic gases to fill the tunnel. The Black Swan aspect of this event is that "Since such foodstuffs are not 'dangerous goods' they do not fall within the Sixth Issue of the ADR Directive Road Tunnels controls but, in reality, they can present a far more likely tunnel fire hazard than do petrochemical fuels." Kjell Bjoern Vinje, of the Norwegian Public Roads Administration, said it was the first time he could remember cheese catching fire on Norwegian roads. The accident occurred near one of the tunnel exits and no one was injured.

* * * * *

We now discuss three 'modern' underground rock engineering projects: shale gas extraction by hydraulic fracturing; underground carbon dioxide sequestration (storage); and radioactive waste disposal.

## 8.6 EXTRACTION OF SHALE GAS BY HYDRAULIC FRACTURING — FRACKING

Shale gas, i.e., methane, is sometimes present within shale deposits — in the natural fractures, within the rock pores, and adsorbed onto organic material. Some of this gas will have already escaped but much of it remains in the shale strata which have a relatively low permeability. In order to

HYDRAULIC **FRACTURING**

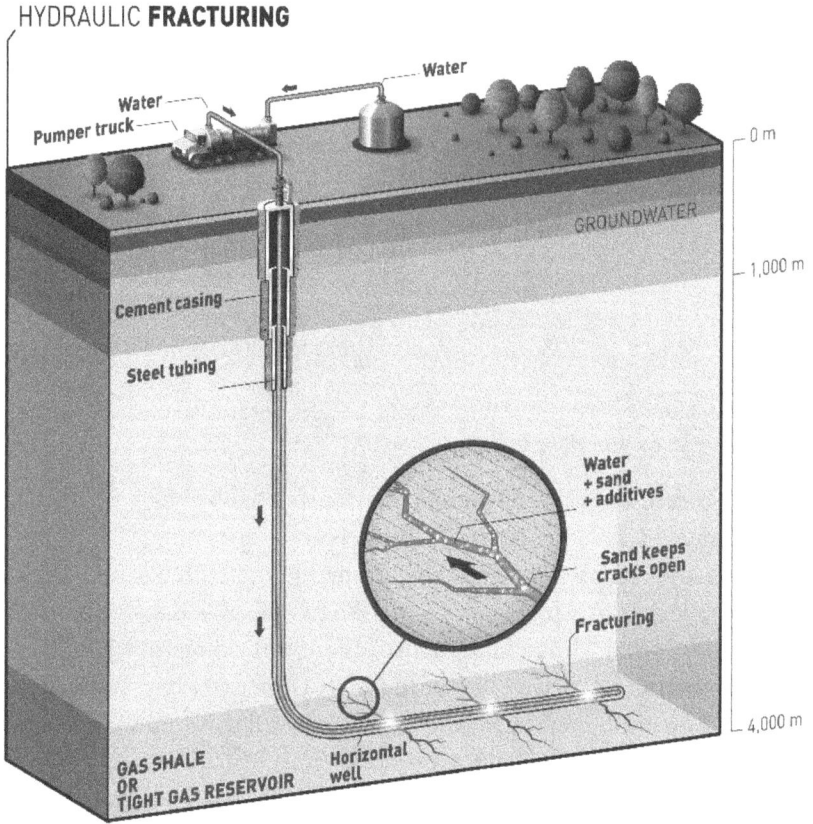

Figure 8.3. The process of extracting shale gas at depth via hydraulic fracturing of the strata (from http://www.skifergas.dk/en/technical-guide/what-is-hydraulic-fracturing.aspx).

extract the methane, this permeability has to be increased by the creation of additional fractures formed by hydraulic fracturing, a process known as 'fracking', Figure 8.3.

The use of hydraulic fracturing is well known in rock mechanics studies through its use in rock stress measurement, both for fracturing intact rock and for opening pre-existing fractures which enables estimates of the six stress tensor components. However, in the present context, the purpose is to provide a set of fractures within the rock mass to enable the methane to escape and be collected, hopefully under controlled engineering circumstances. Water, sand and chemicals are pumped under high pressure into the production well creating fractures which can extend hundreds of metres into the shale formation, with the injected sand holding the fractures open. Depressurising the well then allows the shale gas to travel to the surface.

There are large resources of shale gas in the world, with, in order of the magnitude of their resources, China, USA, Argentina, Mexico, South Africa, Australia and Canada. The UK does not have a large resource in comparison to these countries (although its value is estimated to be £1.5 trillion at the time of writing), but this is more than enough for economic production to be considered and to have stimulated a joint Royal Society/Royal Academy of Engineering RS/RAE report (2012). In the UK, the gas belongs to the Crown Estate.

## 8.6.1  Risks associated with shale gas extraction

Environmental concerns relating to shale gas production have been expressed by many organisations, especially with regard to global warming (because methane has a greater global warming potential than carbon dioxide), contamination of ground water and hence drinking supplies and seismic events. Malin (2012) in a Talking Point feature in the magazine 'Ground Engineering' considers the uncertainties and risks involved in 'fracking' and notes that controlling the impacts is a function of the predictability of an individual 'fracking' activity. He notes that, "…fracking is controlled primarily by natural fractures and these do not lend themselves to any type of average description. As a consequence, local physical property measurements at a given site do not predict anything about a nearby one, even though the two sites might be geologically and mechanically connected." (This description is reminiscent of the opencast mining project in the previous Chapter which was badly affected by glacial action whereas a nearby site was not.) In other words, this is both aleatory and epistemic uncertainty. He goes on to say that, "Even taking into account obvious factors such as sub-surface depth and rock type, the erratic nature of fracture flow paths renders frack-drive flow improvement and leak avoidance unpredictable."

This means that the actual extraction operation needs to be adequately controlled. For example, the surface operation should ensure that there are no significant spills of the water-based fracturing fluid (which contains additives). This is accomplished by the use of an impermeable site lining. Similarly, the series of casings in the well itself must be sufficient for the purpose (the conductor casing near the ground surface, the 'surface' casing that isolates the well from any fresh groundwater, the intermediate casing that isolates the well from non-fresh water, and the production casing itself which is installed into the production strata). The risk of blowouts is rare for shale gas wells because the shale strata have a low permeability.

The development and orientation of rock fractures initiated by the hydraulic fracturing process (and which can be seismically monitored) depend to a large extent on the rock stress at the shale strata location. Near the surface, rocks tend to have the greatest stress component acting horizontally but, as the depth increases, so does the overburden and hence the vertical stress component, so that eventually the vertical stress component exceeds the horizontal stress component (Figure 8.4).

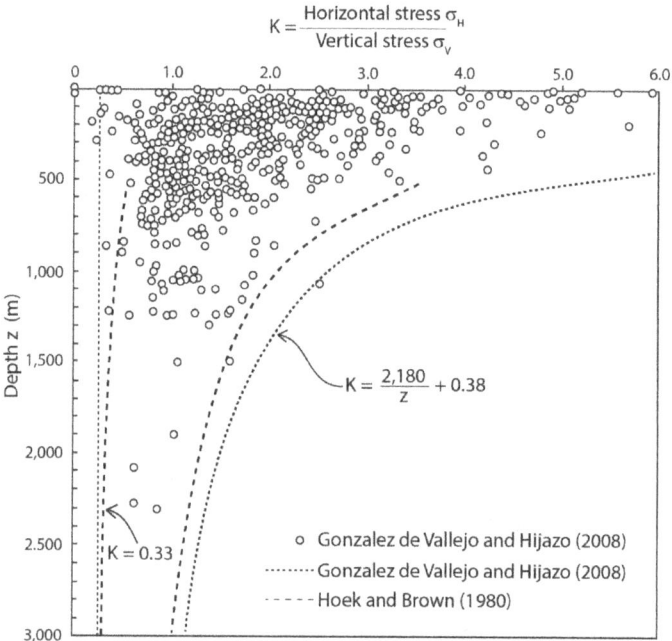

Figure 8.4. Variation in the relation between the mean of the horizontal stress components and the vertical stress component (after Gonzalez de Vallejo and Ferrer, 2011).

From Figure 8.4, and assuming that fractures propagate perpendicular to the least principal stress (following the least work principle), we would expect horizontal fractures to develop near the surface, where the horizontal stresses tend to be high compared to the vertical stress, and vertical fractures to develop below that. However, it should be remembered that the magnitude and orientation of the *in situ* stresses are a function of the local geology, especially the presence of any major discontinuities such as faults, see Figure 8.5. Although the modelling illustration in Figure 8.5 refers to a depth of only 150 m, such stress perturbations can occur at any depth depending on the nature of the geological discontinuities. Fracturing at one site in the UK (Bowland trough) occurred at 1,700–3,100 m where we could expect the vertical and horizontal principal stresses to be similar, see Figure 8.4.

The fractures produced by the fracking process may be irregular because of the rock mass inhomogeneity, but the idea is to maximise the fracture surface area. Various techniques are used to maximise this area,

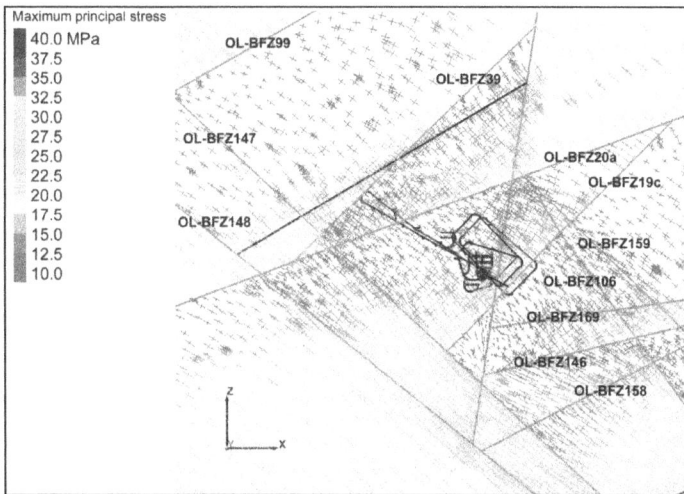

Figure 8.5. Computer simulation of the variation in the orientation and magnitude of the maximum principal stress at a depth of 150 m in the gneissic rocks at Olkiluoto, western Finland, as affected by large, brittle fault zones, the OL-BFZ features in the display (from Valli *et al.*, 2011).

given the nature of the shale formation. At present, the environmental regulator does not permit fracturing below freshwater aquifers. The other concern relating to hydraulic fracturing is that it could cause seismicity. Naturally, the fracturing process itself will induce low intensity seismicity; the danger is that it will reactivate an existing fault causing an earthquake, as is thought to have occurred during exploitation of the Bowland shale in the UK. However, and in the context of hydraulic fracturing in shale formations, there are several factors mitigating against the stimulation of a large earthquake — mainly that the shale is a relatively weak rock, the volume of rock involved is relatively small, and the pressure of the hydraulic fracturing fluid is limited, controllable, and only sustained for a relatively short time.

## 8.6.2 The approach to risk management of shale gas extraction

In 2012, a study of the safety of shale gas extraction using hydraulic fracturing was completed jointly by the Royal Society and the Royal Academy of Engineering (RS/RAE, 2012). The report, stimulated by the potential risks associated with this engineering process, discusses the many aspects of the subject and specifically addresses "the major risks associated with hydraulic fracturing as a means to extract shale gas in the UK, including geological risks, such as seismicity, and environmental risks, such as groundwater contamination." The report also considers whether and how these risks can be effectively managed. Although this report was written in the UK context, its explanations and conclusions apply generally to the fracking process.

The RS/RAE (2012) report highlights the Offshore Installations and Wells Regulations 1996 as follows: they state that wells should be designed and constructed so that "as far as is reasonably practicable, there can be no unplanned escape of fluids from the well; and risks to the health and safety of persons from it or anything in it, or in strata to which it is connected, are as low as is reasonably practicable". The report also mentions the need to develop a more goal-based approach to environmental protection: "Operators are forced to identify and assess risks in a way that fosters innovation and continuous improvements in risk management". This has advantages over a prescriptive approach based upon standards which "tends to support routine practices and limit innovation in risk management". The report also states that, "Operators should also ensure mechanisms are put in place to audit their risk management processes" — which is directly in line with the auditing procedure recommended by Feng and Hudson (2011) for all rock engineering design and construction work.

The RS/RAE report (2012) makes two important recommendations: "To manage environmental risks:

— an Environmental Risk Assessment (ERA) should be mandatory for all shale gas operations, involving the participation of local communities at the earliest possible opportunity; and
— the ERA should assess risks across the entire lifecycle of shale gas extraction, including the disposal of wastes and well abandonment. Seismic risks should also feature as part of the ERA."

The first recommendation, to involve the participation of local communities at the earliest possible opportunity, which might not be the first aspect being considered during a technical evaluation, is particularly important — as radioactive waste disposal implementers have found to their cost in different countries over the years. The second recommendation, to consider the entire lifecycle, is also important as this then includes all environmental factors.

The RS/RAE report (2012) goes on to recommend that, "Co-ordination of the numerous bodies with regulatory responsibilities for shale gas extraction should be maintained. A single body should take the lead. Consideration should be given to: clarity on roles and responsibilities; mechanisms to support integrated ways of working; more formal mechanisms to share information; joined-up engagement of local communities; and mechanisms to learn from operational and regulatory best practice internationally." The report concludes that more research is required to address the uncertainties involved in both exploratory and production activities, and discusses research funding options.

\* \* \* \* \*

This discussion of shale gas extraction has also highlighted the need in a risk assessment for a 'corporate memory' of existing data and experiences as baseline information for the design of extraction. The example further reinforces the need for an integration of the relevant technical subjects, e.g., structural geology, rock mechanics, fluid flow, coupled numerical modelling, etc., when developing the design. In terms of the practicalities of risk evaluation, a more 'goal-based approach to environmental protection' has been recommended. Additionally, a technical and operational auditing procedure should be implemented concurrently with the various stages of the process, from concept to abandonment. All these aspects assist in reducing the uncertainty and hence risk.

## 8.7 UNDERGROUND CARBON DIOXIDE STORAGE

'*Sequester*': 1. To cause to withdraw into seclusion
　　　　　2. To remove or set apart; segregate

### 8.7.1 The purpose of underground carbon dioxide storage

The purpose of storing carbon dioxide in the ground is to 'create a closed carbon loop', i.e., to return the carbon generated by the use of fossil fuels back into rock masses. Whilst the exact reasons for climate change are subject to debate, there is no question that the release of significant quantities of carbon dioxide into the atmosphere and subsequently the oceans must have some effects. For this reason, the underground sequestration of carbon dioxide is being promoted and researched. The proposal is that the $CO_2$ should be captured at coal- and gas-fired power stations and other industrial facilities such as cement plants, transported to a storage location, and then injected into a rock mass for long-term storage. Considerable research has already been undertaken in assessing the viability of the concept in different geological formations, in identifying suitable formations in different countries, in computer modelling of such storage, and in pilot storage schemes. In addition, and because $CO_2$ emission is of global concern, networks have been established to co-ordinate related information and research results, e.g., the European Commission's $CO_2$ GeoNet European Network of Excellence.

Because the proposal is to store the $CO_2$ underground and hence perturb the rock mass, there are risks associated with such a storage concept — in a similar manner to shale gas extraction, except that, for $CO_2$ storage, a material is being injected into the rock mass rather than being extracted from it. In both cases, the rock stresses are being anthropogenically altered and the overall entropy level is being increased with the associated risks. In this Section on $CO_2$ storage, the principle will be explained first and then the potential risks discussed.

## 8.7.2 The principle of underground carbon dioxide storage

To store the $CO_2$ underground, it is cheaper to utilise naturally occurring containment such as a depleted gas/oilfield, a saline aquifer, or another type of natural trap such as a dome formed by folded rock strata having a relatively impermeable cap rock and an absence of faulting, see Figures 8.6 and 8.8, with the link between structural geology and rock engineering again being reinforced.

As illustrated in Figure 8.6, the relatively impermeable cap rock traps the $CO_2$ within the rock mass. There are also subsidiary effects, such as

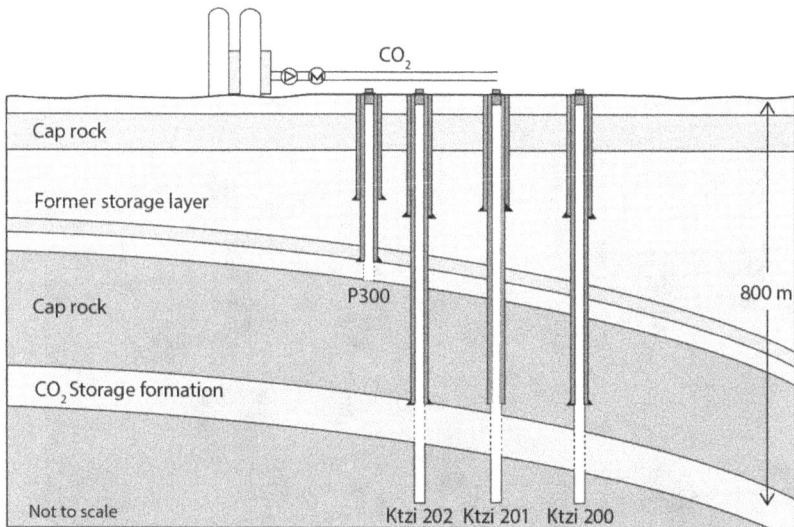

Figure 8.6. Underground storage of $CO_2$ beneath an impermeable cap rock stratum (from GFZ German Research Centre for Geosciences). The two deep boreholes to the left and right of the central injection borehole (Ktzi 201) are observation boreholes.

trapping of the $CO_2$ in small pores, dissolution in which some $CO_2$ becomes dissolved in water which sinks, and mineralisation in which the presence of the $CO_2$ can lead to precipitation of calcite.

### 8.7.3 The Ketzin pilot site for $CO_2$ disposal

At Ketzin, near Potsdam west of Berlin in Germany, an *in situ* pilot scale experiment operated by the GFZ German Research Centre for Geosciences has been underway since 2008. The purposes of the experiment have been to "improve the scientific understanding of the geological storage of $CO_2$, and to study the sub-surface processes of the $CO_2$ injection and distribution". The $CO_2$ has been injected at 650 m into porous sandstone of the Stuttgart formation which is locally in the shape of a dome and below a 210 m thick shaly cap rock, see Figure 8.6. More than 50,000 tonnes have been injected. The $CO_2$ storage tanks can be seen in Figure 8.7. Seismic measurements and numerical computer simulation models support the pilot scale experiment and enable assessments of the distribution of the injected $CO_2$.

Figure 8.7. Large cylindrical $CO_2$ storage tank (visit to the Ketzin $CO_2$ pilot site by members of the DECOVALEX project, 2012).

## 8.7.4 The risks in storing carbon dioxide

There are two main categories of risk associated with the general underground storage of $CO_2$: the $CO_2$ could leak from the site; and the alteration of the pressure within the reservoir and hence the stress state within the surrounding rock mass could lead to earthquakes (Figures 8.8 and 8.9). Leakage from the reservoir requires a pathway — which could be natural, through one or more major fractures in the overlying strata, or man-made through the injection and other wells. Such leakage could be locally dangerous to humans if the resultant $CO_2$ concentrations are sufficiently high. Naturally, the prime $CO_2$ storage function of the reservoir could be prejudiced if such leakage occurs. However, leakage should not occur if the site is sufficiently well geologically understood and if care has been taken with the engineering procedures.

The injection of $CO_2$ and other fluids into rock formations during fracking and sequestration processes and the consequential alteration of the natural rock stresses can lead to 'earth tremors', which have been experienced in Lancashire in the UK. In a paper on "Earthquake triggering and

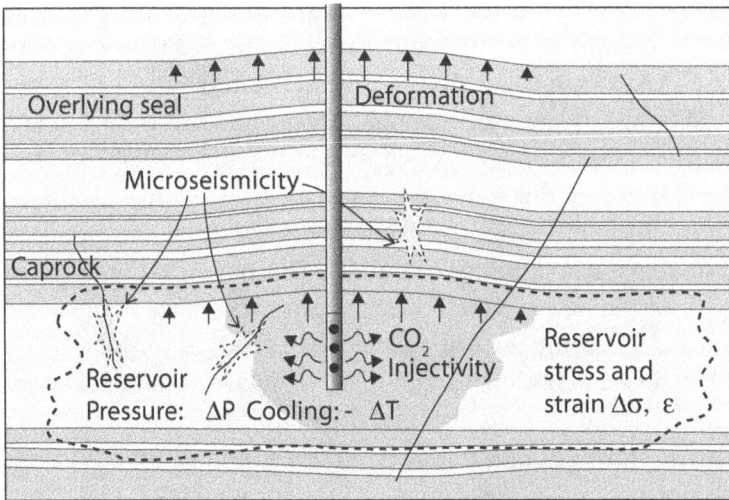

Figure 8.8.   Injection-induced stress, strain and deformation resulting from $CO_2$ injection into rock strata which could lead to loss of integrity of the cap rock and seal causing leakage (after Rutqvist, 2012).

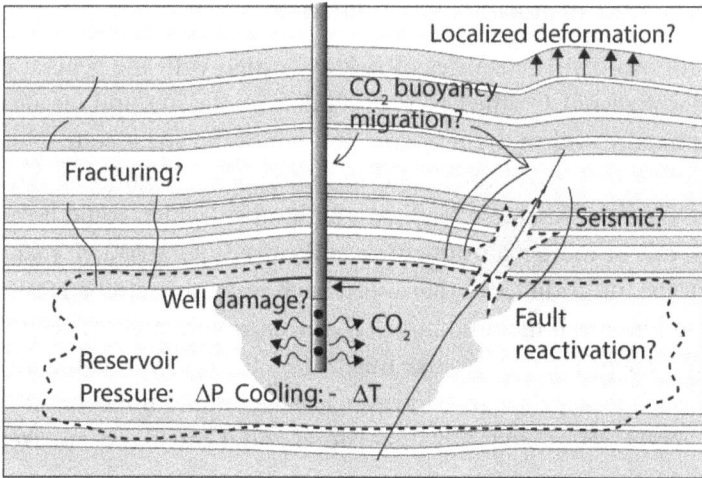

Figure 8.9. Potential unwanted mechanical changes resulting from $CO_2$ injection into rock strata (after Rutqvist, 2012).

large-scale geologic storage of carbon dioxide", Zoback and Gorelick (2012) argue that, "…there is a high probability that earthquakes will be triggered by injection of large volumes of $CO_2$ into the brittle rocks commonly found in continental interiors. Because even small- to moderate-sized earthquakes threaten the seal integrity of $CO_2$ repositories, in this context, large-scale carbon capture and storage is a risky, and likely unsuccessful strategy for significantly reducing greenhouse gas emissions." However, this view has been challenged on the basis that experience from injection activities worldwide related to oil and gas, geothermal, waste disposal and storage operations does not show the triggering of large scale earthquakes or the breach of cap rock integrity, and that the possibility of breaching the seal by injection of $CO_2$ is a risk factor that is routinely assessed in both the preliminary characterisation of storage sites and in the risk management plan that drives the lifecycle of a storage project. Thus, a potential injection site that was found to have a significant risk will not proceed to development in any case. However, in Europe, where country sizes are smaller than the USA and the areal population densities much higher, any earth tremors are more likely to be felt and have a strong psychological effect on the population — and hence on the granting of any future permissions for fracking or $CO_2$ storage.

In terms of the design life of a $CO_2$ storage reservoir, we note that the design life of rock engineering projects in general varies within the spectrum of a few days to a million years, depending on the project. In a longwall coal mine where the coal is being extracted and in which an advancing face moves forward, the superincumbent strata behind the advancing face are designed to collapse and hence provide support so that the region in the immediate vicinity of the face can remain open and move forward with the advancing face. Thus, at any particular time and location, the opening at the coal face should only remain clear for a few days. At the other end of the spectrum, the design life of a repository for underground disposal of radioactive waste is based on the criterion that unacceptable quantities of radionuclides should not return to the biosphere and the regulators of such a facility can recommend a design life of a million years (see Section 8.8). Within this extremely wide time spectrum, the question arises as to what should be the design life of a $CO_2$ storage reservoir?

The main steps in such a storage project (see the $CO_2$GeoNet) are site selection, site characterisation, storage design and construction, injection operations and site closure — a period of about 45 years — but how long should the monitored post-closure design phase be? In the case of radioactive waste disposal, the objective is to ensure safe disposal, i.e., no monitoring should be required. Can we apply this philosophy to the underground disposal of $CO_2$ or do we need to continuously monitor the disposal facility for escaping $CO_2$ via one of the potential escape mechanisms illustrated in Figure 8.10?

\* \* \* \* \*

This case example of $CO_2$ storage, like the previous case example of shale gas extraction, has highlighted the risks involved particularly in terms of whether the project objective can be successfully achieved. There are powerful arguments both for and against the ability to successfully store $CO_2$ underground, but the way forward must be to enhance the understanding of the geology, the risks and their mitigation.

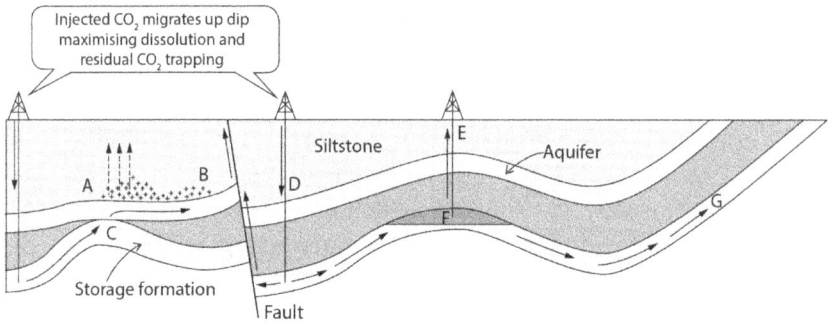

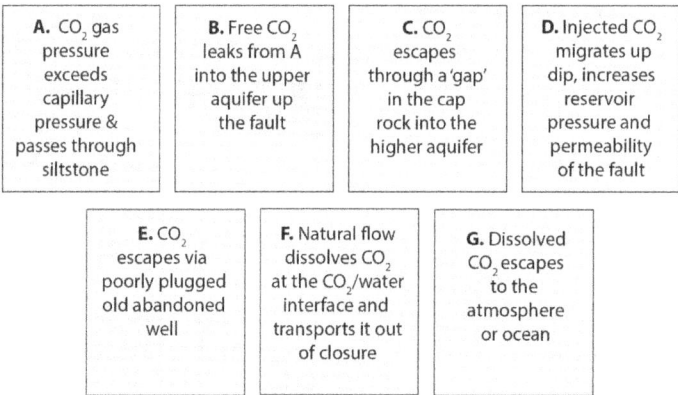

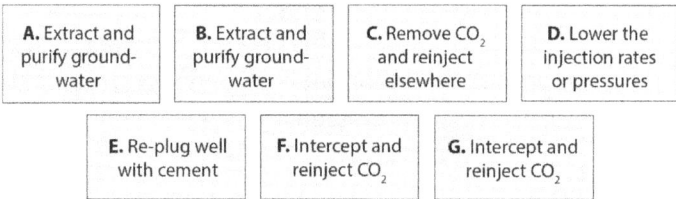

Figure 8.10. Some potential escape routes for $CO_2$ injected into saline formations and the associated remedial measures (from Metz *et al.*, 2005).

## 8.8 DESIGN OF AN UNDERGROUND REPOSITORY FOR RADIOACTIVE WASTE DISPOSAL

### 8.8.1 The unique combination of characteristics of an underground radioactive waste repository

The ultimate example of a long design life in underground rock engineering projects and the need for structural geology–rock engineering co-operation is a repository for the disposal of radioactive waste. This is because the design of such a repository involves an unusual combination of factors, as follows:

— from a technical point of view, the repository can be located anywhere suitable, as opposed to other structures such as transport tunnels and mines which have to be positioned in specific localities,
— the function of the repository is that of isolation of the waste, i.e., there is no active function such as, for example, transport of people or ore extraction,
— the design involves understanding the coupled geological-thermal-hydrological-mechanical-chemical-biological processes to a much greater degree than for other rock engineering projects, and
— the design life of the repository is measured in hundreds of thousands of years.

Within a presentation of some of the special conditions that can be involved in different rock engineering designs, some of the special conditions for the case of radioactive waste disposal are highlighted by the dark squares in Figure 8.11.

The key to the design of such a repository is that unacceptable quantities of radionuclides should not migrate from the repository to the biosphere. This means that, in the case of a repository for high-level waste, the regulators insist on designing for a very long period, i.e., of the order of one million years. There are many ramifications of this, not least is that all the possible features, events and processes (FEPs) must be evaluated, much more site investigation is required (e.g., 50 km of cored boreholes at a potential site), and the supporting design programme from concept to approval can take 20–30 years.

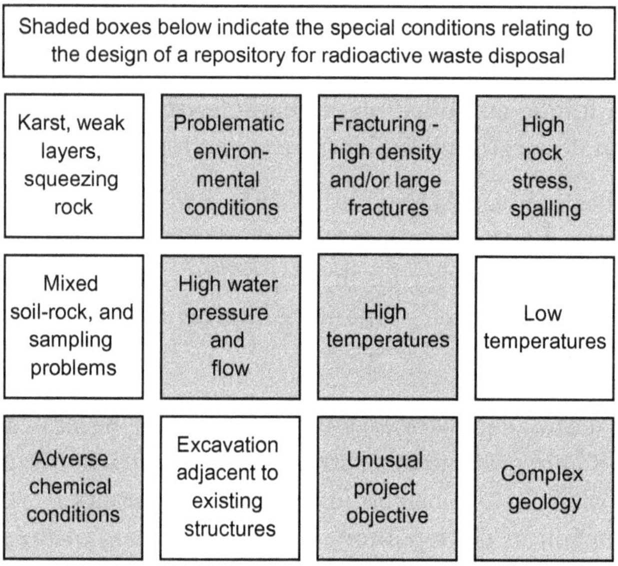

| | | | |
|---|---|---|---|
| Shaded boxes below indicate the special conditions relating to the design of a repository for radioactive waste disposal | | | |
| Karst, weak layers, squeezing rock | Problematic environmental conditions | Fracturing - high density and/or large fractures | High rock stress, spalling |
| Mixed soil-rock, and sampling problems | High water pressure and flow | High temperatures | Low temperatures |
| Adverse chemical conditions | Excavation adjacent to existing structures | Unusual project objective | Complex geology |

Figure 8.11.   Some of the rock engineering special conditions (shaded) which are associated with designing a repository for radioactive waste disposal.

In order to be able to predict the potential migration of radionuclides over such a long time period, it is necessary to understand and be able to characterise the hydrogeologically connected fracture network. The repositories are anticipated to be situated at depths of ~ 400–500 m (e.g., Figure 8.12) which means that there will be a high rock stress compared to other civil engineering projects and rock spalling then becomes a possibility. The same applies to the water pressure at these depths. High-level waste is significantly heat emitting so that, after a period of ~50–100 years, the maximum thermal load can induce additional rock stresses of the order of 25 MPa and hence further spalling of the excavations. Also, given the long design life, it is necessary to be able to model long-term effects, which then include chemical effects and, at the Finnish site (Figure 8.12), the potential impact of several ice ages with the attendant loading, unloading and erosion of the crust as the glaciers advance and retreat.

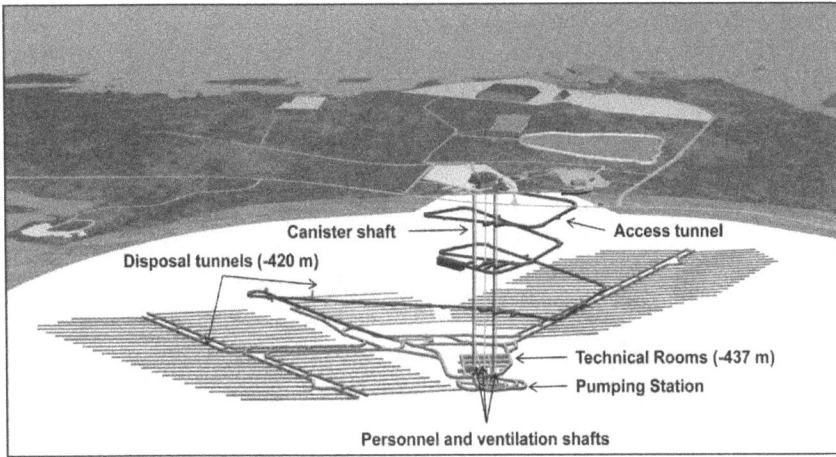

Figure 8.12. The layout of a radioactive waste repository, as illustrated by one of Posiva Oy's schemes at the Olkiluoto site in western Finland (from Posiva Oy, Finland).

## 8.8.2 The design process

Needless to say, these aspects cannot all be simulated in one computer model; nor can the results of individual sub-models be validated (i.e., confirmed by the actual rock behaviour) because of the long design life required. Accordingly, the repository design has to be based on a systems overview methodology. There are two main rock engineering design aspects: developing a design that will ensure safe disposal; and developing the construction operations for excavation and disposal. In terms of the contributions from structural geology and rock mechanics, the subject of this book, these aspects include the following:

— obtaining the rock properties at the chosen site (*in situ* stress, intact rock, fractures, faults, thermal properties, etc.);

— using/developing coupled numerical models to assess the effects of different design options;

— designing the underground excavations, given the characteristics of the site and the required openings;

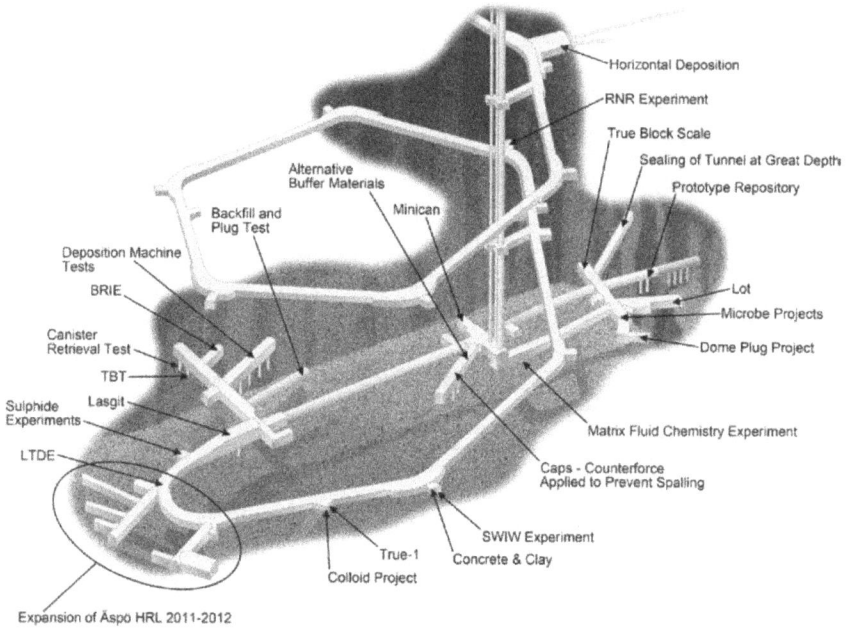

Figure 8.13.   The Äspö Hard Rock Laboratory in Sweden and the many experiments conducted therein (from SKB, Sweden).

— understanding the short- and long-term results of excavation; and
— understanding the short- and long-term effects of the disposal of radioactive waste at elevated temperatures, i.e., up to 100°C.

There will have to be extensive *in situ* and laboratory tests. Additionally, an underground research laboratory (URL) is generally required so that tests can be conducted on a representative scale and under appropriate and representative environmental conditions, see Figure 8.13.

### 8.8.3 Validation of numerical models

Also, and because much of the support for the design of an underground radioactive waste repository is provided through thermo-hydro-mechanical-chemical coupled numerical modelling, it is crucial to ensure that the modelling is reproducible and has been technically audited. This is to check that all the necessary factors are included and that the technical work is correct within the limits of the site information and current scientific knowledge.

An interesting approach to establish the reproducibility of numerical modelling was conducted within the 2007–2011 phase of the DECOVALEX research programme (DEvelopment of COupled models and their VALidation against EXperiment). In this Test Case example, the principal stress alterations caused by excavation of a small diameter shaft at the Äspö Hard Rock Laboratory in Sweden (Figure 8.14) were numerically modelled by separate teams from China, Japan, Korea, Finland, Sweden, Czech Republic, and another from China. As is evident from Figure 8.14, the stress path results obtained using different numerical models are approaching sufficiently close reproducibility for the results to be considered adequate in terms of verification, i.e., comparison of different modelling techniques. Continuing studies through the DECOVALEX project, Hudson and Jing (2013), and www.decovalex.org,

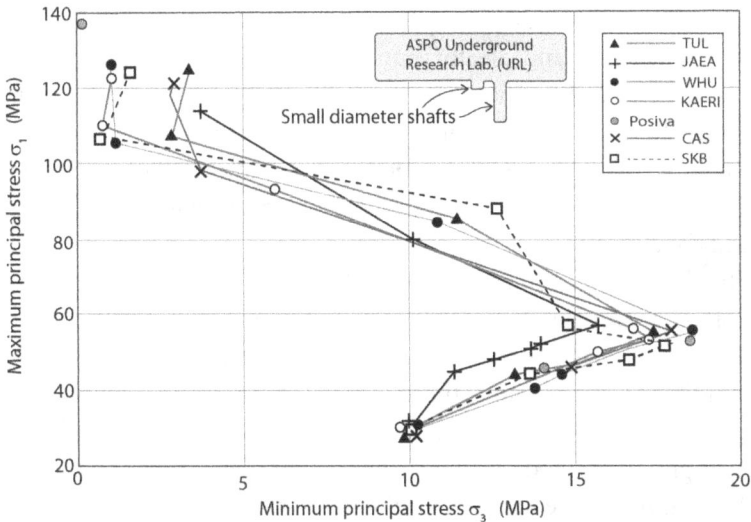

Figure 8.14. Numerical modelling of the stress path during excavation of a shaft in the Äspö Hard Rock Laboratory in Sweden — by seven different international teams from China, Japan, Korea, Finland, Sweden, Czech Republic, and another from China (from the DECOVALEX programme, see www.decovalex.org).

will improve the agreement and validation of the modelling, i.e., comparison with *in situ* data.

\* \* \* \* \*

From the general discussion on uncertainty and risk, plus the three rock engineering application case examples presented in this Chapter, a variety of conclusions is evident, as follows.

1. The subject of uncertainty and risk requires much more development so that the concepts can be built into all the aspects of rock mechanics and rock engineering, i.e., through rock characterisation, modelling and design. There is considerable literature available on the philosophies and techniques to support such development, both in general and related to rock engineering, as summarised by Brown (2012b).

2. In particular, it is helpful to study uncertainty through the two concepts of epistemic and aleatory uncertainties. These two concepts (the former dealing with lack of knowledge but reducible uncertainty, and the latter dealing with intrinsic randomness) should be implemented in design procedures.

3. Both epistemic and aleatory uncertainty reduce as excavation proceeds — indicating that, where possible, the rock engineering design should have built-in flexibility so that it can be adapted to the continually improving knowledge and understanding. This supports the inclusion of design flexibility and the 'observational method' in design and construction.

4. The three rock engineering application examples, of shale gas extraction, $CO_2$ storage and radioactive waste disposal, are all cases where the overall project objective could be prejudiced by a lack of understanding of, for example, the structural geology, the fluid flow, or the long-term behaviour of the facilities. Thus, for all three of the case examples presented in this Chapter, the risks should be assessed through the entire life cycle of the project.

5. Such comprehensive risk assessment has been extensively undertaken by many countries for the case of radioactive waste disposal — and the associated techniques could be most usefully applied when suitably streamlined to all major rock engineering projects.

6. Rock engineering design should specifically include consideration of potential 'black swan' events. Application of Conclusion 5 above should assist in this process, even if such events are mainly aleatory in nature.

7. Further work is required to establish the validity of thermo-hydro-mechanical-chemical coupled numerical models in order to reduce the risks of incomplete conceptual understanding and/or inaccuracy in the predicted values of physical variables as excavation proceeds.

8. Additionally, more information about the coupled numerical models should be made available and procedures established to make their wider use more practical.

9. An international, corporate knowledge database should be established for rock engineering projects to enhance the information available for all projects and hence reduce the uncertainty and thus the risks.

10. There is a conflict between the enhanced study of uncertainty and risk as recommended in this Chapter and the necessity to follow a code of engineering practice, such as EUROCODE7, which may not be appropriate, either for rock engineering or for the particular project in hand. Therefore, pressure should be applied to modify such codes of practice so that they are indeed suitable and do include the structural geology aspects and the concepts of uncertainty and risk in an appropriate manner for rock engineering.

# CHAPTER 9

# HISTORICAL MONUMENTS AND STONE BUILDINGS

In the preceding Chapters 4–8, we have discussed a variety of engineering applications of structural geology and we have emphasised the advantages of geologists and engineers working together on construction sites. In this Chapter, and in the same 'structural geology–rock engineering synthesis' mode, we consider a variety of historical monuments and stone buildings in the context of their overall mechanical stability and their degradation. We discuss the stability of a necropolis in Egypt, obelisks in and from Egypt, runestones in Sweden, cave temples in India, Easter Island statues, the mechanics of building stones, the use of building stones, the conservation of ancient monuments and rock used directly for ornamental purposes.

## 9.1 ROCK STABILITY AT THE THEBAN NECROPOLIS, LUXOR, EGYPT

> *"A palace of the god, wrought with gold and silver,*
> *it illuminated the faces of the people with its brightness."*
> Tomb inscription of Hatshepsut's official, Djehuty.

In this first example, the mechanical stability of the Theban Necropolis on the west bank of the Nile at Luxor is explained where some of the most important and spectacular monuments of Ancient Egypt are located. The geological and geomorphological settings of the site are considered together with the rock types and their deformational history in order to understand and predict the potential instabilities that might occur in the rock mass in which and on which the tombs and temples have been constructed.

## 9.1.1 The geological and geomorphological setting

From south to north across Egypt, the river Nile winds along a flat, fertile valley, Figure 9.1(a). On each side are pink cliffs of weathered Eocene limestone; the scenery is colourful, yet it varies little. An exception is near Luxor where the Nile follows a U-turn around the arid Theban Highlands and the cliffs rise to 500 m. In this area lies the Theban Necropolis — perhaps the foremost archaeological site on Earth, Figure 9.1(b). The pharaohs of the New Kingdom (1570–1070 BC) including the famous boy-king, Tutankhamun, built their tombs in the hidden Valley of Kings. Nearer the

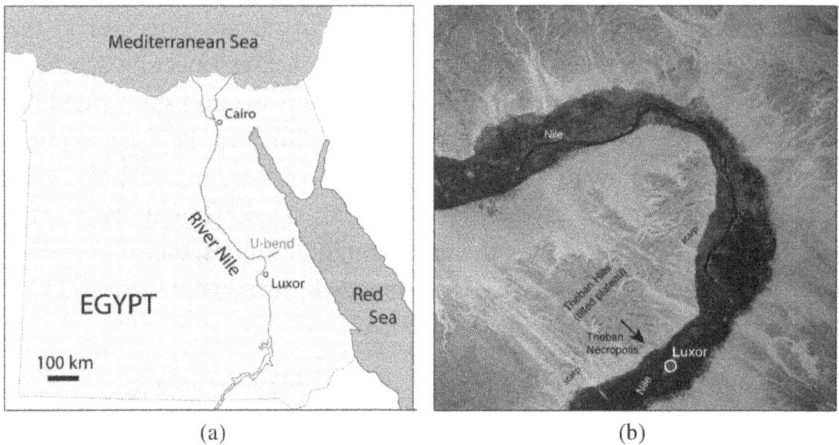

(a)                                        (b)

Figure 9.1.    (a) Schematic map of Egypt. Notice the U-bend in the River Nile, near Luxor. (b) A satellite image of the Theban Hills and of the U-bend in the River Nile, near Luxor. The limestone plateau (Thebes Formation) dips gently to the NNW, away from the escarpment defined by the Theban cliffs. The Theban Necropolis (arrow) is within the incised Valley of Kings.
*Source of image*: Google Earth.

Nile, amongst low rolling hills, their consorts had another burial ground, the Valley of Queens, while officers of the realm occupied the Tombs of the Nobles. Finally, at the edge of the Nile floodplain, the pharaohs built immense mortuary temples.

Why the rulers chose these sites for their resting places is not clear: it may be that they had mystical significance. However, a strong argument can be made that the sites were chosen on the basis of their ability to offer a solid and stable rock mass within which secure tombs could be excavated capable of keeping the pharaohs safe 'for all time'. The following discussion of the site indicates that safety and stability were probably of major concern.

## 9.1.2 Rock type

The main rock types at the site are shown in the stratigraphic column of the Theban Hills in Figure 9.2. At the base is the weak Esna shale which is beneath the Thebes Formation, a succession of thick limestones, shales and marls of Lower Eocene age.

The Nile valley has been incised into this stratigraphy and a schematic section through the Theban cliffs, from the Nile valley into the Valley of Kings is shown in Figure 9.3. The base of the cliff is comprised of the weak Esna shale on top of which are thick limestone units, members of the Thebes Formation.

The tombs of the Pharaohs have been excavated in gently dipping rocks which are away from the cliff edge. The tombs of the Queens are in rock which is closer to the cliff face and which contains numerous normal faults linked to landslide formation. These faults have a listric geometry, Figure 9.3, because of the low friction of the underlying shale. Thus, the faults are similar, although on a much larger scale, to those described in Chapter 7 above a coal seam, Figure 7.9. The tombs of the Nobles are built in the low hills at the foot of the cliff which are made up of eroded old landslides, rotated blocks that slid on the underlying Esna shales.

In terms of overall mechanical stability, it is apparent from the schematic profile shown in Figure 9.3 that the Valley of Kings is in the most stable region of the rock mass. The Valley of Queens is nearer to the cliff edge and therefore more susceptible to potential landslides and toppling.

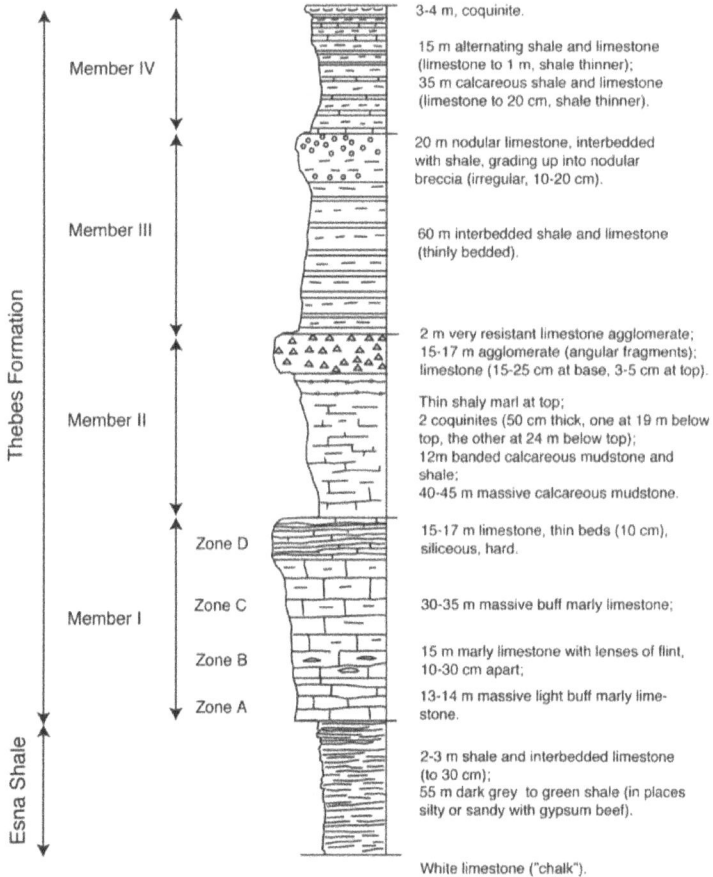

Figure 9.2. Stratigraphic column for the Valley of Kings. Egypt. Notice four members of the Thebes Formation (dominantly limestone), overlying the Esna Shale.

*Source of data*: Theban Mapping Project.

Finally, the nobles whose tombs were built in the rotated blocks are situated in front of the main scarp in a region that is the result of landslides.

### 9.1.3 Deformation history and the generation of fractures

The rocks of the Theban Highlands generally dip very gently to the NNW but their dip is locally disturbed by normal faulting. The faults can be divided into two types: those whose origin is tectonic; and those that are

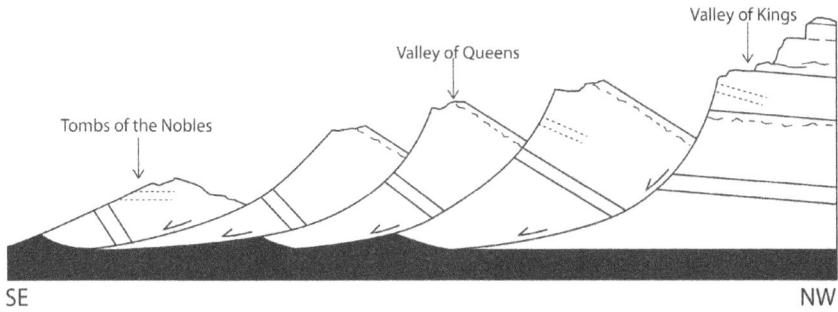

Figure 9.3.   Schematic section through the Theban cliffs showing the position of the Valley of Kings, Valley of Queens and the tombs of the Nobles with respect to the cliff face. The Esna shale is represented by the black unit at the base of the cliff. Note that the galleries leading to the tombs of the nobles (dotted lines) are still horizontal indicating that no movement of the blocks has occurred since the tombs were excavated.

induced by geomorphological processes, Figure 9.3. The tectonic normal faults are often characterised by crystal fibres of calcite that form sub-parallel to the fault plane and declare both the direction and sense of shear (see Sections 2.7.4 and 2.7.5). The orientation of these faults has been determined by the tectonic stress fields linked to the opening of the Red Sea and the collision of Africa with Europe, and therefore bears no relation to geomorphological features such as cliffs. In contrast, there are a number of normal faults more recent in age. They are related to landslides which occur along the steep scarp of the NE–SW trending Theban cliffs which in the vicinity of the Necropolis define the NW boundary of the Nile Valley, Figure 9.1(b).

These faults do not contain veins of calcite, are localised along the cliff line, and trend parallel to the cliff face — observations compatible with them being related to landslides. Both types of normal fault detach into the weak Esna shale and have a listric profile. Slip on such faults results in a rotation of the fault block (the hanging wall) with the result that the bedding locally becomes steeply inclined, Figure 9.3.

In addition to the faults (shear fractures) noted above, the rocks of the region are cut by several sets of important, often through-going, joints (extensional fractures). Most of them are normal to bedding and two of the most important in terms of the stability of the rock mass in the vicinity of the Necropolis trend ~ NE–SE and NW–SE, i.e., approximately parallel and normal to the cliffs.

### 9.1.4 The impact of the rock fractures, rock type and geomorphological processes on stability

**The Valley of Kings**

The Valley of Kings and the numerous tombs it contains is situated in the Theban Hills away from the cliff face (Figures 9.3 and 9.4). The ancient Egyptian tomb builders came across tabular, flat-lying beds of Eocene limestone of the Thebes formation (Figures 9.2 and 9.3). These were able to support galleries, shafts and the roofs of burial chambers. The Valley of Kings was remote and easy to guard; it was deeply incised, providing good opportunities for horizontal and inclined tunnelling into the hillside. However, although the site is some distance from the face of the Theban cliffs and is therefore not prone to landsliding, there were — and still are — two major problems relating to the stability and integrity of the tombs.

— Firstly, the Esna shale is prone to swelling when it becomes damp, and this damages the surrounding rock. Some of the tombs cut into this rock are particularly susceptible to damage, both structurally as a result of the local swelling and artistically in that the decorated plaster which is used to line parts of the tomb becomes detached as the shale hydrates.
— Secondly, there are some large tectonically induced faults, which could be reactivated in response to seismic activity.

Since the Eocene (50 mya) which was when the limestones accumulated, the opening of the Red Sea and the collision of Africa and Eurasia have occurred. Despite the fact that the Necropolis is situated some distance from these two plate margins, it is possible that these tectonic events have been subjecting, and are still subjecting, the region to tectonic activity and deformation. Evidence that middle Egypt is currently seismically active has been reported by Badawy *et al.* (2006).

**The Valley of Queens**

The schematic cross-section shown in Figure 9.3 is based on the part in Figure 9.4(a) which shows a panoramic view from the Theban cliffs capped

Figure 9.4. (a) Panoramic view over the Theban lowlands. The view is towards the SW. The highest peak (al-Qurn, (6)) is at the top right and the River Nile is at the top left (1). The black symbols indicate dips of bedding along two ridges. Notice the steep dips in and above the Valley of Queens (2). (3) Workers village, (4) tombs of the nobles, (5) the Theban cliffs. (b) Listric normal faults (dashed white traces) and tilted bedding (dashed black traces) above the Valley of Queens. The view is to the SW. The field of view is ~300 m along the skyline. After Cobbold *et al.* (2008).

by al-Qurn, the highest peak, over the Theban lowlands and onto the flood plain of the Nile valley. The main rock mass on the right of the picture contains the tombs of the pharaohs in the Valley of Kings and has sub-horizontal bedding.

It can be seen from Figure 9.4(a) that, for the low-lying hills between the horizontally-bedded limestones and the Nile flood plain, the geological context is very different. Here the bedding in the Eocene limestone is no longer flat lying but dips generally to the northwest at 45° or more. These Theban lowlands represent an area of large landslips. The arcuate form of the Theban Cliffs, concave toward the Nile, is typical of an upper landslip scar. Exposures in rocks above the Valley of Queens (Figure 9.4(b)) are cut by normal faults which dip to the south–east, i.e., normal to the cliff face, an orientation characteristic of landslide-related faults. The faults flatten at depth: in other words, they are listric (smoothly curving) and appear to link into a flat-lying slip surface (detachment) near the top of the Esna shale. Motion on the listric faults has caused blocks in their hanging walls to tilt, accounting for the steep bedding.

As one moves away from the cliffs towards the flood plain, the foothills become lower and normal faults are less visible, Figure 9.4(a). Instead, one finds a few low-angle thrust faults and folds, indicating horizontal shortening. Such structures are typical of toe thrusts and associated folds encountered at the leading edge of landslide blocks, Figure 9.5. Note that it can be seen from Figures 9.3 and 9.4 that the Valley of Queens lies within a large rotated fault block.

Thus, the tombs in the Valley of Queens are in a less stable setting than those in the Valley of Kings. In addition to the possibility of damage from swelling of the Esna shales and slip on tectonically induced normal faults as a result of current seismic activity, there is also the possibility of movement on the much more prevalent, landslide-induced, normal faults.

## Tombs of the Nobles and Mortuary Temples

The tombs of the Nobles were excavated in the oldest, most eroded and weathered landslide blocks, i.e., those furthest away from the current cliff

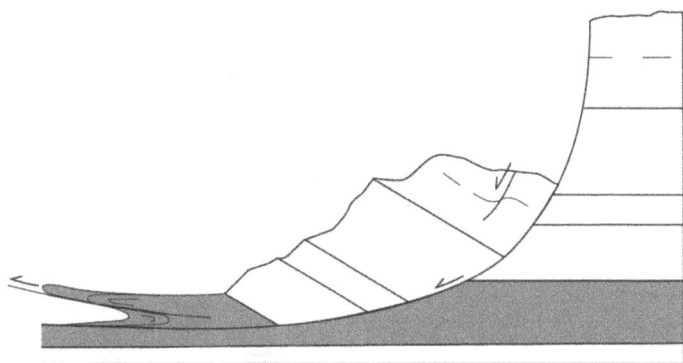

Figure 9.5. A schematic section through a landslip block showing extensional features at the rear and compressional features, folding, and thrusting, near the toe.

line, Figures 9.3 and 9.4(a). As noted above, the pharaohs built immense mortuary temples at the edge of the Nile flood plain. One of the most famous of these is the temple of Queen Hatshepsut, Figures 9.6 and 9.7, generally regarded by Egyptologists as one of the most successful pharaohs.

The temple is built at the foot of a vertical cliff of Eocene limestones which belong to Member I of the Theban formation, see Figure 9.2. The limestones are characterised by closely spaced, through-going joints (see Section 2.8.5) and, as noted in the Section above on the deformational history of the site, there are two particularly important joint sets, one normal and the other parallel to the cliff face. These two sets can be seen in Figures 9.6(a), 9.6(b), and 9.7. A close-up of the joints parallel to the cliff face and exposed in the Valley of Kings is shown in Figure 9.6(b). These joint sets result in the cliff being susceptible to toppling (Principle 37). The fractured limestones that make up the cliff face rest on the thick bed of Esna Shale on which the Temple is built and which can be seen in Figures 9.6 and 9.7. It is the erosion of this weak layer at the base of the cliff, probably by the Nile, that is responsible for toppling instabilities occurring in the overlying limestones and the consequent formation and persistence of the vertical cliff profile. As noted earlier, in addition to the likelihood of

Figure 9.6.   (a) Queen Hatshepsut's mortuary temple situated at the foot of the Theban cliffs. Two vertical joint sets occur in the limestone of the cliff, one normal to and the other parallel to the cliff face. A close-up of the cliff-parallel joints exposed in the Valley of Kings is shown in (b). These joint sets and the weak shale at the base of the cliff together increase the possibility of toppling.

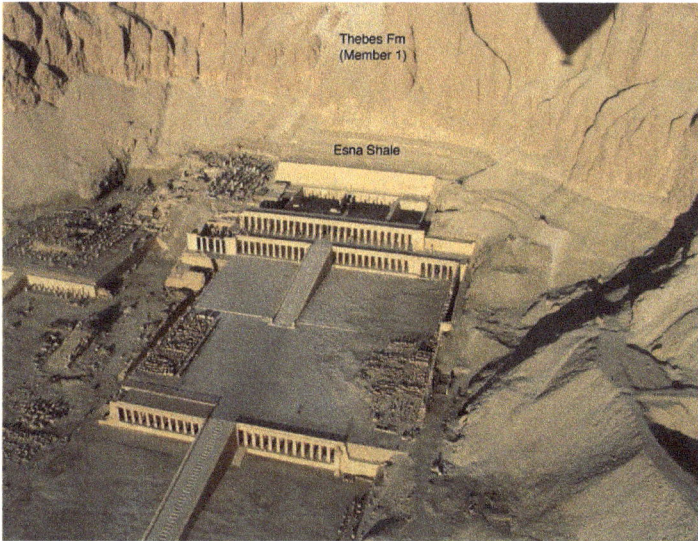

Figure 9.7.   Oblique view northward over the Theban Cliffs and Temple of Hatshepsut. The Temple lies on Esna Shale which can be clearly seen outcropping at the base of the cliff, the main face of which is made up of limestone of the Thebes Formation (Member I). Note the trace of joints sub-parallel to the cliff face on the cliffs that tower above the Temple.

toppling, the vertical cliffs are also prone to erosion by landslides — which slide on and now rest on the Esna shale, Figures 9.2 and 9.3.

### 9.1.5 Current threats to stability

The two main factors controlling the stability of the rock mass within the Necropolis are:

— seismicity; and
— the undercutting of the Esna shales.

Both may trigger the toppling of the tall, joint-bounded blocks along the cliff face and slip on the normal faults associated with the landslides. However, despite the threat of toppling on the cliff face immediately behind and above the Temple, it appears that no rock blocks of any significance have fallen since the Temple was built around 3,500 years ago. Similarly, observation shows that the galleries of the tombs of the nobles, which were driven into the inclined beds of the tilted landslide blocks, Figures 9.3 and 9.4, are still horizontal, testifying to the fact that no further rotation has occurred on these blocks since the construction of the tombs.

The main risk to these monuments is that posed by seismicity as the chances of significant undercutting of the cliff by the Nile (the other risk) has been considerably reduced by the building of the Aswan dams. The first of these, the Aswan Low Dam, was built at the beginning of the 20th century and the Aswan High Dam was completed in 1970. Since then, the annual inundation of the Nile and the flooding of the valley has not occurred.

This protection of the Theban cliffs is one of the few benefits resulting from the construction of the dams. Regrettably, these are outweighed by the negative impact that this project has had. Prior to their construction, the Nile valley was flooded annually as a result of rain in the mountains of the countries to the south. This resulted in the water level at Luxor rising around 12 m and the yearly flood brought silt-laden waters which covered the Nile Valley. When the waters receded, the silt would stay behind, fertilising the land for growing crops. In addition, it helped to stem the relentless encroaching of the desert across the valley. The result of preventing the inundation is that the desert continues to encroach onto the fertile flood plain and the land has to be fertilised by the addition of nitrogen rich fertilisers.

The construction of Lake Nasser behind the current dam has also had some unfortunate side effects. As a result of international collaboration and some remarkable engineering projects, many of the monuments built along the Nile in the area where the Lake was to be formed were rescued by raising them to positions above the level of the lake. However, the formation of this huge body of water (5,250 km$^2$) has created a microclimate in the desert. Rain is now more frequent and the monuments which survived for over 3,000 years in their arid environment are now being rapidly eroded.

## 9.2 EGYPTIAN OBELISKS

In the previous Section on the Theban necropolis, the negative impact of natural fractures on the stability of the tombs and temples of the pharaohs and their entourage was considered. However, natural fractures in rocks, such as bedding planes and joints, have also been exploited by masons of past civilisations in their efforts to build monumental structures in and from rocks. In this Section, we describe two further examples where ancient mankind has successfully exploited fractures in rocks, one from Egypt and the other from Easter Island. An example will be discussed later, in Section 9.4, where the role of the original, horizontal layering in the layered basalt flows that make up the Deccan Traps in India has affected the construction of the monolithic Buddhist temples at the Ellora site.

### 9.2.1 An unfinished obelisk in Egypt

The Egyptians are renowned for their production of monolithic granite obelisks — tall, four-sided, narrow tapering monuments which end in a pyramid-like shape or pyramidion at the top. Their symbolism is still unclear but they were usually erected in pairs at the entrance to temples and their tapered geometry is thought to reflect the rays of the Sun, which increase in width as they reach the Earth.

One of the largest ever to be excavated still lies in an abandoned quarry in Aswan in Upper Egypt, Figure 9.8(a). The quarry is sited in the Pre-Cambrian shield of the African plate and, although the rock type is commonly referred to as a granite, it is in fact a syenite. Like granite, syenite is a coarse grained, intrusive igneous rock, with a similar composition to granite but deficient in quartz. The quarries in Aswan are the type localities for syenite which is named after Syene, the ancient name of Aswan.

The obelisk in the quarry is over 40 m in length and weighs over 1,000 tonnes (a million kilograms). It was abandoned because of the formation of cracks and it has provided a remarkable insight into the techniques used by the ancient masons to construct these monuments. The most striking geomechanical feature of the granite surrounding the obelisk is the occurrence of a regular set of vertical fractures. The obelisk lies parallel to these joints which were clearly exploited by the Egyptian masons. Evidence for the widening of these joints to release the monolith from the rock by

Figure 9.8(a). Aerial view of the 'unfinished obelisk' in a syenite quarry in Aswan. Its orientation is controlled by the main set of extensional joints. Note the fracture that caused the project to be abandoned running through the upper portion of the monolith.

Figure 9.8(b).   Left photo — view of the obelisk showing the fracture that developed during construction and which caused the project to be abandoned. Evidence for the technique used to free the obelisk from the granite block when a joint was not available, using the abrasive properties of small dolerite builders, can be clearly seen. Right photo — shows a series of small holes probably made with copper tools in a line at regular intervals into which sun dried wooden wedges would be hammered. These were then repeatedly soaked with water and would gradually expand over time, eventually fracturing the rock. See Figures 5.11 & 5.12.

the patient process of abrasion with small dolerite boulders can be clearly seen in Figure 9.8(b), which also shows a detail of the fatal fracture that brought the endeavour to a halt.

## 9.2.2 Cleopatra's Needle

Cleopatra's Needle (Figure 9.9) on the Thames Embankment in London, is one of three ancient Egyptian obelisks created during the reign of the 18th Dynasty Pharaoh Thutmose III (approx. 1450 BC) and which were installed in London, Paris, and New York City during the nineteenth century. Transporting the obelisk from Egypt to London was no easy matter. Firstly, the obelisk was moved from its original site to Alexandria where it remained for 60 years due to lack of funding. However, it eventually arrived in London in 1878 and was then erected in its current position.

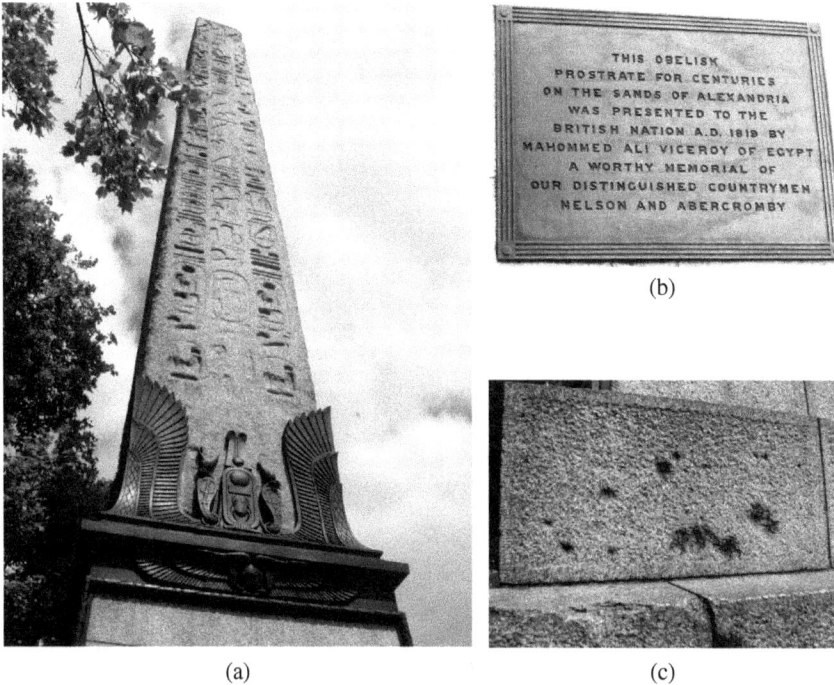

(b)

(a)

(c)

Figure 9.9. (a) Cleopatra's Needle, Thames Embankment, London. (b) Plaque commemorating the gift of Cleopatra's needle to the British nation in 1819. (c) Shrapnel holes caused by bomb damage to the granite base during the First World War.

Such obelisks are dramatic rock monuments, especially given their age and the fact that they were formed directly from monolithic rock masses using early stone engineering techniques. The methods that the Egyptians used, especially in forming, releasing and undercutting, and finally transporting and erecting such obelisks are described in Dieter (1991).

## 9.3 RUNESTONES

An example of deterioration in historical monuments can be directly seen in some runestones. Sawyer (2000) notes that in Scandinavia there are about 3000 runestones which date from the $10^{th}$ and $11^{th}$ centuries. Most

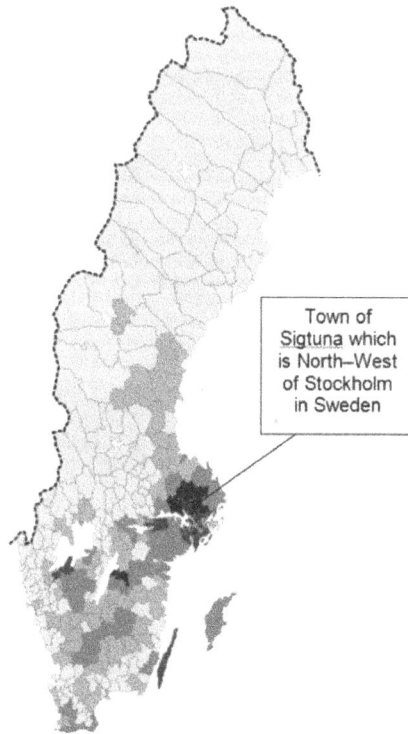

Figure 9.10.    Map of Sweden with the location of the town of Sigtuna indicated. The shading of the map represents the density of runestones, the darkest area where Sigtuna is located having greater than 10 runestones/square km. Map from Wikipedia.

of these are memorials to local people who lived and died in the vicinity and most of the late Viking Age runestones were upright. The town of Sigtuna in Sweden is located in the darkest part of the runestone density map in Figure 9.10 which indicates that there are greater than 10 runestones per square kilometre in that region.

The runestone in Sigtuna shown in Figure 9.11 is remarkably well preserved, being made of a Pre-Cambrian granite gneiss from the Scandinavian Shield. However, many of the runestones have deteriorated, mainly due to the weather but also because of aggravating environments such as their location by a road and the effects of vegetation. If a runestone originally contained pre-existing fractures (which may not have been

Figure 9.11.    Intact runestone in Sigtuna, Sweden, showing little deterioration with age.

visibly evident), the continuing effects of rain associated with significant daily and yearly temperature changes over a thousand years are likely to cause an incipient fracture to extend and split the runestone. We show photographs of a cracked Sigtuna runestone in Figure 9.12 and enlargements of the two fractures in Figures 9.13 and 9.14.

The fractures shown in Figures 9.13 and 9.14 are irregular and discontinuous extensional fractures (see Chapter 2, Section 2.3.3) and are probably weathering-induced, i.e., caused by freeze-thawing cycles through the Swedish seasons. The irregularity is caused by variations in the microstructure; for example, there is evidence of the crack in Figure 9.14 being made up of two sections, each being arrested at a thin vertical vein.

Figure 9.12.    Runestone from Sigtuna, Sweden, with two fractures traversing its complete width.

Figure 9.13.    Closer view of the upper fracture in Figure 9.12.

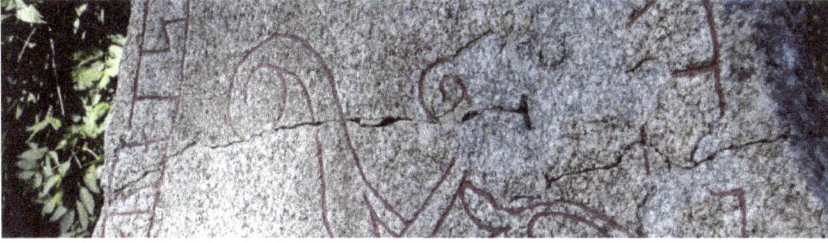

Figure 9.14.   Closer view of the lower fracture in Figure 9.12.

## 9.4 THE STABILITY OF MONOLITHIC BUDDHIST CAVE TEMPLES AT ELLORA, INDIA

### 9.4.1 The temples and the case example

Cave temples at the Ellora World Heritage site in India have been excavated since the 6th century in basalt cliff faces of the Deccan traps, Figure 9.15. These monolithic temples are situated about 250 km NE of Bombay in the Aurangabad region of India. The monuments span a period of about

Figure 9.15.   9th century monolithic Buddhist temple excavated in the Deccan basalts in India.

350 years from the 6$^{th}$ to the 10$^{th}$ centuries AD and are excavated out of the vertical face of the Khuldabad escarpment to face westwards across the Deccan plain.

Most of the temples are in the form of sculpted caves but in several cases some of the surrounding rock has also been removed creating a monolithic free standing structure. The prime example of this is the Kailasa Temple, Figure 9.16, which is considered to be one of the wonders of the ancient world because of the scale and extent of the rock-cut carving. However, these rock structures have deteriorated as a result of the presence of fractures and stress. In some locations, the primary layering of the basalt flows and the tectonically-induced jointing have progressively

Figure 9.16. The Kailasa Temple in India — a monolithic temple, i.e., carved directly from the *in situ* rock.

opened. Also, the orientation of the present day regional stress field and the local stress concentrations caused by the topography and excavation geometry have exacerbated the deterioration.

A basic knowledge of the geology of the site and the present day tectonic setting of the Deccan region provides sufficient information for a structural geologist to be able to predict the type and orientation of potential fractures that might be expected to be found in the area. Thus, with this information, and with an understanding of the effects of excavation on the host rock mass, the potential instabilities can be forecast with the additional knowledge of rock mechanics principles — thus again illustrating the value of the synthesis of structural geology and rock engineering.

### 9.4.2 The geological setting of the Ellora cave temples

Throughout the geological history of the Earth, repeated eruptions of flood basalt lavas have taken place, burying the land surface or sea bottom and spreading out to form nearly level plains. This fine grained rock, when unweathered, has a high strength. If the lava flows are at high altitudes, they are termed 'lava plateaux' and the basaltic lava flows that form them are called 'plateau basalts'. They represent the accumulation of numerous individual lava flows, some of which followed each other with remarkable rapidity, whilst others are separated from the underlying lava by a considerable time interval.

The Deccan traps represent the largest lava plateau in the world, covering over half a million square kilometres, and were originally probably significantly larger before erosion. The origin of the word 'traps' is from the Swedish 'trappa' meaning 'stair' or 'step' in reference to the stepped appearance created by the abrupt termination of successive flows. The total mean thickness of the lava flows is at least 1 km, and locally reaches thicknesses greater than 2 km. Most of these flows are thought to have erupted around 65 million years ago at the end of the Cretaceous period over a period of less than half a million years. These flat, relatively unstructured lavas lie with a marked angular unconformity on the highly structured and complexly deformed Pre-Cambrian basement of the Indian Shield.

To understand the fractures within the basalt including the primary layering, it is necessary to consider their complete geological history. Flood basalts form during periods of important crustal extension and are generally fed from dyke swarms which intrude along major, sub-vertical, extensional fractures in the Earth's crust and which tap magma sources in the lower crust and upper mantle. Because the lavas are extruded during a time of regional extension, once they have cooled and solidified they too become subjected to the regional stress field and often develop extensional fractures parallel or sub-parallel to the dykes that fed them. Also, because of the rapid cooling the lavas experience on extrusion, they may develop cooling fractures which form normal to the cooling surface and result in vertical columnar joints. Interestingly, most of the lava flows of the Deccan traps lack such joints. In places, vertical fractures and shear zones have off-set the lavas indicating post-volcanic faulting and tectonism of the trappean rocks. It is during these events that the basement lineaments induce associated fracturing in the overlying basalts.

In addition to the vertical cooling and tectonically-induced fractures mentioned above, the lavas also contain an important horizontal planar anisotropy resulting from the bedded nature of the flows. The planes of weakness separating successive lava flows will also vary in strength depending on the times between depositions. Junctions between lavas extruded almost simultaneously would be welded and would be likely to have a relatively high adhesion and cohesion. Conversely, large time intervals between flows would allow weathering of the basalt surface and, in due course, the formation of soils. Under these circumstances, the adhesion between two such adjacent basalt flows would be extremely low and there would be an associated high permeability along the weathered horizon. One can anticipate that this low adhesion would have been further reduced by the action of fluids channelled along these high permeability zones.

Each flow will have an upper scoriaceous (rough and irregular) part which, may be characterised by horizons of vesicles — small, almond-shaped holes caused as a result of the entrapment of gases beneath a chilled crust which forms on top of the flow. The vesicles usually become infilled with minerals, typically zeolites which are hydrous alumino-silicates. Because these minerals weather and erode easily, the vesicular horizons often become planes of low strength. The two processes of weakening of the vesicular horizons and the formation of soil layers impart important variations in the anisotropy of the lava pile.

Thus, the lavas have an important primary anisotropy resulting from their bedded nature and a secondary anisotropy resulting from vertical fractures caused by cooling and/or by syn- and post-volcanic tectonism. Subsequently, new fractures or the enhancement of old fractures may occur during the exhumation of the lavas when the removal of the over-burden causes a change in the stress field. In addition, when the lavas are in the weathering zone, corrosion and removal of the zeolites further weakens certain bedding planes and dramatically increases their porosity and permeability.

Having understood the processes likely to generate fractures within the rock mass during its geological evolution, it is now possible to pre-dict some key features of the rock mass properties and to foresee the type of rock engineering problems that might be encountered during and subsequent to the construction of the temples. Additionally, it is important to consider the present-day stress field acting on the lavas. The stress field relates to their plate-tectonic setting which is dominated by the continent–continent collision between the Indian and Asian plates and the associated north–south compression which is generating the Himalayas. Major thrust movements dominate the tectonics of northern India, indicating that at present the maximum principal stress, $\sigma_1$, is acting horizontally N–S and the minimum principal stress, $\sigma_3$, is acting vertically.

So, if new fractures were to form in response to the current plate stresses, their orientations can be predicted, see Section 2.3, Figure 2.14(c) and the related discussion. If they were shear fractures, they would strike E–W and dip at 30° north or south; if they were extensional fractures, they would form normal to $\sigma_3$ and hence would be horizontal. Because of the existence of a planar anisotropy in the lavas, i.e., the bedding, it is likely that the expression of extensional failure would be to further weaken existing fractures rather than generate new failure surfaces. Thus, these subsequent fractures would enhance the existing anisotropy within the lava pile by enhancing the existing planes of weakness parallel to bedding or by generating new, low angle shear fractures within the lavas. Both these effects would impact directly on the stability of the structures at Ellora. Note that the major principal stress, acting horizontally N–S, is sub-parallel to the cliff-face in Figures 9.17–9.20.

### 9.4.3 The cave temples

The architectural style of the temples is shown in Figures 9.17–9.19. The interaction between the primary basaltic layering and cave temple excavation is shown in Figure 9.20. This is a more complex veranda collapse caused by a combination of the pre-existing fractures and the excavation-induced fractures. Note that the temple has been excavated in harmony with the layering because the roofs and floors are defined by separate bedding planes. The lava layer between the two verandas has been unloaded and, as a result, sub-horizontal, excavation–induced fractures have formed. These, together with jointing, have caused the formation of rock blocks, and associated collapse. In turn, the failure has induced fall-out from the rock face above.

Internally, the temples are often decorated with carved motifs, as is illustrated in Figures 9.21–9.23. The internal carvings are an integral part of the *in situ* rock mass having been sculpted during excavation of the

Figure 9.17. Cave temple at the Ellora site, India. Note that the roof of the temple entrance has exploited a bedding plane between lava flows and that, at the top left, differential weathering of the lava flows can be seen.

Figure 9.18.   Cave temple at the Kailasa complex at the Ellora site in India. Note the rock failure above the second storey. Four major lava flows can be seen and the bedding planes between them form the roofs and floors of the temple storeys.

temple. Thus, all the cave temple deterioration effects, internally and externally, relate to the basaltic rock mass deterioration.

## 9.4.4  Remedial measures

The Figures in Section 9.4.3 illustrate the variety of degradation effects suffered by the monolithic cave temples with their external and internal decorations. Naturally, the implementation of any remedial measures should be sympathetic to the purpose and nature of these Buddhist temples, i.e., both with the intentions of the original master builders and the nature of the temples' status as World Heritage sites. In other words, it should

Figure 9.19.    Cave temple at the Ellora site in India. Note the rock failure above the entrance induced by pre-existing, tectonically-induced fractures. Again, a bedding plane has been used as the roof of the temple with failure facilitated by this and interaction with other fractures in the lava flow.

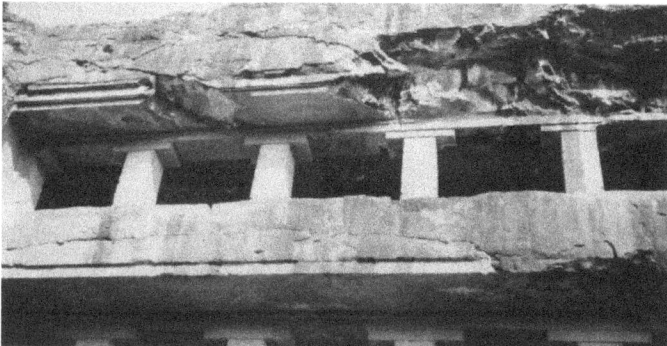

Figure 9.20.    Cave temple at the Ellora site in India, Note that sub-horizontal curved fracture surfaces within the lava flow have exacerbated the rock deterioration.

Figure 9.21. Interior of cave temple at the Ellora site in India showing sculpted supporting pillars. Note that the builders realised the importance of internal pillars; without them, the expanse of a separated bedding plane would lead to collapse of the roof.

Figure 9.22. Partial failure of a pillar caused by the intersection of sub-vertical pre-existing joints in a cave temple at the Ellora site in India.

Figure 9.23.   Monolithic elephant carving inside a temple at the Ellora site in India. Note the primary horizontal layering highlighted by the accumulation of vesicular horizons now infilled with zeolite. (Zeolites, microporous, aluminosilicate minerals, form where volcanic rocks and ash layers react with alkaline groundwater.)

be as discreet as possible to avoid spoiling the appearance and ambiance of the rock temples, whilst avoiding any further failure in the future. Thus, application of shotcrete (sprayed cement) or the installation of supporting steel structures is inappropriate in the case of these temples, which are works of art. However, the insertion of rockbolts (i.e., reinforcement) would be an ideal method of avoiding degradation in both the short- and long-terms. In many cases, the length and orientation of such bolts can be determined directly by observing the existing failure, *cf.* Figure 9.20.  In other cases, the structural geology knowledge of the rock layering and the directions of the primary fractures enables the optimal choice of preventative support measures.

Let us consider what occurs when a cave temple is excavated in these basaltic lavas. The three primary generic geotechnical effects of excavation will be as according to Principle 31 in Chapter 3:

— at the free surfaces, i.e., all the excavation surfaces inside the temple, there is no direct resistance to inward movement of the rock mass after the rock has been excavated;
— at these free surfaces, no normal or shear stresses act and so the surfaces are all principal stress planes with the principal stresses oriented normal and parallel to the surfaces; and
— any hydraulic pressure in the rock mass has been reduced to atmospheric in the boundary rock, causing the temples to act as hydraulic sinks.

From the geological analysis of the temple sites, further specific geotechnical effects (bearing in mind the essentially vertical walls and horizontal roofs of the temples) will be:

— opening of both horizontal and vertical fractures;
— concentration of the maximum principal stress in the roof, floor and N–S orientated walls; and
— exacerbation of water flow through high permeability horizontal fracture zones.

Also, these effects can occur on all surfaces simultaneously and in combination. For example, if there are two adjacent perpendicular free rock surfaces, such as at the entrance to a temple, the effects will occur at both surfaces and hence potentially cause a greater disturbance, depending on the temple orientation; or the stress could enhance fracturing in a temple roof, thus exacerbating roof movement. If water is also entering the temple area through a high permeability zone in the roof, instability will be even more likely. Furthermore, the access to the temples is via the vertical rock faces of river valleys which themselves are influenced by the geometry of the rock structure. As a general principle, we might therefore expect the temples' walls to be parallel and perpendicular to the fracture sets.

The predicted deterioration mechanisms are therefore as follows:

— rock blocks becoming detached and falling where there are perpendicular surfaces, especially at convex corners in portals, pillars, and balconies;

— rock blocks becoming detached and falling where the temples' walls and fracture sets are sub-parallel;
— in sculptures traversed by the horizontal and vertical fracturing, rock blocks becoming detached and falling;
— damaged zones occurring in the temple walls with fractures developing parallel to the walls and at the corners of temples where there is an especially adverse concentration of stress;
— damaged zones in pillars where the vertical component of stress has been significantly concentrated (and has become the major principal stress) by the rock extraction; and
— softened zones occurring in the vicinity of fractures conducting water into the temples.

The geology, the regional stress field, the orientation of the temples, and their internal architecture cannot be altered. Thus, the only main remaining option is to improve the strength of the fractures and to provide mechanical support through the use of rockbolts which can be installed such that they are hardly visible.

*****

This monolithic cave temple case example again illustrates the value of a combined structural geology–rock engineering approach. In particular in this case, the understanding of the mode of formation of the fractures gives greater confidence in the rockbolting reinforcement strategy, especially deciding on the required lengths, orientations and locations of the rockbolts.

## 9.5 EASTER ISLAND STATUES

### 9.5.1 The carved heads (moai) of Easter Island

Easter Island is one of the most remote inhabited islands in the world. The nearest continental point lies in central Chile, 3,512 km (2,182 miles) away. The name "Easter Island" was given by the island's first recorded European visitor, the Dutch explorer Jacob Roggeveen who encountered it on Easter Sunday (5 April) in 1722. The Island is famous for its 887 extant

Figure 9.24.   Easter Island heads (moai) carved from tuff. The bedding in the tuff has been picked out by weathering. The exposed parts of the figures are ~3 m high.

monumental statues, called moai, Figure 9.24, created by the early Rapa Nui people whose civilisation flourished on Easter Island between 1250 and 1500 AD.

Easter Island is a volcanic island 2,000 m above the sea bed. Its maximum height above sea level is 507 m and it is made up of three extinct, coalesced volcanoes which give the island its roughly triangular shape. The island is dominated by basalt flows but a variety of other volcanic rocks occur, including tuffs (a type of rock made of volcanic ash ejected from a vent during a volcanic eruption). Following ejection and deposition, the ash is compacted into a solid rock by the overburden load of later tuffs and lavas. Most moai are made of tuff (Figure 9.25) because it is a relatively soft material, although a few were carved from basalt. Because tuff erodes easily, few of the moai's original designs remain. However, some moai that were buried retain original markings.

Like the Egyptian obelisk which has been left in the quarry at Aswan, Section 9.2, there are several carved figures still *in situ* in the quarries on Easter Island, Figure 9.25. They are thought to have been abandoned when

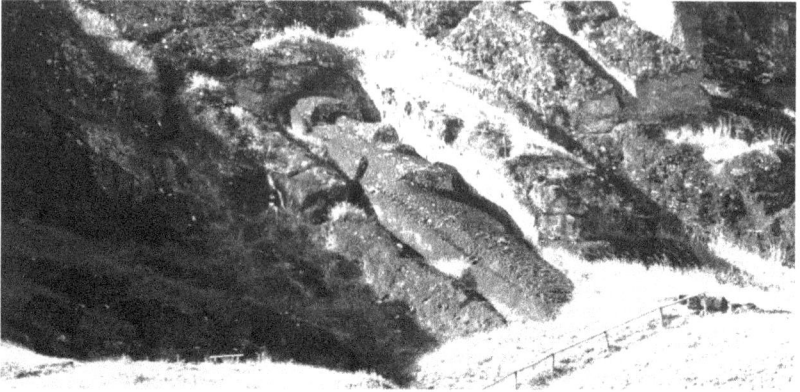

Figure 9.25.  A recumbent statue still *in situ* in the quarry and lying parallel to the bedding which is dipping at ~30°. Note the vertical, bedding-normal, sub-vertical planar joint at the back of the figure.

Figure 9.26.  Partially toppled heads. Weathering has highlighted the bedding and has left the relatively resistant blocks entrapped within the tuff standing out as positive features. These give an estimate of the amount of weathering that has occurred since the figures were carved.

the civilisation collapsed, an event believed to be the result of total defor-
estation of the island.

Tuffs are bedded and can contain several types of joints. These include
columnar jointing which is generally normal to the bedding and forms as a
result of cooling, long, regular bedding-normal joints and sheeting jointing
(shallow, surface-parallel, opening-mode rock fractures) sub-parallel to bedding. It is clear from Figure 9.25, that the sculptors exploited both the bedding planes and the regular, bedding-normal joints when carving the statues.
The bedding is dipping at approximately 30° and the joints can be clearly
seen, behind the recumbent figure and coated in a light coloured material,
the product of the combined action of volcanic fluids and weathering.

The effect of weathering on the statues is shown in Figure 9.26, especially the way in which weathering accentuates the bedding. Resistant
clasts within the softer tuff now stand proud of the surface of the statue as
the result of weathering. Studies of such features can assist in estimating
the relative ages of the statues.

## 9.6  BUILDING STONES AND STONE BUILDINGS

*"That", said the quarry foreman, "is where St. Paul's came from!" I looked
down towards the sea from a high cliff on the eastern side of the Isle of
Portland. I saw a valley gouged out of the hillside: a dead, desolate wilder-
ness; a cutting away of high stone cliffs as if some race of giants had scooped
out the stone to its bed. I realise now that no one understands London until
they have explored the significant chasms of the Portland stone quarry.
Somerset House, St Paul's, the Bank, the Royal Exchange, the Mansion
House, the Law Courts, the British Museum, all the Wren churches have left
their caves, gullies, and gaps in the Isle of Portland. I thought, not only of the
buildings which Portland has already given to London, but of the London to
be that slumbers still in darkness in the womb of this pregnant Isle.*

From a description of a visit to the limestone quarry at the Isle of
Portland on the south coast of England by H.V. Morton, in his book
"In Search of England", Methuen 1927, Penguin Books, 1960.

We include the quotation above not only because it relates to the most
well-known building stone in England, Portland stone, but also

Figure 9.27.   The Royal School of Mines building of Imperial College London, UK, made of Portland stone and where this book originated.

because it is the stone used for the Royal School of Mines building of Imperial College London (Figure 9.27) where this book originated.

Four main topics are highlighted and discussed in this Section:

— the origin of building stones,
— the rock mechanics properties of different building stones,
— examples of the use of building stones,
— understanding structural problems in stone buildings, and
— the repair of deteriorated building stones in old structures.

The examples will cover cases chosen from across the world, but there will also be a focus on UK examples.

## 9.6.1 The origin of building stones

In the same way that the landscape of a country reflects the underlying geology, so the geology affects the appearance of local buildings in the countryside (the vernacular architecture) and indeed many of the town buildings because of the availability of buildings stones from local outcrops and quarries. Despite this general principle, there are many exceptions due to the transport of stones, sometimes from distant sources. For example, after William I from France conquered England in 1066 AD, he built a series of castles and cathedrals using Caen stone which was imported from France. Nowadays, it is much easier to import building stones from different overseas sources, especially when there are sufficient financial resources for prestige buildings.

In the UK, we are lucky to have rocks from essentially all geological ages and so there are many different types of building stone. Figure 9.28 illustrates a couple of readily identifiable types. Two major collections of the UK's building stones were made at the beginning of the 20th century (Howe, 1910; Watson, 1911). The Howe collection is on permanent loan from the Natural History Museum in London to the British Geological Survey in Keyworth; the Watson collection is located in the Sedgwick Museum of Earth Sciences, which is part of the Department of Earth Sciences of the University of Cambridge.

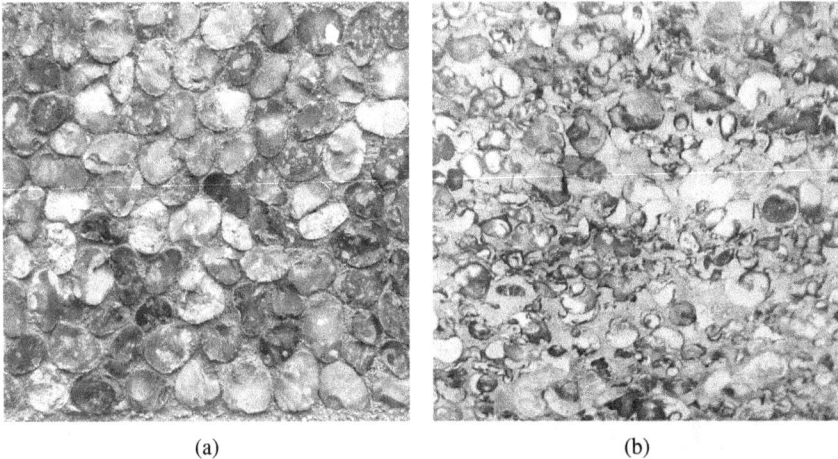

(a)                                        (b)

Figure 9.28.   Two examples of readily identifiable UK building stones. (a) Image ~ 400 mm across: flint pebbles, Norfolk; flint is composed of cryptocrystalline silica and occurs in the upper part of the Chalk in the UK. (b) Image ~ 60 mm across: Purbeck 'marble', Temple Church, London; although this stone is known as a marble because it can be well polished, it is actually a freshwater limestone consisting mainly of fossils of the snail *Viviparus*. See also Figure 9.38.

## 9.6.2 The rock mechanics properties of different building stones

*[Note: In this Section, the building stone data from two old books are used, one book dated 1887 and the other 1911. Both these books use Imperial units, e.g., for density, lbs/ft³. These units have been retained in the text and Figures here; however, the equivalent SI units are also given, e.g., kg/m³.]*

The main rock mechanics properties of building stones are the density and the crushing strength (the latter being the compressive strength, see Principle 3 in Chapter 3). Because building stones have been used for centuries, there has been a continuing study of their properties, well before modern rock mechanics was developed in the 1960s. In particular, there is a wealth of information in the 1887 book by W.J.M. Rankine titled "A Manual of Civil Engineering". Rankine is well known in the UK because, *inter alia*, of the Rankine Lecture which is organised by the British Geotechnical Society and held annually at Imperial College London.

Figure 9.29 shows an image of the Rankine 1887 book together with page 348 listing the 'heaviness' (i.e., density) of a sample of different rock types, ranging from 117 lbs/ft³ for low strength chalk to 187 lbs/ft³ for basalt. [Note that 100 lbs/ft³ ≅ 1.28 kg/m³.]

348     MATERIALS AND STRUCTURES.

TABLE OF THE HEAVINESS OF ROCK.

| | Lbs. in one Cubic Foot. | | Lbs. in one Cubic Yard. | | Cubic Feet to a Ton. |
|---|---|---|---|---|---|
| Basalt, | 187 | ... | 5060 | ... | 12 |
| Chalk, | 117 to 174 | ... | 3160 to 4730 | ... | 19·1 to 12·9 |
| Felspar, | 162 | ... | 4370 | ... | 13·8 |
| Flint, | 164 | ... | 4430 | ... | 13·6 |
| Granite, | 164 to 172 | ... | 4430 to 4640 | ... | 13·6 to 13 |
| Limestone, | 169 to 175 | ... | 4560 to 4720 | ... | 13·2 to 12·8 |
| „ magnesian, | 178 | ... | 4810 | ... | 12·6 |
| Quartz, | 165 | ... | 4450 | ... | 13·6 |
| Sandstone, average, | 144 | ... | 3890 | ... | 15·6 |
| „ different kinds, | 130 to 157 | ... | 3510 to 4240 | ... | 17·2 to 14·3 |
| Shale, | 162 | ... | 4370 | ... | 13·8 |
| Slate (Clay), | 175 to 181 | ... | 4720 to 4890 | ... | 12·8 to 12·4 |
| Trap, | 170 | ... | 4590 | ... | 13·2 |

W.J.M. Rankine 1887 "A Manual of Civil Engineering" 16th Ed., C. Griffin and Co., London, 812p.

Figure 9.29.   Left: Rankine's 1887 book; Right: Part of page 348 with physical data for different minerals and rock types.

STRENGTH OF STONES—TESTING DURABILITY. 361

| | Crushing Stress, in lbs. on the Square Inch. |
|---|---|
| Grauwacke from Penmaenmaur,...................... | 16,893 |
| Basalt, Whinstone,.......................................... | 11,970 |
| Granite (Mount Sorrel),................................. | 12,861 |
| ,, (Argyllshire), ................................. | 10,917 |
| Syenite, (Mount Sorrel),................................. | 11,820 |
| Sandstone (Strong Yorkshire, mean of 9 experiments), ..................................... | 9,824 |
| ,, (weak specimens, locality not stated), | 3,000 to 3,500 |
| Limestone, compact (strong),........................... | 8,528 |
| ,, magnesian (strong), ........................ | 7,098 |
| ,, ,, (weak),....................... | 3,050 |

Mr. Fairbairn's experiments further show, that the resistance of strong sandstone to crushing in a direction parallel to the layers, is only *six-sevenths* of the resistance to crushing in a direction perpendicular to the layers.

The hardest stones alone give way to crushing at once, without previous warning. All others begin to crack or split under a load less than that which finally crushes them, in a proportion which ranges from a fraction little less than unity in the harder stones, down to about *one-half* in the softest.

The mode in which stone gives way to a crushing load is in general by *shearing*. (Article 157, pp. 235, 236; Article 158, p. 237.)

Experiments on the strength of stones have hitherto been made almost universally on cubical specimens. It is desirable that they should be made on prismatic specimens, whose heights are at least once and a-half their diameters; for an experiment made by crushing a cube indicates somewhat more than the real strength of the material.

Figure 9.30. Page 361 of Rankine's 1887 book providing example data for the compressive strength of various building stones.

Page 361 of Rankine's 1887 book (Figure 9.30) lists example values of the 'crushing stress', i.e., compressive strength, of sample rock types [10,000 psi $\cong$ 69 MPa]. The two paragraphs of text below the table are particularly interesting: that the rock strength is less when loaded parallel to the bedding as opposed to perpendicular to the bedding; and that the rock samples begin to crack before the compressive strength "in a proportion which ranges from a fraction little less than unity in the harder stones to about *one-half* in the softest". We know that cracking in a compressive strength test is initiated by the development of axial cracks, see Figures 2.110 and 2.113 and related text, so we would expect that a rock sample loaded parallel to the bedding would exhibit a lower strength than one loaded perpendicular to the bedding, but we would not necessarily expect the rock sample

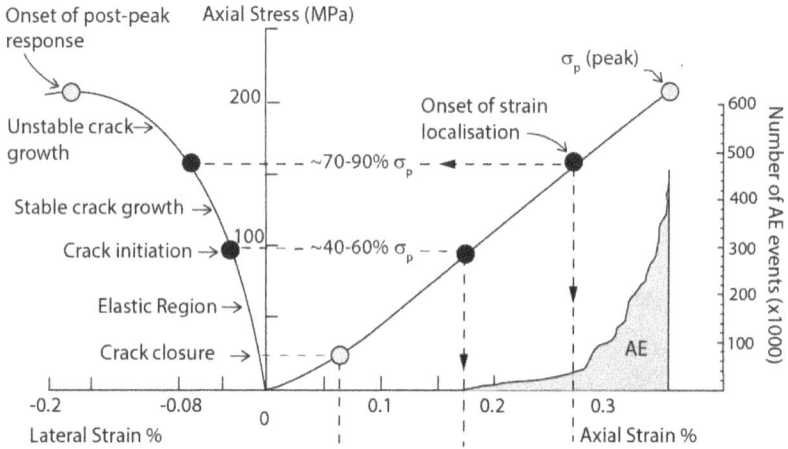

Figure 9.31. Difference between peak strength and crack initiation strength in a compressive strength test on rock (from Hoek and Martin 2014).

to start cracking or splitting at only half of the compressive strength during loading. However, this is precisely what is found to happen when modern testing results are analysed, see Figure 9.31, where the crack initiation strength is 40–60% of the compressive strength, and the paper by Siren *et al.* (2015) where the crack initiation strength of the Finnish migmatitic gneiss is reported as 40 MPa and the rock mass strength is about 90 MPa, i.e., cracking starts at 44% of the compressive strength.

The book by Watson (1911) provides a unique listing of the density for 412 British building stones and the compressive strengths for many of these. The densities are shown in histogram form in Figure 9.32, exhibiting a normal type distribution ranging from 105–205 lbs/ft$^3$, i.e., 1,682–3,284 kg/m$^3$.

The histogram in Figure 9.33 is of the mean densities of the Watson stones according to geological period, i.e., their age. This has the definite and satisfying trend that the older rocks have higher densities, with the mean densities dropping from just less than 170 lbs/ft$^3$ to just less than 140 lbs/ft$^3$ travelling upwards through the geological timescale.

The distribution of compressive strength data shown in Figure 9.34 is of a log-normal type with many rock types in the 200–700 tons/ft$^2$ (20–70 MPa) range and then more scattered values from 800–2,400 tons/ft$^2$ (80–240 MPa). Note that the data in Figures 9.33 and 9.34 illustrate the

**Distribution of Building Stone Densities**
**(all geological periods - British Isles)**

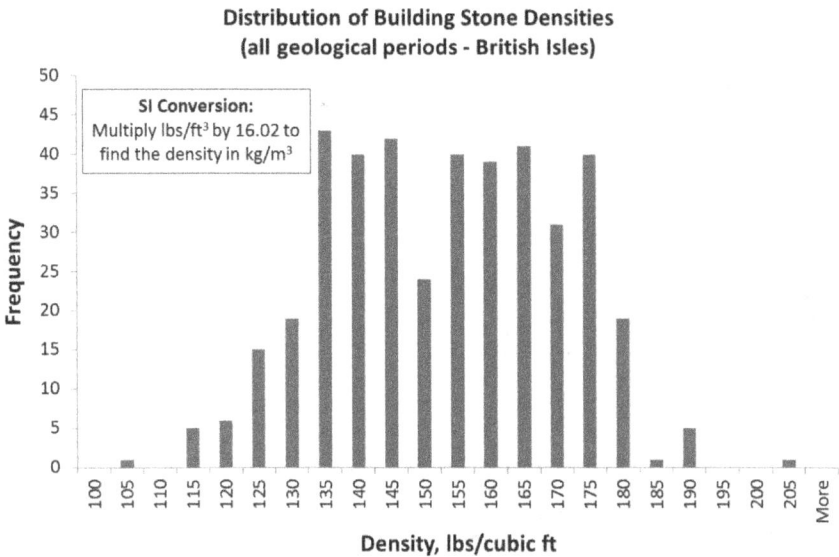

SI Conversion:
Multiply lbs/ft³ by 16.02 to
find the density in kg/m³

Figure 9.32.  Summary of the density data for 412 British building stones listed in Watson (1911).

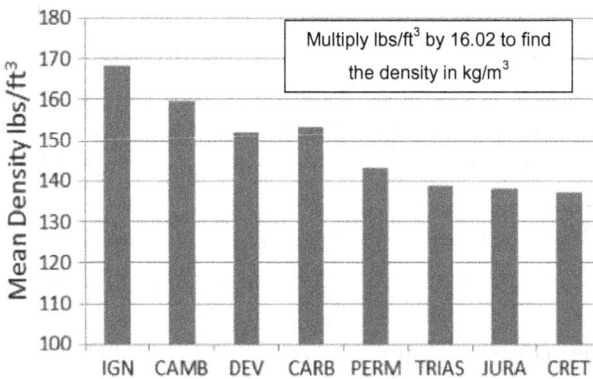

Multiply lbs/ft³ by 16.02 to find
the density in kg/m³

Figure 9.33. Summary of the mean density data according to geological period (age) for the 412 British building stones listed in Watson (1911). The abbreviations along the x-axis refer to Igneous and, older to younger, Cambrian, Devonian, Carboniferous, Permian, Triassic, Jurassic, and Cretaceous.

**Compressive strength**

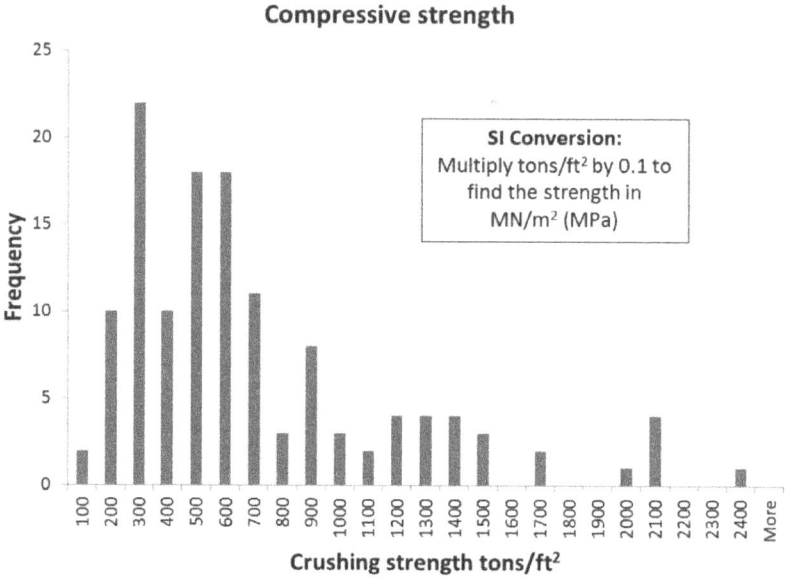

Figure 9.34. Summary of the compressive strength data for 412 British building stones listed in Watson (1911).

nature of the wide spectra of building stone densities and compressive strengths, but modern testing is now required to obtain these and other properties for specific building stones, e.g., from, the Building Research Establishment in the UK and the stone suppliers.

An interesting question is "How high can a pillar of rock blocks be before the bottom stone collapses due to the weight of the blocks above?" This can be calculated directly from the rock density and compressive strength as follows (using the Imperial units as in the data in the Figures 9.32–9.34).

— Take a mean density (Figure 9.32) of, say, 150 lbs/ft³ ($\cong$150 × 16.02 kg/m³).
— This exerts 150 lbs on each square foot of the base for each foot of pillar height.
— Assume the compressive strength of the rock is, say, 500 tons/ft² (Figure 9.34).
— Then the pillar height before the base block collapses is 500 × 2240/150 ft high.
— Which is 7,467 ft $\cong$ 1.4 miles = 2.3 km.

Of course, this assumes that there is a good pillar foundation, the blocks are stacked with perfect alignment, there is no effect from the wind, etc. In fact, and as we shall see in the following sections of this Chapter, it is not the original stone strength that is the main factor governing the stability and longevity of stone buildings but the adverse quality of stone building foundations, block alignments in the buildings, degradation due to the weather, human activities such as military attack, etc.

### 9.6.3 Examples of the use of building stones

In Figures 9.35 and 9.36, there are just some examples of the spectrum of building stone uses and types. Some of the building stones are easily recognised, especially if they are in common use, such as Portland stone, or have a characteristic appearance, such as Radyr conglomerate (Figure 9.36, upper right); many others are not so easily recognised and recourse needs to be made to descriptive literature about building stones.

An interesting example relating to paving stones on the volcanic island of Jeju situated off the southern coast of the Korean Peninsula is shown in Figure 9.37. The island was formed by monogenetic basaltic volcanism which produced many small vents and vesicle tracks in the resultant rocks. A cross-section though a small vesicle track is shown in Figure 9.37(a) and a series of orthogonal tracks are evident in Figure 9.37(b).

| (a) Dry stone wall using glaciated boulders and pebbles | (b) Dry stone wall using non-glaciated stones |
|---|---|

Figure 9.35.   Dry stone walls using (a) rounded, glaciated stones and (b) angular, non-glaciated stones.

Lloyds Bank exterior
facade, Salisbury,
England: Ham Hill Stone,
a Jurassic limestone

Paving stones in the Pohjoisesplanadi Street,
Helsinki, Finland:
foliated (high-grade) mica gneiss
with granitic interlayers

Llandaff Cathedral,
Wales:
Eroded Radyr stone
pillar; Triassic breccia

Caerphilly Castle, Wales:
Carboniferous Pennant
sandstone with white
Jurassic (Sutton stone)
limestone window
dressing

Ilfracombe, England: Illustrating how an interesting
juxtaposition of stone textures and sizes can make
a most attractive wall

King Street,
Nottingham, England:
Millstone grit,
Carboniferous
sandstone

Figure 9.36. Some examples of the wide variety and uses of building stones.

As the basalt cools it generates cooling fractures which act as conduits along which gases can escape. Two dominant sets of cooling fractures are formed and their abutting relationships suggest that they are contemporaneous, see Section 2.12.

## 9.6.4 Understanding structural problems in stone buildings

The compressive strength of a building stone is generally sufficient to withstand the stress induced by the weight of the rock above in most structures. In the example calculation at the end of Section 9.6.2, it was

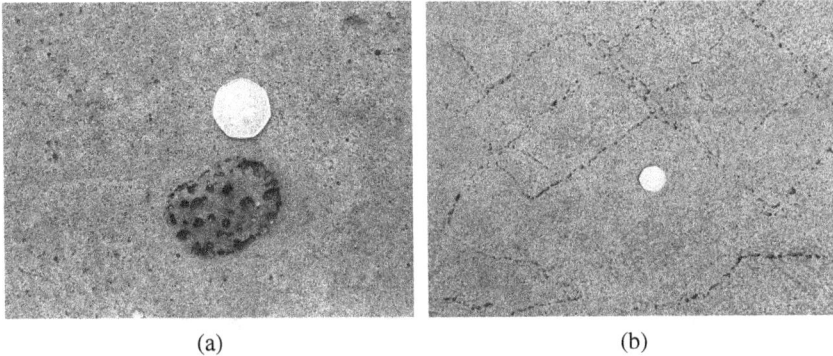

(a)                      (b)

Figure 9.37.   Basalt paving stones with vesicule tracks on Jeju Island, South Korea.

shown that, for typical stone values, a stone column could be built 1.4 miles high before the lowest stone collapses. Whilst this calculation is technically correct, in practice there are many other factors that are involved in stone buildings, not least whether the load is being uniformly applied to the stone, the nature of the foundations, the orientation of any bedding planes in the stone, whether any tensile stresses are present, deterioration with time, etc. One of the interesting cases in the UK is that of the pillars at the east end of the nave in Salisbury Cathedral, Figure 9.38.

The pillars in Figure 9.38 are of Purbeck marble — which is in fact a bedded Lower Cretaceous limestone but has been used as a 'marble' because of its attractive surface when polished. In order to obtain the lengths of columns shown in Figure 9.38 it is necessary to lathe the pillars out — with the bedding along the length of the pillar. Thus, the pillar is transversely isotropic (see Principle 16 in Chapter 3), and so is susceptible to splitting, Figure 9.39(b), especially if the loading is misaligned as in Figure 9.39(d). However, it is difficult to evaluate the safety of the pillars in Figure 9.38 without a detailed study of whether these are structural and hence load bearing or simply decorative only supporting their own weight. The severely split Purbeck marble pillar shown in Figure 9.40 does not provide confidence, although this pillar has been subject to the ravages of the outside environment for centuries.

(a) East end of the nave, North side      (b) East end of the nave, South side

Salisbury Cathedral, England

Figure 9.38.   Bending of the Purbeck 'marble' pillars in the nave of Salisbury Cathedral, England.

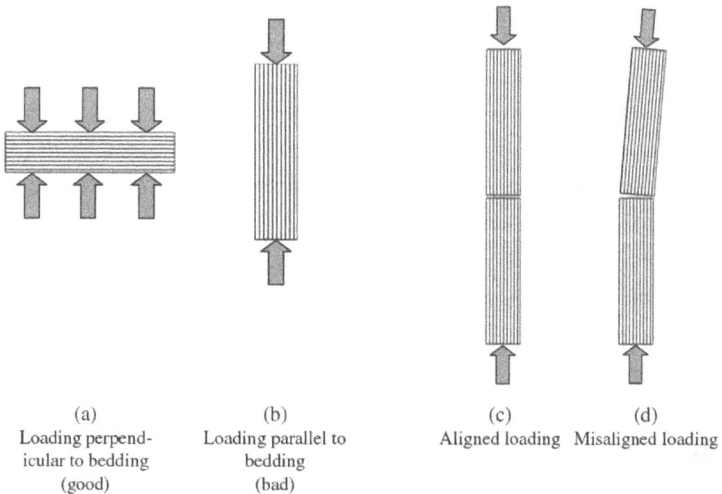

|            (a)              |          (b)              |           (c)            |            (d)            |
| Loading perpend-           | Loading parallel to       | Aligned loading          | Misaligned loading        |
| icular to bedding          | bedding                   |                          |                           |
| (good)                     | (bad)                     |                          |                           |

Figure 9.39.   (a) Loading perpendicular to the bedding is good. However, loading parallel to the bedding as shown in (b) is likely to result in splitting. However, if this is necessary, (c), then the load should not be misaligned (d) as this will exacerbate the likelihood of failure because the stress is highly magnified at the reduced pillar loading area.

Figure 9.40.   Severely split Purbeck marble pillar in the cloisters of Salisbury Cathedral —
loaded and split along the bedding.

Indeed, weathering is a major problem with building stones, as
shown in Figure 9.41 by the deterioration of decorative limestone pillars
outside a building in Montréal. The first two examples (a) and (b), which
are replacement pillars above the pedestal, show pillars in an unweathered state. The remaining four examples, (c)–(f), show various stages of
degradation. The annual temperature variation in Montréal is large, the
lowest recorded temperature having been −38°C and the highest
recorded temperature having been +38°C, so, together with the ice,
snow, and rain, the environmental conditions are not agreeable for some
building stones.

We finish this Section with a relevant extract from an 1842 paper
"Observations on stone used for building" by C.H. Smith published in the
Transactions of the Royal Institute of British Architects, 1(2), and as

Figure 9.41. Weathering of decorative limestone pillars outside a building in downtown Montréal, Canada. (a) Pillar in good condition showing the decorative features; (b) weathering of the vermiculated decoration of the pillar's support base; (c) early pillar degradation; (d) severe pillar deterioration; (e) advanced deterioration of part of a pillar; and (f) advanced deterioration of a whole pillar.

quoted in the "Conservation of Building and Decorative Stone" book by Ashurst and Dimes (1990):

> "Whoever expects to find a stone that will stand from century to century, deriding alike the frigid rains and scorching solar rays, without need of reparation, will indeed search for the philosopher's stone."

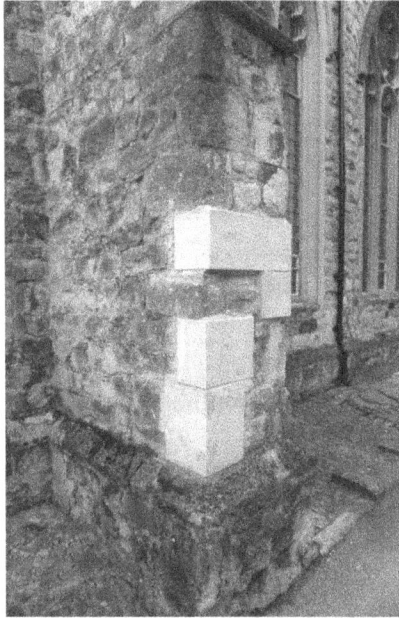

Figure 9.42.    Replacement of weathered blocks in a Kentish ragstone wall, Rochester, UK.

### 9.6.5 The repair of deteriorated building stones in old structures

The repair of deteriorated stones is not an easy task. In the previous Section, we included a sequence of illustrations showing the deterioration of limestone pillars (Figure 9.41), but these could be conveniently replaced with new pillars as required. However, this is more difficult in the case of a building stone that is an integral part of a structure (Figure 9.42).

The bright, white replacement stones shown in Figure 9.42 certainly look out of place, but hopefully they will eventually weather down to a less incongruous state. We deal next with the wider aspects of the conservation of ancient monuments.

## 9.7 THE CONSERVATION OF ANCIENT MONUMENTS

*"They came to a grizzled church, whose massive square tower rose unbroken into the darkening sky, the lower parts being illuminated by the nearest*

> *lamps sufficiently to show how completely the mortar from the joints of the stonework had been nibbled out by time and weather, which had planted in the crevices thus made little tufts of stone-crop and grass almost as far up as the very battlements."*
>
> from "The Mayor of Casterbridge" by Thomas Hardy, 1902,
> Macmillan and Co., London

The subject of the conservation of ancient monuments and sites is an especially interesting one for two main reasons:

— the technical aspects of structural geology and rock mechanics have to be considered in association with many other subjects, e.g., chemistry, archaeology, long-term monitoring, management and artistic factors; and
— there are many issues associated with the subject that are controversial and often intractable, e.g., To what extent is it necessary to conserve? Should the same stone be used? Who is going to pay for the conservation? Does the conservation have a long life? Should the same architectural style be used in the conservation?

So, in addition to the more direct geological and engineering aspects discussed in other Chapters, it is instructive to consider how the synthesis of the two subjects is helpful in the context of ancient monuments, i.e., structures made in rock masses or with rock material many years ago. Can the degradation be modelled and understood and can suitable remedial measures be recommended? In his paper on "Understanding the Earth scientist's role in the pre-restoration research of monuments: an overview", Přikryl (2007) noted that, "Pre-restoration research into building materials is generally conducted to provide information on types of material, their damage and repair. Although the technologist and restorer must manage the practical aspects of repair, the earth scientist can make a significant contribution in terms of material research. First, he or she can answer questions on the nature of the stone(s) used, their provenance (location of the quarry), and their weathering characteristics in terms of the deterioration of physical and mechanical properties and destruction of rock fabric. Second, the earth scientist can research the physical and mechanical properties of new stone proposed as a replacement for decayed stonework, including recommendations for alternative materials where stone from the original quarry is no longer available." It should be added in the context of this book that there is also a role for the rock engineer in considering how remedial measures can be effected, e.g., the use of rock-bolts in supporting unstable monolithic and other stone structures.

Two of the seminal books on the conservation subject are:

1.  "Conservation of Historic Buildings" by B.M. Feilden, first published in 1982 with a 3$^{rd}$ edition in 2003 published by the Architectural Press, an imprint of Elsevier. Sir Bernard was the Director Emeritus of the International Centre for the Study of the Preservation and Restoration of Cultural Property, based in Rome. In the author's obituary (2008), the author is himself described as "a monument to conservation" and "the list of monuments with which he was involved is a roll-call of the world's most precious and   important cultural sites, including the Great Wall of China". The book covers structural aspects of historic buildings, causes of decay in materials and structure, and the work of the conservation architect.

2.  "Conservation of Building and Decorative Stone" edited by J. Ashurst and F.G. Dimes, first published in 1990 with a paperback edition in 1998 in the Butterworth–Heinemann Series in Conservation and Museology. This book covers an introduction to the restoration, conservation and repair of stone, the nature of building and decorative stones, igneous rocks, sedimentary rocks, metamorphic rocks, and the weathering and decay of masonry.

These two books together provide a superb background to the philosophy and technical aspects of such conservation, the content being applicable to all ancient sites and monuments. Some of the major issues can be illustrated with reference to UK monuments and buildings. For example, most readers will be familiar with the Stonehenge monument in the UK which has been resistant to weathering over the last ~5000 years (because the stones are sandstone with a silica cement) but may not be so resilient to tourists over far fewer years. To what extent should the site be isolated from people? Similarly, Hadrian's Wall in the UK, was built to

Figure 9.43.   Salisbury Cathedral in the UK, built of sandy limestone starting in the year 1220 AD.

protect the northern boundary of the Roman Empire 2000 years ago but, given that this wall is ~100 km long, should it be protected from damage by sightseers and, if so, how can it be protected from such damage?

In England, there are 44 cathedrals, many of which are medieval, having been built about 900 years ago, and are jewels of Britain's architectural heritage as exemplified by Salisbury Cathedral illustrated in Figure 9.43. The cathedrals were built with different types of stone, some even with stone imported from Caen in France, as has been mentioned earlier. However, the stone has deteriorated — mainly because of the weather: sun, rain, and frost. Some of the cathedrals have experienced major structural problems, often related to insufficient recognition of the need for suitable foundations. This was the case at Wells Cathedral, Figure 9.44. About 200 years after construction was completed in 1230 AD, the central tower began to collapse and scissor arches were added to stabilise the structure, Figure 9.45. This demonstrates the skill of the stonemasons in

Figure 9.44.    Wells Cathedral in the UK built in 1230+ AD with coarse-grained limestone.

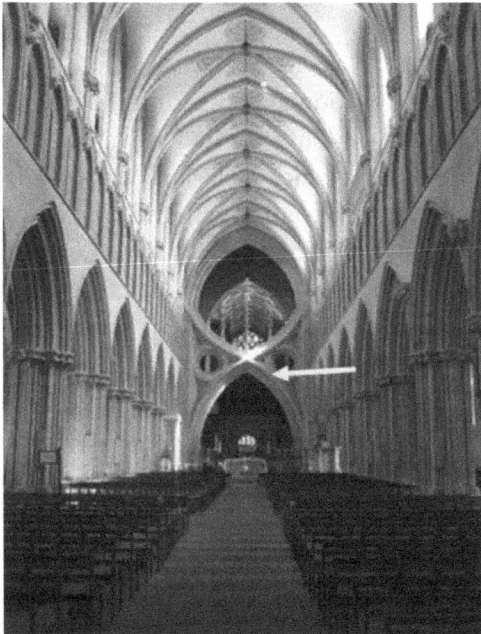

Figure 9.45.    The scissor arches in Wells Cathedral, UK, (indicated by the white arrow) added in the 15th century about 200 years after the original construction in order to stabilise the collapsing central tower above.

Figure 9.46. Caerphilly Castle in Wales, UK.

the 15th century in conservation and ensuring the life of the structure to the present day. The stonemasons devised a brilliant structural solution to the instability problem — which is elegant, aesthetically impressive and which uses material totally in keeping with the original structure.

Coming now to the questions raised by conservation and restoration, consider the state of the partially ruined Caerphilly Castle shown in Figure 9.46. Is restoration necessary, or should the castle be left like this?

Similarly, architectural styles within the same cathedral can have changed during construction of a building. At Norwich Cathedral, Figure 9.47, the building work was held up by the Black Death (plague, year 1340 AD) for 40 years — during which the architectural style changed from the Decorated style to the Perpendicular style, Figure 9.48. Then the question facing modern day restorers is to what extent should the restoration be in exactly the same architectural style as the original? Also, should exactly the same stone, or even the same material, be used for the restoration?

In Figure 9.49 are examples of the surface spalling and general deterioration of the Caen stone at Norwich cathedral. It is a mammoth, and

Figure 9.47. Norwich Cathedral, UK, 900 years old, built with Caen limestone bought from France in barges.

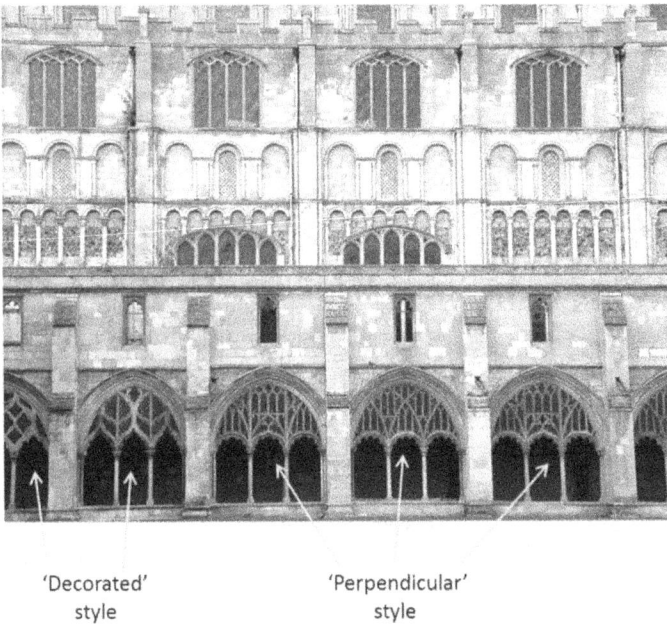

'Decorated'
style

'Perpendicular'
style

Figure 9.48. Different architectural styles at Norwich Cathedral which changed during a long break in construction caused by the Black Death; early building work was held up by the plague for 40 years — during which the architectural style changed.

Figure 9.49.    Examples of the deterioration of the stone at Norwich Cathedral, UK.

indeed, in some cases, impractical task to attempt to replace all the deteriorating stone with fresh stone of the same origin and exactly the same geometry.

An example of partial restoration of a Norman arch is shown in Figure 9.50. This kind of work is time consuming and expensive, and the question arises as to who will pay for the restoration?

Some of the questions listed earlier have been illustrated: To what extent is it necessary to conserve? Should the same stone be used? Who is going to pay for the conservation? Does the conservation have a long life? Should the same architectural style be used in the conservation? There are no easy answers, but we hope that knowledge of the geological sources of building stones and the application of rock mechanics knowledge for the structural considerations as described in this book will assist in the development of the answers.

Figure 9.50.    Partial restoration of a Norman arch at Norwich castle originally built in the year 1100 AD. Note the restoration to the left of the dashed line.

## 9.8 ORNAMENTAL ROCKS

In this Chapter on historical monuments and stone building, we have discussed the stability of a necropolis in Egypt, obelisks from Egypt, rune-stones in Sweden, cave temples in India, Easter Island statues, the mechanics of building stones, the use of building stones, and the conservation of ancient monuments. We conclude by showing that unaltered samples of rock masses can be used directly for ornamental purposes. The following illustrative examples can be found in front of the Wakkanai City Hall in the north of Hokkaido Island, Japan.

Figure 9.51.  Layered, unfolded, fine-grained, marine sediments of variable hardness with strata-bound and through-going fractures.

Figure 9.52.  Portion of a fine-grained layered marine sediment showing evidence of folding linked to the subduction process that dominates the geology of Japan. Hotel keycard scale: 85 mm long.

Figure 9.53.   Deep water, marine, metasediments showing alternations of jasper-rich and jasper-poor bands with layer-normal extensional fractures infilled with quartz.

Figure 9.54.   Inter-bedded layers of mudstone and jasper rich horizons, now metamorphosed. The jasper layers are relatively brittle and frequently display intense brecciation as a result of deformation (as shown in Figure 9.55 below). Hotel keycard scale: 85 mm long.

Figure 9.55. An originally layered sedimentary rock in which deformation has caused intense brecciation of the brittle jasper-rich horizons. Hotel keycard scale: 85 mm long.

Figure 9.56. A veined boulder showing both randomly orientated veins that have led to brecciation, together with a systematically orientated, infilled fracture set.

# CHAPTER 10

# CONCLUDING SUMMARY

We began this book by explaining that our purpose was to:

(a) Outline the principles of both structural geology and rock engineering (the latter via engineering rock mechanics); and
(b) to explain with illustrative examples why the synthesis of the two subjects is mutually beneficial.

Accordingly, the book consists of two main parts, with Chapters 2 and 3 outlining respectively the principles of structural geology and rock mechanics, and then Chapters 3–9 covering a variety of rock engineering projects where structural geology is the key to understanding the site and hence optimising the engineering.

It has been our teaching experience that the concept of 'stress' together with its mathematical ramifications is the most difficult subject of those that we have covered. This is because 'stress', as a second order tensor quantity and all that implies, is not consciously experienced in daily life, unlike scalar and vector quantities such as temperature and force which are much easier to understand. For this reason, and because the concept of stress is potentially important in all the rock engineering applications, following Section 2.1, an introduction to structural geology, we have explained the concept of stress in some detail in Section 2.2. The two most significant properties of stress are (a) that six components are required to specify the state of stress at a point and (b) that there are no normal and shear stress components on a free surface, e.g., on the surfaces of an open fracture or the unsupported walls, roof and floor of a rock engineering excavation.

The remaining Sections in Chapter 2 outlined a series of important structural geology subjects, i.e., brittle rock failure, the natural state of stress in rocks, fluid-induced failure, small scale fracture processes, recognising fracture types in the field, fracture spacing, fracture orientation, the brittle-ductile transition, complex fracturing, fracture networks, bulk properties of fracture networks and the influence of both regional and more local stresses on rock fracturing. This comprehensive set of structural geology subjects was then followed in Chapter 3 by an explanation of the rock mechanics aspects in the engineering context — via a compact set of 50 basic rock mechanics principles. The intimate link between structural geology and basic rock engineering was immediately evident via these rock mechanics principles because they are primarily related to the influence of rock stress and rock fractures on engineered structures.

Armed with both the structural geology principles and the rock mechanics principles, a set of engineering applications was then described, thus providing examples of the necessary linkage between the two subjects. These examples included the foundation of the Clifton suspension bridge in the UK (Chapter 4), the rock engineering aspects of four quarries (Chapter 5), three cases of difficulties with dam stability (Chapter 6), opencast coal mining in glacially disturbed strata (Chapter 7), three examples of different underground rock engineering objectives together with the related factors (Chapter 8), and a wide ranging set of examples concerning historic monuments and stone buildings (Chapter 9). Note that the two subjects of rock stress and rock fractures pervade all the examples in Chapters 4–9.

Hopefully, you have enjoyed reading this book and now have a greater knowledge of the individual subjects of structural geology and rock mechanics. Recalling that the dictionary definition of synthesis is "the combination of components or elements to form a connected whole", we also hope that the need for the two subjects to be intimately linked in applied rock engineering has been demonstrated.

We finish with the following Chapter 11 Epilogue which describes a very small scale application of the structural geology–rock mechanics approach in solving a difficult problem.

# CHAPTER 11

# EPILOGUE

This is the concluding example of the synthesis of structural geology and rock mechanics/rock engineering — and it is a story about a fossil, a very important fossil, some say the most important fossil.

## 11.1 THE AUTHENTICITY OF THE FOSSIL ARCHAEOPTERYX LITHOGRAPHICA — UNNATURAL SELECTION?

The *Archaeopteryx lithographica* fossil is a transitional species between dinosaurs and birds. The generic name, *Archaeopteryx*, is developed from the Greek words 'ancient' and 'wing'; the specific name, *lithographica*, alludes to the fact that the fossil was found in the Solnhofen limestone in Bavaria. The rock formation, and hence the fossil, is of Late Jurassic age, i.e., around 150 million years old. The fossil is shown in Figure 11.1, the use of the Solnhofen limestone for printing purposes in Figure 11.2, and an artist's impression of the dinosaur/bird in Figure 11.3.

The fossil was purchased in Victorian times by the British Museum in London, UK, because, as noted by Charig *et al.* (1986), it is "an obvious and comprehensible example of organic evolution". In Darwin's theory of evolution, if lizards had evolved into birds, there should be fossil evidence

Figure 11.1.   The *Archaeopteryx lithographica* fossil in the Natural History Museum in London, U.K. Note the tail feather to the left. The specimen is of the order of 0.5 m across.

of intermediate species, and *Archaeopteryx* was considered by the British Museum to be such a link as a transitional form. On the one hand, *Archaeopteryx* had teeth and hind limbs like reptiles, but it also had feathers and a wishbone like birds.

However, in the 1980s, the authenticity of the fossil was challenged by N. C. Wickramasinghe and F. Hoyle on the basis that the fossil had been forged in the nineteenth century by impressing a modern day feather into a layer of fresh plaster spread over part of the fossil specimen. This accusation had some credibility because (a) both Wickramasinghe and Hoyle were well known scientists and (b) the specimen does superficially look as though it is made of dental plaster because of the fine grained uniform nature of the limestone. Indeed, the Solnhofen limestone is used as a type rock for rock mechanics testing because of its uniformity and fine grain, Figure 11.4.

In addition to the British Museum's own investigations, one of the authors of this book (JAH) was asked to study the specimen from a rock

Figure 11.2.    Use of the Solnhofen limestone for printing purposes.

Figure 11.3.    Artist's impression of the *Archaeopteryx lithographica* dinosaur/bird.

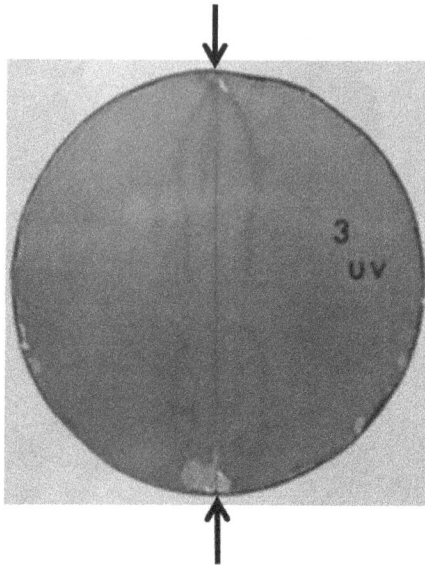

Figure 11.4.   Indirect tensile strength test on a disc of Solnhofen limestone. The applied axial compressive loading generates a lateral tensile stress which splits the specimen. Note the linearity of the crack formed between the two loading points — which is an indication of the extreme uniformity of this type of limestone (see Principle 17 in Chapter 3). The fainter secondary cracks are caused by the subsequent loading of the two half-discs.

mechanics point of view. Did a part of the fossil appear to have been manufactured or not, especially the surface layer? Given that this *Archaeopteryx lithographica* fossil is thought to be one of the most important in the world and that its authenticity is critically important for evolutionary studies, the task was somewhat daunting. In other circumstances, a quick way to establish whether a layer of plaster had been added to the fossil would have been to drill through the fossil and examine the rock core; needless to say, this was definitely not in line with paleontological protocol! So, an examination was made of the fractures evident at the fossil surface, see Figures 11.5(a)–11.5(c).

In Figure 11.5(a), a small rock fracture can be seen traversing the barbs of the large feather. The calcite crystals that have grown in this fracture are shown in Figure 11.5(b) and see also Figures 4.15(b) and 4.15(c). A Victorian forger could not have artificially created such a persuasive mineralised infill of the fracture. In addition, in Figure 11.5(c), the *en echelon* nature of

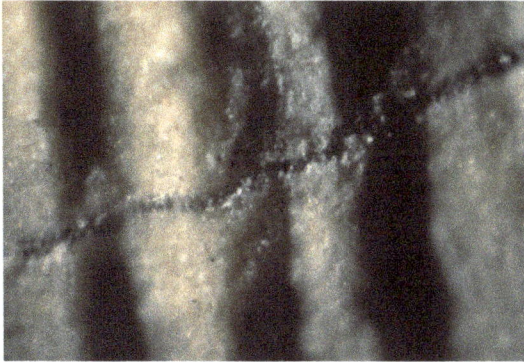

(a)

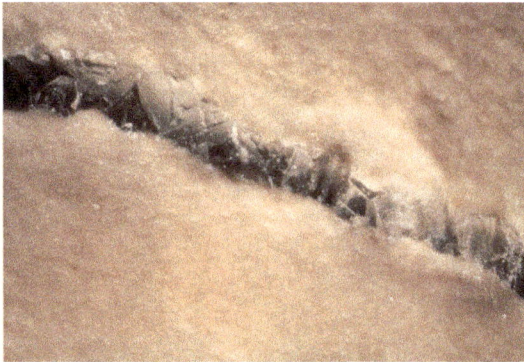

(b)

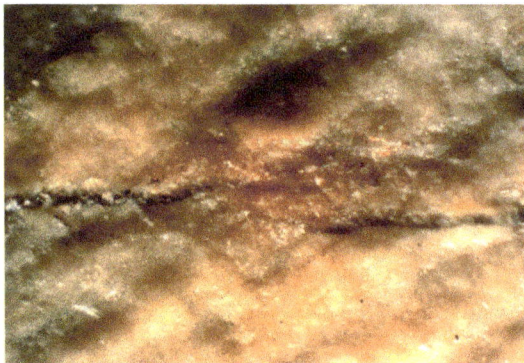

(c)

Figure 11.5. (a) Small rock fracture traversing the feather barbs. (b) Close-up of calcite crystals in the fracture. (c) *En echelon* crack development showing staining region around the tips of the fractures.

the fracture formation is illustrated with a zone of staining around the overlap, which is a phenomenon often observed on the larger rock outcrop scale.

Charig *et al.* (1986) justified the fossil's authenticity by "exactly matching hairline cracks and dendrites on the feathered areas of the opposing slabs, which show the absence of the artificial cement layer into which modern feathers could have been pressed by a forger". That conclusion was further supported by the fracture mineralisation shown in Figure 11.5(b).

This concluding example is an unusual application of rock fracturing knowledge, but one that once again demonstrates the value of understanding the combination of structural geology and rock mechanics — in this case, the formation and characteristics of rock fractures.

# REFERENCES AND BIBLIOGRAPHY

Abraham, T.J. & Sloan, R.C. (1978) TVA cuts deep slot in dam, ends cracking problem. *Civil Engineering — ASCE*, 48, 66–70.

Adams, J. (2001) *Risk*. London: Routledge, Taylor & Francis Group, p. 228.

Anderson, E.M. (1942) *The Dynamics of Faulting and Dyke Formation with Applications to Britain*. Edinburgh: Oliver & Boyd, p. 191.

Albertoni, S.C. (2009) The Itapebi hydroelectric development in the Jequitinhonha river. In: Corrado Piasentin (ed.) *Main Brazilian Dams III: Design Construction and Performance*. pp. 222–229. Published by Brazilian Committee on Dams — www.cbdb.org.br/documentos/mbdiii/Itapebi.pdf

Ashurst, J. & Dimes, F.G. (eds.) (1990) *Conservation of Building and Decorative Stone*. Oxford: Butterworth-Heinemann Series in Conservation and Museology. Paperback edition in 1998.

Atkinson, B.K. (1984) Subcritical crack growth in geological materials. *J Geophys Res*, 89(6), 4077–4114.

Atkinson, B.K. (ed.) (1987) *Fracture Mechanics of Rock*. London, UK: Academic Press, p. 534.

Badawy, A., Abdel-Monem, S.M., Sakr, K. & Ali, Sh.M. (2006) Seismicity and kinematic evolution of middle Egypt. *J Geodyn*, 42, 28–37.

Bai, T. & D.D. Pollard (2000a) Closely spaced fractures in layered rocks; initiation mechanism and propagation kinematics. *J Struct Geol*, 22(10), 1409–425.

Bai, T. & D.D. Pollard (2000b) Fracture spacing in layered rocks; a new explanation based on the stress transition. *J Struct Geol*, 22(1), 43–57.

Bankwitz, P., Bankwitz, E., Thomas, R., Wemmer, K. & Kampf, H. (2004) Age and depth evidence for pre-exhumation joints in granite plutons: fracturing during the early cooling stage of felsic rock. In: Cosgrove, J.W. & Engelder, T. (eds.). *The Initiation, Propagation and Arrest of Joints and Other Fractures.* Geological Society of London, Special Publications, 231, 25–47.

Barnes, M.P. (2012) *Runes: A Handbook.* Woodbridge, UK: The Boydell Press.

Barton, N.R. (1976) The shear strength of rock and rock joints. *Int J Rock Mech Min Sci*, 13, 255–279. DOI: 10.1016/0148-9062(76)90003-6.

Barton, N.R. (2000) *TBM Tunnelling in Jointed and Faulted Rock.* London, UK: CRC Press, Taylor & Francis Group, p. 184.

Barton, N.R. (2007) *Rock Quality, Seismic Velocity, Attenuation and Anisotropy.* London, UK: Taylor & Francis Group, p. 729.

Barton, N.R. (2012) Reducing risk in long deep tunnels by using TBM and drill-and-blast methods in the same project — the hybrid solution. *J Rock Mech Geotech Eng*, 4, 115–126.

Barton, N.R. (2013) Shear strength criteria for rock, rock joints, rockfill and rock masses: Problems and some solutions. *J Rock Mech Geotech Eng*, 5, 249–261. DOI:10.1016/j.jrmge.2013.05.008.

Barton, N.R. (2014) Cover picture. In: Ulusay, R. (ed.) *The ISRM Suggested Methods for Rock Characterization, Testing and Monitoring: 2007–2014.* London: Springer Science, p. 293.

Barton, N.R., Lien, R. & Lunde, J. (1974) Engineering classification of rock masses for the design of tunnel support. *Rock Mech Rock Eng*, 6, 189–236.

Bedi, A. (2013) *A proposed framework for characterising uncertainty and variability in rock mechanics and rock engineering.* PhD thesis, Imperial College London, p. 223.

Bogdanov, A.A. (1947) The intensity of cleavage as related to the thickness of the bed. (Russian text), *Soviet Geology*, 16.

Brown, E.T. (2012a) Fifty years of the ISRM and associated progress in rock mechanics. In: Qian Q. & Zhou Y. (eds.) *Harmonising Rock Engineering and the Environment: Proc. 12th ISRM International Congress on Rock Mechanics 18–21 October 2011, Beijing, China.* London: Taylor & Francis Group. pp. 29–45.

Brown, E.T. (2012b) Risk assessment and management in underground rock engineering — an overview. *J Rock Mech Geotech Eng* 4, 193–204.

Burford, R.O. & Harsh, P.W. (1980) Slip on the San Andreas fault in central California from alignment array surveys, *B Seismol Soc Am* 70, 1233–1261.

Campos e Matos, A., Ribeiro e Sousa, L., Kleberger, J. & Lopes Pinto, P. (eds.) (2006) *Geotechnical Risk in Rock Tunnels.* London: Taylor & Francis, p. 191.

Cartwright, J.A., Trudgill, B.D. & Mansfield, C.S. (1995) Fault growth by segment linkage; an explanation for scatter in maximum displacement and trace length data from the Canyonlands Grabens of SE Utah. *J Struct Geol*, 17, 1319–1326.

Cartwright, J.A., Mansfield, C. & Trudgill, B. D. (1996) The growth of normal faults by segment linkage. In: Buchanan, P.G. & Nieuwland, D.A. (eds.) *Modern Developments in Structural Interpretation, Validation and Modelling.* Geological Society of London, Special Publications, 99, 163–177.

Cartwright, J., James, D. & Bolton, A. (2003) The genesis of polygonal fault systems: a review. In: Van Rensbergen, P., Hillis, R.R., Maltman, A.J. & Morley, C.K. (eds.) *Subsurface Sediment Mobilisation.* Geological Society of London, Special Publications, 216, 223–243.

Casabianca, D. & Cosgrove, J.W. (2012) A new method for top seals predictions in high-pressure hydrocarbon plays. *Petrol Geosci*, 18, 43–57.

Charig, A.J., Greenaway, F., Milner, A.N., Walker, C.A. & Whybrow, P.J. (1986). *Archaeopteryx* is not a forgery. *Science* 232(4750), 622–626.

Charles, J.A., Tedd, P. & Warren, A. (2011) *Lessons from historical dam incidents.* Bristol, UK: Environment Agency, UK, p. 159. ISBN: 978-1-84911-232-1.

Ciftci, N.B. & Bozkurt, E. (2007) Anomalous stress field and active breaching at relay ramps: A field example from Gediz Graben, SW Turkey. *Geol Mag*, 144, 687–699.

Clayton, C.R.I. (2001) *Managing Geotechnical Risk.* London: Thomas Telford. p. 80.

Cobbold, P.R., Watkinson, A.J. & Cosgrove, J.W. (2008) Faults of the Pharaohs. *Geoscientist*, 18, 18–22.

Coleman, T.S. (2011) *A Practical Guide to Risk Management.* The Research Foundation of CFA Institute, p. 212. ISBN-10: 1934667412; ISBN-13: 978-1934667415.

Coli, M., Livi, E., Berry, P., Bandini, A., He, M. & Jia, X. (2010) Studies for rockburst prediction in the Carrara Marble (Italy). In: Xie, F. (ed.) *Rock Stress and Earthquakes*, Beijing: CRC Press, Taylor & Francis Group, pp. 367–373.

Cosgrove, J.W. (1976) The formation of crenulation cleavage. *J Geol Soc London*, 132, 155–178.

Cosgrove J.W. (1997) Hydraulic fractures and their implications regarding the state of stress in a sedimentary sequence during burial. In Sengupta, S. (ed.) *Evolution of Geological Structures in Micro- to Macro-scales.* London: Chapman & Hall, pp. 1–25.

Cosgrove, J.W. (2004) The role of stress, anisotropy and fluid pressure in controlling the movement of magmas and mineralizing fluids through the crust. *Z Geol Wiss*, 32(1), 1–8.

Cosgrove, J.W. (2015) The association of folds and fractures and the link between folding, fracturing and fluid flow during the evolution of a fold-thrust belt: a brief review. In: Richards, F.L., Richardson, N.J., Rippington, S.J., Wilson, R.W. & Bond, C.E. (eds.) *Industrial Structural Geology: Principles, Techniques and Integration.* Geological Society of London, Special Publications, 421, first published online May 12, 2015, http://doi.org/10.1144/SP421.11.

Cosgrove, J.W. & Sedman J.H.F. (2005) A review of the geological setting and structural geology of Tytherington Quarry and its relevance to quarry design. In: Walton, G. (ed.) *Proceedings of the 13th Extractive Industry Geology Conference, University of Leeds, 2 — 4 September* 2004. Birmingham: Minerals Industry Research Organisation, pp. 10–26.

Coulomb, C.A. (1776) Sur une application des règles maximis et minimis à quelques problèmes de statique, relatifs à l'architecture. *Acad Sci Paris Mem Math Phys*, 7, 343–382.

Cretu, O., Stewart, R. & Berends, T. (2011) *Risk Management for Design and Construction.* Hoboken, N.J., USA: John Wiley, p. 261.

Cruikshank, K.M., Zhao, G. & Johnson, A.M. (1991) Analysis of minor fractures associated with joints and faulted joints. *J Struct Geol*, 13, 865–886.

Delaney, P.T., Pollard, D.D., Ziony, J.I. & McKee, E.H. (1986) Field relations between dikes and joints: Emplacement processes and paleostress analysis. *J Geophys Res*, 91, 4920–4938.

Dillon, P. (2009) *The Mournes Walks.* O'Brien Walks series, O'Brien Press. Dublin, Ireland: The O'Brien Press. ISBN: 9781847171412. (2015) edition: p. 160.

Eichhubl, P. (2004) Growth of ductile opening-mode fractures in geomaterials. In: *The Initiation, Propagation, and Arrest of Joints and Other Fractures.* Geological Society of London, Special Publications, 231, 11–24.

Engelder, T. (1993) *Stress Regimes in the Lithosphere.* New Jersey, USA: Princeton University Press, p. 492.

Essex, R.J. (2007) *Geotechnical Baseline Reports for Construction: Suggested Guidelines.* Reston, Virginia, USA: ASCE, American Society of Civil Engineers. ISBN: 9780784409305.

Evans, A.G. & Blumenthal, W. (1984) High temperature failure mechanisms in ceramic polycrystals. In: Tressler, R.E. & Bradt, R.C. (eds.) *Deformation of Ceramic Materials II.* New York: Plenum Press, pp. 487–505.

Feilden, B.M. (1982) *Conservation of Historic Buildings.* 3rd Edn. (2003) Oxford: Architectural Press, Elsevier. ISBN-10: 0750658630; ISBN-13: 978-0750658638

Feng, X.-T. & Hudson, J.A. (2011) *Rock Engineering Design.* London: CRC Press, Taylor & Francis Group, p. 468.

Friedman, M. (1968) X-ray analysis of residual elastic strain in quartzose rocks. In: Gray, K.E (ed.) *Basic and Applied Rock Mechanics: Proceedings of 10$^{th}$ U.S. Symposium on Rock Mechanics, 20–22 May, 1968, Austin, Texas, USA.* A.A. Balkema, pp. 573–595.

Fujii. Y., Takemura, T., Takahashi, M. & Lin, W (2007) Surface features of uniaxial tensile fractures and their relation to rock anisotropy in Inada granite. *Int J Rock Mech Min Sci,* 44, 98–107.

Gonzalez de Vallejo, L.I. & Ferrer, M. (2011) *Geological Engineering.* London: CRC Press, Taylor & Francis Group, p. 678. ISBN 9780415413527.

Goodman, R.E. & Shi, Gen-Hua. (1985) *Block Theory and Its Application to Rock Engineering.* New Jersey, USA: Prentice-Hall International Series in Civil Engineering and Engineering Mechanics, Hall, W. J., (ed.), p. 338. ISBN: 0130781894; 9780130781895.

Gosschalk. E.M., Hinks, J.L., Johnson, F.G. & Jarvis, R.M. (1991). Overcoming the build-up of stresses, cracking and leakage in Mullardoch dam, Scotland. In: *Transactions 17$^{th}$ International Congress on Large Dams, Vienna, 17–21 June 1991.* Vienna, Austria: International Commission on Large Dams, Volume 2, pp. 475–498.

Griffith, A.A. (1920) The phenomena of rupture and flow in solids. *Philos Trans R Soc Lond Ser A,* 221,163–197.

Griffith, A.A. (1924) Theory of rupture. In: *Proceedings of the First International Conference of Applied Mechanics, Delft, Netherlands,* pp. 55–63.

Griggs, D.T. & Handin, J. (1960) Observations on fracture and a hypothesis of earthquakes. In: Griggs, D. & Handin, J. (eds.) *Rock Deformation. (A Symposium).* Memoir 79 of the Geological Society of America. New York, USA: Waverly Press, Baltimore, Maryland, pp. 347–73. ISBN-10: 1258352109; ISBN-13: 978-1258352103 (2012)

Hakala, M., Hudson, J.A., Harrison, J.P. & Johansson, E. (2008) *Assessment of the potential for rock spalling at the Olkiluoto site.* Posiva Oy, Finland. Working Report Posiva 2008-83.

Hallbauer, D.K., Wagner, H. & Cook, N.G.W. (1973) Some observations concerning the microscopic and mechanical behaviour of quartzite specimens in stiff, triaxial compression tests, *Int J Rock Mech Min,* 10,713–726.DOI:10.1016/0148-9062(73)90015-6.

Harrison, J.P. (2012) Rock engineering, uncertainty and Eurocode 7: implications for rock mass characterisation. In: Barla, G., Barla, M., Ferrero, A.M. & Rotonda, T. (eds.) *Nuovi metodi di indagine monitoraggio e modellazione degli ammassi rocciosi: Proc. XIV Ciclo di conferenze di meccanica e ingegneria delle rocce. Torino, Italy, 21–21 November 2012.* Torino, Italy: Celid.

Harrison, J.P. & Hadjigeorgiou, J. (2012) Challenges in selecting appropriate input parameters for numerical models. In: *Proc. 46th US Rock Mech. & Geomech. Symp, Chicago, 24–27 June* 2012. ARMA.

Harrison, J.P. & Hudson, J.A. (2000) *Engineering Rock Mechanics: Part 2 — Illustrative Worked Examples.* Oxford: Pergamon Press, Elsevier Science, p. 506.

Hayward, D. (1990). Mullardoch dam. *New Civil Engineer,* 25 October, p. 12.

Helgeson, D.E. & Aydin, A. (1991) Characteristics of joint propagation across layer interfaces in sedimentary rocks. *J Struct Geol,* 13, 897–911.

Hinks, J.L., Burton I.W., Peacock A.R. & Gosschalk, E.M. (1990) Post-tensioning Mullardoch Dam in Scotland. *Water Power and Dam Engineering,* 42, 12–15.

Hoagland, R.G., Hahn, G.T. & Rosenfield, A.R. (1973) Influence of microstructure on fracture propagation in rock. *Rock Mech,* 5, 77–106.

Hobbs, D.W. (1967) The formation of tension joints in sedimentary rocks. *Geol Mag,* 550–556.

Hodgson, R.A. (1961) Classification of structures on joint surfaces. *Am J Sci,* 259, 493–502.

Hoek, E. & Brown, E.T. (1997) Practical estimates of rock mass strength. *Int J Rock Mech Min Sci,* 34, 1165–1186.

Hoek, E. & Martin, C.D. (2014) Fracture initiation and propagation in intact rock — A review. *J Rock Mech Geotech Eng,* 6, 287–300.

Hopkin, P. (2012) *Fundamentals of Risk Management.* (2nd Edn.) London: Kogan Page Ltd, p. 419.

Hoshino, K., Koide, H., Inami, K. Iwamura, S. & Mitsui, S. (1972) *Mechanical properties of Japanese tertiary sedimentary rocks under high confining pressures.* Report 244, Geological Survey of Japan.

Howe, J.A. (1910) *The Geology of Building Stones.* London: Edward Arnold, p. 455.

Hudson, J.A. (1992) *Rock Engineering Systems: Theory and Practice.* Chichester, UK: Ellis Horwood Ltd., UK, p. 135.

Hudson, J.A. & Priest, S.D. (1983) Discontinuity frequency in rock masses. *Int J Rock Mech Min,* 20, 73–89.

Hudson, J.A. & Harrison, J.P. (1997) *Engineering Rock Mechanics: An Introduction to the Principles.* Oxford. UK: Pergamon Press, Elsevier Science, p. 444.

Hudson, J.A. & Jing, L. (2013) Demonstration of coupled models and their validation against experiment: the current phase — DECOVALEX 2015. In: Feng, X.T., Hudson, J.A. & Tan, F (eds.) *SINOROCK 2013: Rock Characterization, Modelling and Engineering Design Methods: Proc. ISRM SINOROCK2013 Symp, 18–20 June 2013, Shanghai, China.* London: Taylor & Francis Group.

Hudson, J.A. & Feng, X.-T. (2015) *Rock Engineering Risk.* London: CRC Press, Taylor & Francis Group, p. 572.

Inglis, C.E. (1913) Stresses in a plate due to the presence of cracks and sharp corners. *Proc Inst Naval Arch*, 55, 219–230.

Ingram, G.M. & Urai, J.L. (1999) Top-seal leakage through faults and fractures: the role of mudrock properties. In: *Muds and Mudstones: Physical and Fluid-Flow Properties*. Geological Society of London, Special Publications, 158, 125–135. DOI:10.1144/GSL.SP.1999.158.01.10.

ISRM (2012) International Society for Rock Mechanics (ISRM): Ten Suggested Methods. *Rock Mech Rock Eng*, 45(6), 955–1022.

ISRM 'Orange Book' (2014) Ulusay, R. (ed.) *The ISRM Suggested Methods for Rock Characterization, Testing and Monitoring*: 2007–2014. London: Springer Science, p. 293.

Jaeger, J.C., Cook, N.G.W. & Zimmerman, R.W. (2007) *Fundamentals of Rock Mechanics*. 4th Edn. revised by R.W. Zimmerman. Oxford: Blackwell Publishing Ltd, p. 475.

Jahns, R.H. (1943) Sheet structure in granites: its origin and use as a measure of glacial erosion in New England. *J Geol*, 51, 71–98.

Johansson, E., Äikäs, K., Autio, J., Hagros, A., Malmlund, H., Rautakorpi, J., Sievänen, U., Wanne, T., Anttila, P. & Raiko, H. (2002) *Preliminary KBS-3H layout adaptation for the Olkiluoto site. Analysis of rock factors affecting the orientation of a KBS-3H deposition hole*. Working Report 2002-57. Posiva Oy, Olkiluoto.

Johnson, F.G. (1986) Appendix to the 1986 BNCOLD Lecture: recent events at Mullardoch dam. In: *Proceedings of BNCOLD Conference on Reservoirs, Edinburgh*, 1986. Discussion volume, pp. a–f.

Jolly, R.J H. & Cosgrove, J.W. (2003) Geological evidence of patterns of fluid flow through fracture network; examination using random realizations and connectivity and analysis. In: *Fracture and in-situ Stress Characterization of Hydrocarbon Reservoirs*. Geological Society of London, Special Publications, 209, 177–186.

Kahyaoglu, H. (2012) The profit in risk. *Tunnels & Tunnelling*, 16 Oct., 45–51.

von *Kármán*, T. (1911) Festigkeitsversuche unter allseitigem Druck. *(in German) Z Ver Dtsch Ing*, 55, 1749–1757.

Kidan T.W. & Cosgrove J.W. (1996) The deformation of multilayers by layer-normal compression; an experimental investigation. *J Struct Geol*, 18, 461–474.

Kirsch, G. (1898) Die Theorie der Elastizität und die Bedürfnisse der Festigkeitslehre. *Z Ver Dtsch Ing*, 42, 797–807.

Kutter, H.K. & Fairhurst, C. (1967) The roles of stress wave and gas pressure in presplitting. In: *Status of Practical Rock Mechanics: Proceedings of the 9th*

*Symposium of Rock Mechanics, Colorado School of Mines, April* 1967. New York: AIME, 1968, pp. 265–284.

Kutter, H.K. & Fairhurst, C. (1971) On the fracture process in blasting. *Int J Rock Mech Min*, 8, 181–202.

La Pointe, P.R. & Hudson, J.A. (1985) Characterization and interpretation of rock mass jointing patterns. *Special paper* 199 *of the Geological Society of America*, p. 37, (presented as a University of Wisconsin-Madison, Engineering Experiment Station Report, June 1981). ISBN 0-8137-2199-7.

La Pointe, P., Wallmann, P., Thomas, A. & Follin, S. (1997) *A methodology to estimate earthquake effects on fractures intersecting canister holes*. SKB Technical Report: TR 97-07, Svensk Kärnbränslhantering AB, p. 97.

Ladeira, F.L. & Price, N. J. (1980) Relationship between fracture spacing and bed thickness. *J Struct Geol*, 3, 179–184.

Lawn, B.R. & Wilshaw, T.R. (1975) *Fracture of brittle solids*. Cambridge: Cambridge University Press, UK, p. 204.

Lisowski, M. & Prescott, W.H. (1981) Short-range distance measurements along the San Andreas fault system in central California, 1975 to 1979. *B Seismol Soc Am*, 71, 1607–1624.

Lopez Jimeno, C., Lopez Jimeno, E. & Ayala Carcedo, F.J. (1995) *Drilling and Blasting of Rocks*. London: Balkema, Taylor & Francis Group, p. 391.

Malin, P. (2012) Talking Point. *Ground Engineering*, UK, September, p. 7.

Mandl, G. (2000) *Faulting in Brittle Rocks — An Introduction to the Mechanics of Tectonic Faults*. Berlin: Springer Verlag, p. 434. DOI: 10.1007/978-3-662-04262-5.

Mandl, G. (2005) *Rock Joints; the Mechanical Genesis*. Berlin: Springer, p. 232.

Matheson, G.D. (1983) *Presplit blasting for highway road excavation*. Transport and Road Research Laboratory Report LR 1039, Crowthorne, Berks, UK.

McMahon, C.J. & Graham, C.D. (1992) *Introduction to Engineering Materials*. Merion Books. 386p. ISBN 0964659808; 9780964659803

Metz, B., Davidson, O., de Coninck, H., Loos, M. & Meyer, L. (eds.) (2005) *Carbon dioxide capture and storage. Report of the intergovernmental panel on climate change*. UK: Cambridge University Press, p. 431.

Milnes, A.G., Hudson, J.A., Wickström, L. & Aaltonen, I. (2006) *Foliation: geological background, rock mechanics significance, and preliminary investigations at Olkiluoto*. Posiva Work Report 2006-3, p. 87 Available on: www.posiva.fi.

Mohr, O. (1900) Welche Umstände bedingen die Elastizitätsgrenze und den Bruch eines Materiales? *Z Ver Dtsch Ing*, 44, 1524–1530; 1572–1577.

Mollema, P.N. & Antonellini, M. (1999) Development of strike-slip faults in the dolomites of the Sella Group, Northern Italy. *J Struct Geol*, 21, 273–292.

Murrell, S.A.F. (1958) Strength of coal under triaxial compression. In: Walton, W.H. (ed.) *Mechanical Properties of Non-Metallic Brittle Materials.* London: Butterworth, pp. 123–145.

Navier, M. (1833) *Résumé des leçons données à l'École des ponts et chaussées.* Part 1. Paris: Carilian–Goeury.

Nemcok, M., Gayer, R. & Miliorizos, M. (1995) Structural analysis of the inverted Bristol Channel Basin: implications for the geometry and timing of fracture porosity. In: *Basin Inversion.* Geological Society, London, Special Publications, 88, 355–392.

Ode, H. (1957) Mechanical analysis of the dike pattern of the Spanish Peaks area, Colorado. *Geol Soc Am Bull,* 68, 567–578.

Olson, J.E. (1993) Joint pattern development: Effects of subcritical crack growth and mechanical crack interaction. *J Geophys Res,* 98 (B7), 12251–12265.

Olson, J.E. (2004) Predicting fracture swarms — the influence of subcritical crack growth and the crack-tip process zone on joint spacing in rock. In: *The Initiation, Propagation, and Arrest of Joints and Other Fractures.* Geological Society of London, Special Publications 231, 73–88.

Olson, J.E. & Pollard, D.D. (1988) Inferring stress states from detailed joint geometry. In: Cundall, P.A., Sterling, R.L. & Starfield, A.M. (eds.) *Key Questions in Rock Mechanics: Proceedings of the 29$^{th}$ US Symposium on Rock Mechanics, June 1988, University of Minneapolis, Minnesota, USA.* Rotterdam, Netherlands: A.A. Balkema, (NLD), pp. 159–167. ISBN 9061918359.

Olson, J.E. & Pollard, D.D. (1989) Inferring paleostresses from natural fracture patterns; a new method. *Geology,* 17, 345–348.

Olson, J.E., Holder, J. & Rijken, E. (2002) Quantifying the fracture mechanics properties of rock for fractured reservoir. In: *Proceedings of the SPE/ISRM Rock Mechanics in Petroleum Engineering Conference 2002.* Richardson, TX: Society of Petroleum Engineers, pp. 421–432.

Palmer A. & Yates, N. (2005) *Advanced Geography.* Deddington, UK: Philip Allan Updates, p. 620.

Parry, R.G.H. (2004) *Mohr Circles, Stress Paths and Geotechnics.* 2$^{nd}$ Edn. London: Spon Press, Taylor & Francis Group, p. 264.

Peacock, D.C.P. & Sanderson, D.J. (1994) Geometry and development of relay ramps in normal fault systems. *Bull Am Assoc Petrol Geol,* 78, 147–165.

Peacock, D.C.P. & Sanderson, D.J. (1997) Geometry and development of normal faults. In Sengupta, S. (ed.) *Evolution of Geological Structures in Micro- to Macro-scales.* London: Chapman & Hall, pp. 27–46.

Petružálek, M., Lokajíček, T. & Svitek, T. (2015) Acoustic emission monitoring of the fracturing process of migmatite samples. In: *Innovations in Applied and Theoretical Rock Mechanics: Proc. ISRM 13$^{th}$ International Congress on Rock Mechanics, 10–13 May 2015, Montréal, Canada.* On CD.

Pollard, D.D. & Fletcher, R.C. (2005) *Fundamentals of Structural Geology.* Cambridge, UK: Cambridge University Press.

Price, N.J. (1966) *Fault and Joint Development in Brittle and Semi-Brittle Rock.* Oxord: Pergamon Press, Elsevier, p. 176.

Price, N.J. (1974) The development of stress systems and fracture patterns in undeformed sediments. In: *Advances in Rock Mechanics: Proceedings of the Third Congress of the International Society for Rock Mechanics, Sept 1–7, 1974, Denver, Colorado.* Vol. 1A, 487. Denver, Colorado: National Academy of Sciences.

Price, N.J. & Cosgrove, J.W. (1990) *Analysis of Geological Structures.* Cambridge, UK: Cambridge University Press, p. 502.

Priest, S.D. & Hudson, J.A. (1976) Discontinuity spacings in rock. *Int J Rock Mech Min*, 13,135–148.

Přikryl, R. (2007) Understanding the Earth scientist's role in the pre-restoration research of monuments: an overview. In: Přikryl, R. & Smith, B.J. (eds.) *Building Stone Decay: From Diagnosis to Conservation.* Geological Society of London, Special Publications, 271, 9–21. DOI: 10.1144/GSL.SP.2007. 271.01.02.

Ramsay, J.G. (1967) *Folding and Fracturing of Rocks.* New York: McGraw-Hill, p. 568.

Ramsay, J.G. & Graham, R.H. (1970) Strain variation in shear belts. *Can J Earth Sci*, 7, 786–813.

Rautakorpi J., Johansson, E., Tinucci, J., Palmén, J., Hellä, P., Ahokas, H. & Heikkinen, E. (2003) *Effect of fracturing on tunnel orientation using KBTunnel and 3DEC programmes for the repository of spent nuclear fuel at Olkiluoto.* Posiva Oy, Olkiluoto. Working report 2003-09, p. 38.

Renshaw, C.E. & Pollard, D.D. (1994) Numerical simulation of fracture set formation: a fracture mechanics model consistent with experimental observations. *J Geophys Res*, 99(B5), 9359–9372.

Rives, T., Razack, M., Petit, J-P. & Rawnsley, K.D. (1992) Joint spacing; analogue and numerical simulations. In: Burg, J.-P., Mainprice, D. & Petit, J.-P. (eds.) *Mechanical instabilities in rocks and tectonics: a selection of papers presented at the International Conference on Mechanical Instabilities in Rocks and Tectonics, Montpellier, France, 3–6 September 1991. J Struct Geol*, 14(8–9) 925–937.

Roberts, C. M. (1953) Special features of the Affric Hydro-Electric Scheme (Scotland). ICE Proceedings, *P I Civil Eng*, 2, 520–555. E-ISSN: 1753–7789

Royal Society (1983) *Risk assessment: a study group report*. London.

Royal Society (1992) *Risk: analysis, perception and management*. London.

RS/RAE (2012) *Shale gas extraction in the UK: A review of hydraulic fracturing*. Report by the Royal Society and the Royal Academy of Engineering, p. 75. Available on: www.raeng.org.uk/news/publications/list/reports/Shale_ Gas.pdf.

Rutqvist, J. (2012) The geomechanics of CO2 storage in deep sedimentary formations. *Geotech Geol Eng*, 30, 525–551.

Sawyer, B. (2000) *The Viking-Age Runestones*. UK: Oxford University Press.

Schulz, S.S., Mavko, G.M., Burford, R.O. & Stuart, W.D. (1982) Long-term fault creep observations in central California. *J Geophys Res*, 87, 6977–6982.

Serrano, A. & Olallaa, C. (1998) Ultimate bearing capacity of an anisotropic discontinuous rock mass. Part I: Basic modes of failure. *Int J Rock Mech Min Sci*, 35, 301–324.

Shields, J.G. (1997) Post-tensioning Mullardoch Dam in Scotland. In: Littejohn, G.S. (ed.) *Ground Anchorages and Anchored Structures: Proc. Int. Conf. organized by the Institution of Civil Engineers held 20–21 March 1997, London, UK*. London: Thomas Telford, pp. 205–216. ISBN-10: 0727726072; ISBN-13: 978-0727726070.

Sibson, R.H. (1973) Interaction between temperature and pore fluid pressure during earthquake faulting — a mechanism for partial or total stress relief. *Nature Phys Sci*, 243, 66–68.

Sibson, R.H. (1995) Selective fault reactivation during basin inversion: potential for fluid redistribution through fault-valve action. In: Buchanan, J.G. & Buchanan, P.G. (eds.) *Basin Inversion*, Special Publications Geological Society, London, 88, 3–19.

Sibson, R.H. (2004) Controls on maximum fluid overpressure defining conditions for mesozonal mineralisation. *J Struct Geol*, 26, 1127–1136.

Simonson, E.R., Abou-sayed, A.S. & Clifton, R.J. (1978) Containment of massive hydraulic fractures. Paper SPE 6089, *Soc Petrol Eng J*, 18, 27–32.

Siren, T., Hakala, M., Valli, J., Kantia, P., Hudson, J.A. & Johansson, E. (2015) *In situ* strength and failure mechanisms of migmatitic gneiss and pegmatitic granite at the nuclear waste disposal site in Olkiluoto, Western Finland. *Int J Rock Mech Min*, 79, 135–148.

Skipp, B.O. (ed.) (1994) *Risk and Reliability in Ground Engineering*. London: Thomas Telford, p. 304.

Stearns, D.W., (1964) Macrofracture patterns on Teton Anticline, Northwest Montana. *Transactions of the American Geophysical Union'* 45, 107–108.

Taleb, N.N. (2007 & 2010) *The Black Swan: The Impact of the Highly Improbable.* London: Penguin Books, p. 444.

Tang, A.T. & Hudson, J.A. (2010) *Rock Failure Mechanisms; Explained and Illustrated.* London: CRC Press, Taylor & Francis Group, p. 322.

Terzaghi, R.D. (1965) Sources of error in joint surveys. *Geotechnique,* 15, 287–304.

Trudgill, B. & Cartwright, J. (1994) Relay-ramp forms and normal-fault linkages, Canyonlands National Park, Utah. *Geol Soc Am Bull* 106, 1143–1157.

*Tunnels and Tunnelling* (2013) International Edition. Truckload of burning cheese closes Norway tunnel. March, p. 7.

Ulusay, R. & Hudson, J.A. (eds.) (2007) *The Complete ISRM Suggested Methods for Rock Characterization, Testing and Monitoring:* 1974–2006. 'The ISRM Blue Book'. The compilation prepared by the Turkish National Group on behalf of the ISRM.

Ulusay, R. (ed.) (2014) *The ISRM Suggested Methods for Rock Characterization, Testing and Monitoring:* 2007–2014. London: Springer Science, p. 293.

Valli, J., Kuula, H. & Hakala, M, (2011) *Modelling of the* in situ *stress state at the Olkiluoto site, Western Finland.* Posiva Oy, Finland, Posiva Working Report 2011–34.

Walsh, J.J. & Watterson, J. (1987) Distributions of cumulative displacement and seismic slip on a single normal fault surface. *J Struct Geol,* 9, 1039–1046.

Walsh, J.J. & Watterson, J. (1989) Displacement gradients on fault surfaces. *J Struct Geol,* 11, 307–316.

Walsh, J.J., Nicol, A. & Childs, C. (2002) An alternative model for the growth of faults. *J Struct Geol,* 24, 1669–1675.

Watson, J. (1911) *British and Foreign Building Stones: A Descriptive Catalogue of the Specimens in the Sedgwick Museum Cambridge.* UK: Cambridge University Press, p. 483.

Watterson, J. (1986) Fault dimensions, displacement and growth. *Pure Appl Geophys,* 124, 365–373.

Wei, Z.Q. (1988) *A fundamental study of the deformability of rock masses.* PhD thesis, Imperial College London (previously Imperial College, University of London), p. 268.

White, S.H., Bretan, P.G. & Rutter, E.H. (1986) Fault-zone reactivation: kinematics and mechanics. *Philos T R Soc Lond,* A317, 81–97.

Wilson, E.J., Johnson, T.L. & Keith, D.W. (2003) Regulating the ultimate sink: managing the risks of geologic $CO_2$ storage. *Environ Sci Technol,* 37, 3476–3483.

Wood, D.M. (1990) *Soil Behaviour and Critical State Soil Mechanics.* Cambridge, UK: Cambridge University Press.

Worsey P.N. (1981) *Geotechnical factors affecting the application of pre-split blasting to rock slopes.* PhD thesis, University of Newcastle upon Tyne, p. 515.

Worsey, P.N., Farmer, I.W. & Matheson, G.D. (1981) The mechanics of pre-splitting in discontinuous rock. In: *Proceedings 22nd U.S. Symposium on Rock Mechanics (USRMS),* 29 June–2 July, 1981, Cambridge, Massachusetts, USA. Cambridge, Mass. USA: ARMA, pp. 205–210.

Wu, H. & Pollard, D.D. (1995) An experimental study of the relationship between joint spacing and layer thickness. *J Struct Geol,* 17, 887–905.

Xu, H. & Benmokrane, B. (1996) Strengthening of existing concrete dams using post-tensioned anchors: a state-of-the-art review. *Can J Civil Eng,* 23, 1151–1171. DOI: 10.1139/916-925.

Ziegler, M., Loew, S. & Bahat, D. (2014) Growth of exfoliation joints and near-surface stress orientations inferred from fractographic markings observed in the upper Aar valley (Swiss Alps). *Tectonophysics,* 626, 1–20.

Zoback, M.D. & Gorelick, S.M. (2012) Earthquake triggering and large-scale geologic storage of carbon dioxide. *Proc Nat Acad Sci,* 109, 10164–10168. Available on: www.pnas.org/cgi/doi/10.1073/pnas.1202473109.

# INDEX